Corrosion Properties and Mechanism of Steels

Corrosion Properties and Mechanism of Steels

Editor

Vít Křivý

MDPI • Basel • Beijing • Wuhan • Barcelona • Belgrade • Manchester • Tokyo • Cluj • Tianjin

Editor
Vít Křivý
VSB–Technical University of Ostrava
Czech Republic

Editorial Office
MDPI
St. Alban-Anlage 66
4052 Basel, Switzerland

This is a reprint of articles from the Special Issue published online in the open access journal *Materials* (ISSN 1996-1944) (available at: https://www.mdpi.com/journal/materials/special_issues/Corros_Steels).

For citation purposes, cite each article independently as indicated on the article page online and as indicated below:

LastName, A.A.; LastName, B.B.; LastName, C.C. Article Title. *Journal Name* **Year**, *Volume Number*, Page Range.

ISBN 978-3-0365-6404-3 (Hbk)
ISBN 978-3-0365-6405-0 (PDF)

Cover image courtesy of Vít Křivý

© 2023 by the authors. Articles in this book are Open Access and distributed under the Creative Commons Attribution (CC BY) license, which allows users to download, copy and build upon published articles, as long as the author and publisher are properly credited, which ensures maximum dissemination and a wider impact of our publications.
The book as a whole is distributed by MDPI under the terms and conditions of the Creative Commons license CC BY-NC-ND.

Contents

About the Editor vii

Vít Křivý
Special Issue: Corrosion Properties and Mechanism of Steels
Reprinted from: *Materials* **2022**, *15*, 6796, doi:10.3390/ma15196796 1

Andrés Bonilla, Cristina Argiz, Amparo Moragues and Jaime C. Gálvez
Effect of Sulfate Ions on Galvanized Post-Tensioned Steel Corrosion in Alkaline Solutions and the Interaction with Other Ions
Reprinted from: *Materials* **2022**, *15*, 3950, doi:10.3390/ma15113950 5

Vít Křivý, Zdeněk Vašek, Miroslav Vacek and Lucie Mynarzová
Corrosion Damage to Joints of Lattice Towers Designed from Weathering Steels
Reprinted from: *Materials* **2022**, *15*, 3397, doi:10.3390/ma15093397 19

Yun-Ho Lee, Geon-Il Kim, Kyung-Min Kim, Sang-Jin Ko, Woo-Cheol Kim and Jung-Gu Kim
Localized Corrosion Occurrence in Low-Carbon Steel Pipe Caused by Microstructural Inhomogeneity
Reprinted from: *Materials* **2022**, *15*, 1870, doi:10.3390/ma15051870 45

Sin-Jae Kang, Min-Sung Hong and Jung-Gu Kim
Method for Mitigating Stray Current Corrosion in Buried Pipelines Using Calcareous Deposits
Reprinted from: *Materials* **2021**, *14*, 7905, doi:10.3390/ma14247905 59

Nguyen Thuy Chung, Yoon-Sik So, Woo-Cheol Kim and Jung-Gu Kim
Evaluation of the Influence of the Combination of pH, Chloride, and Sulfate on the Corrosion Behavior of Pipeline Steel in Soil Using Response Surface Methodology
Reprinted from: *Materials* **2021**, *14*, 6596, doi:10.3390/ma14216596 71

Aleksandra Kucharczyk, Lidia Adamczyk and Krzysztof Miecznikowski
The Influence of the Type of Electrolyte in the Modifying Solution on the Protective Properties of Vinyltrimethoysilane/Ethanol-Based Coatings Formed on Stainless Steel X20Cr13
Reprinted from: *Materials* **2021**, *14*, 6209, doi:10.3390/ma14206209 85

Jin Sung Park, Jin Woo Lee and Sung Jin Kim
Hydrogen-Induced Cracking Caused by Galvanic Corrosion of Steel Weld in a Sour Environment
Reprinted from: *Materials* **2021**, *14*, 5282, doi:10.3390/ma14185282 101

Yoon-Sik So, Min-Sung Hong, Jeong-Min Lim, Woo-Cheol Kim and Jung-Gu Kim
Calibrating the Impressed Anodic Current Density for Accelerated Galvanostatic Testing to Simulate the Long-Term Corrosion Behavior of Buried Pipeline
Reprinted from: *Materials* **2021**, *14*, 2100, doi:10.3390/ma14092100 109

Maciej Witek
Structural Integrity of Steel Pipeline with Clusters of Corrosion Defects
Reprinted from: *Materials* **2021**, *14*, 852, doi:10.3390/ma14040852 121

Yuhang Wang, Xian Zhang, Wenzhui Wei, Xiangliang Wan, Jing Liu and Kaiming Wu
Effects of Ti and Cu Addition on Inclusion Modification and Corrosion Behavior in Simulated Coarse-Grained Heat-Affected Zone of Low-Alloy Steels
Reprinted from: *Materials* **2021**, *14*, 791, doi:10.3390/ma14040791 135

Nguyen-Thuy Chung, Min-Sung Hong and Jung-Gu Kim
Optimizing the Required Cathodic Protection Current for Pre-Buried Pipelines Using Electrochemical Acceleration Methods
Reprinted from: *Materials* **2021**, *14*, 579, doi:10.3390/ma14030579 . . . . **151**

Yanru Li, Jiazhao Liu, Zhijun Dong, Shaobang Xing, Yajun Lv and Dawang Li
A Novel Testing Method for Examining Corrosion Behavior of Reinforcing Steel in Simulated Concrete Pore Solutions
Reprinted from: *Materials* **2020**, *13*, 5327, doi:10.3390/ma13235327 . . . . **165**

Yihua Dou, Zhen Li, Jiarui Cheng and Yafei Zhang
Experimental Study on Corrosion Performance of Oil Tubing Steel in HPHT Flowing Media Containing O_2 and CO_2
Reprinted from: *Materials* **2020**, *13*, 5214, doi:10.3390/ma13225214 . . . . **181**

Peng Gong, Guangxu Zhang and Jian Chen
The Corrosion Features of Q235B Steel under Immersion Test and Electrochemical Measurements in Desulfurization Solution
Reprinted from: *Materials* **2020**, *13*, 3783, doi:10.3390/ma13173783 . . . . **197**

About the Editor

Vít Křivý

Vít Křivý, Ph.D. has been an associate professor at the Department of Building Structures, Faculty of Civil Engineering, VSB-Technical University of Ostrava since 2015. His research focuses on the study of corrosion processes of steel structures and the evaluation of corrosion damage on the reliability and durability of building structures. His research pays special attention to weathering steels and the study of environmental effects on these steels. In addition to research, he is also involved in design, especially of steel technological structures.

Editorial

Special Issue: Corrosion Properties and Mechanism of Steels

Vít Křivý

Department of Building Structures, Faculty of Civil Engineering, VSB–Technical University of Ostrava, L. Podeste 1875, 708 00 Ostrava, Czech Republic; vit.krivy@vsb.cz; Tel.: +420-774-442-936

The economic losses caused by corrosion are estimated to be 3–5% of gross domestic product in developed countries. Corrosion losses include the costs of replacing damaged devices, products or constructions, backup solutions, corrosion allowance, corrosion protection systems, loss of productivity, environmental and health damages, etc. Corrosion damage affects products made of various metallic materials, but the main group of products are structures made of steel. In terms of maintaining the required service life of structures or equipment, it is necessary to understand the corrosion damage mechanism, evaluate the impact on reliable services and propose appropriate measures.

The issue of corrosion damage is considered by engineers both in the design phase of a steel structure and during its service life. When designing steel structures, it is in many cases necessary to accurately predict the corrosion damage of the designed structural components. For structures in service, the influence of corrosion damage on the load-bearing capacity and serviceability of the structural component is usually evaluated and then the residual service life is predicted or reconstruction measures are implemented. In all of the above cases, it is important to understand the corrosion mechanisms specific to the material and environmental conditions. Cooperation between corrosion specialists and engineers responsible for the design and operation of a technological equipment or building construction is also very important.

The Special Issue "Corrosion Properties and Mechanism of Steels" has been proposed as a means to present recent developments in the field, and for this reason the fourteen articles included touch different aspects of steel corrosion: a brief summary of the articles' content is given in the following. A significant phenomenon in the focus area of this Special Issue is the research on corrosion properties and mechanisms of steel pipelines. Half of the articles published in the Special Issue are devoted to this area.

The issue of corrosion processes in buried pipelines is covered in two papers prepared by Chung et al. [1,2]. A number of specific corrosion tests were carried out to investigate the synthetic soil corrosion of a pre-buried pipeline in [1]. Following the experimental test results, an empirical equation for the optimized CP current requirement, according to the pipeline service time, was derived. This equation can be applied to any corroded pipeline. External damage to buried pipelines caused by corrosive components in soil solution is studied in [2]. Increased attention is paid to the influence of the combination of pH, chloride and sulfate by using a statistical method according to the design of the experiment. The output of the used statistical methods is an equation that calculates the corrosion current density as a function of pH, chloride and sulfate concentration. Stray current corrosion in buried pipelines is investigated by Kang et al. [3]. In the article, as a countermeasure against stray current corrosion, calcareous depositions were applied to reduce the total amount of current flowing into pipelines and to prevent corrosion. The study examined the reduction of stray current corrosion via the formation of calcareous deposit layers, composed of Ca, Mg and mixed Ca and Mg at the current inflow area. Long-term corrosion mechanisms for steel pipelines in a soil environment are studied by So et al. [4] using electrochemical acceleration methods. Galvanostatic testing allows for accelerating the surface corrosion reactions through controlling the impressed anodic current density. However, a large

Citation: Křivý, V. Special Issue: Corrosion Properties and Mechanism of Steels. *Materials* **2022**, *15*, 6796. https://doi.org/10.3390/ma15196796

Received: 21 September 2022
Accepted: 29 September 2022
Published: 30 September 2022

Publisher's Note: MDPI stays neutral with regard to jurisdictional claims in published maps and institutional affiliations.

Copyright: © 2022 by the author. Licensee MDPI, Basel, Switzerland. This article is an open access article distributed under the terms and conditions of the Creative Commons Attribution (CC BY) license (https://creativecommons.org/licenses/by/4.0/).

deviation from the equilibrium state can induce different corrosion mechanisms to those in actual service. Therefore, applying a suitable anodic current density is important for shortening the test times and maintaining the stable dissolution of steel. To calibrate the anodic current density, galvanostatic tests were performed at four different levels of anodic current density and time to accelerate a one-year corrosion reaction of pipeline steel.

The issue of steel pipe failures is covered in the following two articles [5,6]. Lee et al. [5] investigated the cause of failure of a low-carbon steel pipe meeting standard KS D 3562 (ASTM A135) in a district heating system. In the groundwater environment outside of the pipe, localized corrosion occurred due to crevice corrosion by aluminum inclusions, and localized corrosion was accelerated by the large fraction of pearlite around the aluminum inclusions, leading to pipe failure. Witek [6] investigated the burst pressure and structural integrity of a steel pipeline based on in-line inspection results. Special attention was paid to the evaluation of data provided from the diagnostics using an axial excitation magnetic flux leakage technology in respect to multiple defects grouping. A specific corrosion environment of oilfield tubing and casing was studied by Dou et al. [7]. The high pressure and high temperature flow solution containing various gases and Cl^- ions significantly affects the corrosion processes. The high temperature corrosion conclusions provide references for the anticorrosion construction work of downhole pipe strings.

Several articles published in this Special Issue present detailed studies of specific corrosion processes of steel components used in mechanical engineering or in civil engineering industry. The effect of sulfate ions on galvanized post-tensioned steel corrosion in alkaline solutions is studied by Bonilla et al. [8]. The behavior of galvanized steel exposed to strong alkaline solutions with a fixed concentration of sulfate ions of 0.04 M is studied in detail. The coatings formed on stainless steel X20Cr13 were investigated by Kucharczyk et al. [9]. The article reports the results of the examination of the protective properties of silane coatings based on vinyltrimethoxysilane and ethanol, doped with the following electrolytes: acetic acid, lithium perchlorate $LiClO_4$, sulphuric acid H_2SO_4 and ammonia NH_3. Park et al. [10] examined the hydrogen-induced cracking caused by galvanic corrosion of an ASTM A516-65 steel weld in a wet sour environment using a combination of a standard immersion corrosion test, electrochemical analyses and morphological observation of the corrosion damage. Wang et al. [11] investigated the effects of Ti and Cu addition on inclusion modification and corrosion behavior in the simulated coarse-grained heat-affected zone of low-alloy steels by using an in situ scanning vibration electrode technique, a scanning electron microscope/energy-dispersive X-ray spectroscopy and an electrochemical workstation. The corrosion feature of Q235B steel in desulfurization solution is studied by Gong et al. [12]. The research results presented in the article are related to the tightening of marine diesel engine emission standards. In the article by Li et al. [13], a new mechanical-based experimental method is proposed to determine the corrosion initiation and subsequent corrosion behavior of steel in simulated concrete pore solutions. The proposed experiment is used to investigate the corrosion of the steel wire under various different conditions and to examine the effects of pre-stress level in steel wire, passivation time of steel wire, composition and concentration of simulated concrete pore solution on the corrosion initiation and the subsequent corrosion development in the steel wire. Křivý et al. [14] evaluated the static and corrosion performance of bolted lap joints in long-term operating towers. The article deals with the load-bearing capacity and durability of power line lattice towers designed from weathering steel. The design measures that can be applied in the design of new lattice towers are introduced in the article as well.

Conflicts of Interest: The author declares no conflict of interest.

References

1. Chung, N.-T.; Hong, M.-S.; Kim, J.-G. Optimizing the Required Cathodic Protection Current for Pre-Buried Pipelines Using Electrochemical Acceleration Methods. *Materials* **2021**, *14*, 579. [CrossRef] [PubMed]
2. Chung, N.T.; So, Y.-S.; Kim, W.-C.; Kim, J.-G. Evaluation of the Influence of the Combination of pH, Chloride, and Sulfate on the Corrosion Behavior of Pipeline Steel in Soil Using Response Surface Methodology. *Materials* **2021**, *14*, 6596. [CrossRef] [PubMed]
3. Kang, S.-J.; Hong, M.-S.; Kim, J.-G. Method for Mitigating Stray Current Corrosion in Buried Pipelines Using Calcareous Deposits. *Materials* **2021**, *14*, 7905. [CrossRef] [PubMed]
4. So, Y.-S.; Hong, M.-S.; Lim, J.-M.; Kim, W.-C.; Kim, J.-G. Calibrating the Impressed Anodic Current Density for Accelerated Galvanostatic Testing to Simulate the Long-Term Corrosion Behavior of Buried Pipeline. *Materials* **2021**, *14*, 2100. [CrossRef] [PubMed]
5. Lee, Y.-H.; Kim, G.-I.; Kim, K.-M.; Ko, S.-J.; Kim, W.-C.; Kim, J.-G. Localized Corrosion Occurrence in Low-Carbon Steel Pipe Caused by Microstructural Inhomogeneity. *Materials* **2022**, *15*, 1870. [CrossRef] [PubMed]
6. Witek, M. Structural Integrity of Steel Pipeline with Clusters of Corrosion Defects. *Materials* **2021**, *14*, 852. [CrossRef] [PubMed]
7. Dou, Y.; Li, Z.; Cheng, J.; Zhang, Y. Experimental Study on Corrosion Performance of Oil Tubing Steel in HPHT Flowing Media Containing O_2 and CO_2. *Materials* **2020**, *13*, 5214. [CrossRef] [PubMed]
8. Bonilla, A.; Argiz, C.; Moragues, A.; Gálvez, J.C. Effect of Sulfate Ions on Galvanized Post-Tensioned Steel Corrosion in Alkaline Solutions and the Interaction with Other Ions. *Materials* **2022**, *15*, 3950. [CrossRef] [PubMed]
9. Kucharczyk, A.; Adamczyk, L.; Miecznikowski, K. The Influence of the Type of Electrolyte in the Modifying Solution on the Protective Properties of Vinyltrimethoysilane/Ethanol-Based Coatings Formed on Stainless Steel X20Cr13. *Materials* **2021**, *14*, 6209. [CrossRef]
10. Park, J.S.; Lee, J.W.; Kim, S.J. Hydrogen-Induced Cracking Caused by Galvanic Corrosion of Steel Weld in a Sour Environment. *Materials* **2021**, *14*, 5282. [CrossRef] [PubMed]
11. Wang, Y.; Zhang, X.; Wei, W.; Wan, X.; Liu, J.; Wu, K. Effects of Ti and Cu Addition on Inclusion Modification and Corrosion Behavior in Simulated Coarse-Grained Heat-Affected Zone of Low-Alloy Steels. *Materials* **2021**, *14*, 791. [CrossRef] [PubMed]
12. Gong, P.; Zhang, G.; Chen, J. The Corrosion Features of Q235B Steel under Immersion Test and Electrochemical Measurements in Desulfurization Solution. *Materials* **2020**, *13*, 3783. [CrossRef] [PubMed]
13. Li, Y.; Liu, J.; Dong, Z.; Xing, S.; Lv, Y.; Li, D. A Novel Testing Method for Examining Corrosion Behavior of Reinforcing Steel in Simulated Concrete Pore Solutions. *Materials* **2020**, *13*, 5327. [CrossRef] [PubMed]
14. Křivý, V.; Vašek, Z.; Vacek, M.; Mynarzová, L. Corrosion Damage to Joints of Lattice Towers Designed from Weathering Steels. *Materials* **2022**, *15*, 3397. [CrossRef]

Article

Effect of Sulfate Ions on Galvanized Post-Tensioned Steel Corrosion in Alkaline Solutions and the Interaction with Other Ions

Andrés Bonilla, Cristina Argiz, Amparo Moragues and Jaime C. Gálvez *

Departamento de Ingeniería Civil, Construcción, E.T.S de Ingenieros de Caminos, Canales y Puertos, Universidad Politécnica de Madrid, c/Profesor Aranguren 3, 28040 Madrid, Spain; af.bonilla@alumnos.upm.es (A.B.); cg.argiz@upm.es (C.A.); amparo.moragues@upm.es (A.M.)
* Correspondence: jaime.galvez@upm.es

Abstract: Zinc protection of galvanized steel is initially dissolved in alkaline solutions. However, a passive layer is formed over time which protects the steel from corrosion. The behavior of galvanized steel exposed to strong alkaline solutions (pH values of 12.7) with a fixed concentration of sulfate ions of 0.04 M is studied here. Electrochemical measurement techniques such as corrosion potential, linear polarization resistance and electrochemical impedance spectroscopy are used. Synergistic effects of sulfate ions are also studied together with other anions such as chloride Cl^- or bicarbonate ion HCO_3^- and with other cations such as calcium Ca^{2+}, ammonium NH_4^+ and magnesium Mg^{2+}. The presence of sulfate ions can also depassivate the steel, leading to a corrosion current density of 0.3 µA/cm^2 at the end of the test. The presence of other ions in the solution increases this effect. The increase in corrosion current density caused by cations and anions corresponds to the following orders (greater to lesser influence): $NH_4^+ > Ca^{2+} > Mg^{2+}$ and $HCO_3^- > Cl^- > SO_4^{2-}$.

Keywords: corrosion current density; sulfate; galvanized steel; alkaline solutions; linear polarization resistance; electrochemical impedance spectroscopy

1. Introduction

The phenomenon of corrosion usually causes more severe damage to prestressed steel structures than to conventional reinforced concrete structures. Protection of prestressed galvanized steel wires is ensured by injecting alkaline grout into polyethylene ducts covering the strands. However, in areas not protected by these ducts, in deteriorated places, or in areas with insufficient grout, corrosion phenomena can occur. Accumulation in these areas of water contaminated with aggressive ions coming from the atmosphere, marine environments, with decomposition products of organic matter can cause corrosion of these wires and failures in the post-tensioned strands.

Galvanizing protects steel through two mechanisms. Firstly, it creates a physical barrier that isolates it and acts as a sacrificial anode. In addition, corrosion products create a second protective barrier. The behavior of zinc in alkaline media has already been considered in the literature [1–3]. Zinc in contact with the alkaline matrix of cement in its fresh state shows temporary chemical instability. High pH values of the aqueous phase inside concrete pores, usually above 12.5, cause zinc oxidation. The cathodic reaction is associated with water hydrolysis and generates hydrogen on the galvanized surface, according to Equation (1).

$$2H_2O + 2e^- \rightarrow 2OH^- + H_{2(g)} \qquad (1)$$

The possible transformation of molecular hydrogen into physically adsorbed atomic hydrogen, proposed by Riecke [4], increases the risk of hydrogen embrittlement in galvanized post-tensioned steel.

A Pourbaix diagram of zinc indicates that, at a pH value around 12, it forms an insoluble oxide layer of $ZnO/Zn(OH)_2$ more stable than the oxide layer formed at pH 13. With high alkalinities, Zn has an amphoteric behavior, forming soluble ions $Zn(OH)_3^-$ and $Zn(OH)_4^{2-}$ [5]. Formation of $Zn(OH)_2$ leads to hydrogen formation [6]:

$$Zn + 2H_2O \rightarrow Zn(OH)_2 + H_{2(g)} \qquad (2)$$

The risk of corrosion in an alkaline medium and in the presence of calcium ions can be limited due to the formation of a passive layer of calcium hydroxyzincate $Ca(Zn(OH)_3)_2 \cdot 2H_2O$, which is stable and protective. Some authors [7,8] have identified a value of pH 13.3 ± 0.1 as the limit for the passivation of galvanized steel. At a lower value of pH than 13.3, $Ca(Zn(OH)_3)_2 \cdot 2H_2O$ crystals are small enough to form a thin, homogeneous and stable layer on the surface of the steel capable of keeping it passive. At a higher value of pH than 13.3 and when the calcium content is low, the size of the crystals increases, making it difficult to cover the entire surface of the galvanized steel. In this case, large, isolated $Ca(Zn(OH)_3)_2$ crystals that do not passivate galvanized steel are formed.

Other works [9] have studied the behavior of galvanized steel as a function of pH in the absence of Ca^{2+} ions. In the range of 12 < pH < 12.8, the galvanized layer dissolves at a slow speed. In the range of 12.8 < pH < 13.3, the galvanized layer is capable of being covered with a protective layer that insulates it. However, at pH > 12.8 ± 0.1, hydrogen release occurs. At a value of pH > 13.3, the galvanizing layer dissolves completely. It is worth noting that the role of sulfate ions in the corrosion of galvanized steel has been less studied. Acha [10] studied stress corrosion of prestressed steel immersed in saturated solutions of $Ca(OH)_2$ with five concentrations of sulfate ions at various values of pH (0.01 M SO_4^{2-} at a pH of 12.1; 0.025 M SO_4^{2-} at a pH of 12.2; 0.05 M SO_4^{2-} at a pH of 12.4; 0.1 M and 0.2 M SO_4^{2-} at a pH of 12.85). Results showed a limiting sulfate concentration between 0.025 (pH = 12.2) and 0.05 (pH = 12.4). Above this limit, the steel surface presented severe localized corrosion, and below this concentration limit, the steel remained passive. Liu et al. [11] also showed that a sulfate concentration of between 0.02 and 0.03 mol/L, in a saturated solution of $Ca(OH)_2$, produced steel corrosion.

Therefore, corrosion of prestressed steel in the presence of sulfates depends on the sulphate ion concentration in the solution and on the pH. Carsana and Bertolini [12] identified a pH dependence on the anodic behavior of steel in sulphate solutions for the corrosion of the steel. Acha's thesis also addressed the effect of bicarbonate ions (0.05 M concentration combined with pH = 11 and 0.1 M concentration with pH = 8.2) and of carbonate ions (CO_3^{2-}) in saturated solutions of $Ca(OH)_2$. In both cases, current density reaches values lower than 0.2 µA/cm² after 45 days. These ions do not cause corrosion problems in prestressed steel.

In alkaline media and in the presence of carbonates, the most common corrosion product is hydrozincite ($Zn_5(CO_3)_2(OH)_6$) [13,14], and in the presence of sulfates, zinc hydroxysulfate ($Zn_4(SO_4)(OH)_6 \cdot 3H_2O$). After galvanized steel exposure to marine environments, the formation of a passive layer of hexagonal crystals of simonkolleite $Zn_5Cl_2(OH)_8 \cdot H_2O$, and zincite ZnO has been identified as the main corrosion product. Simonkolleite is formed after hydrozincite, both being white crystalline compounds. Later, a more protective layer of gordaite $NaZn_4Cl(OH)_6SO_4 \cdot 6H_2O$ can be formed by incorporation of sulfate and sodium ions in the crystalline structure of simonkolleite. Soluble compounds such as $ZnCl_2$ and $ZnSO_4$ have also been identified in marine environments [13,15–19].

Xu et al. [20] studied the effect of cations from different sulfate salts ($MgSO_4$, $(NH_4)_2SO_4$, Na_2SO_4, $CaSO_4$) added in a fixed concentration of 0.01 mol/l. The corrosion study was carried out in saturated calcium hydroxide solutions. Solutions of magnesium sulfate and ammonium sulfate showed higher rates of corrosion than solution with sodium sulfate. Solution pH was lowered with the addition of ammonium sulfate and magnesium sulfate. Neupane et al. [21] studied the effect of NH_4^+, Na^+, and Mg^{2+} cations on the corrosion of galvanized steel. Solutions of $(NH_4)_2SO_4$, Na_2SO_4 and $MgSO_4$ 0.5 M were prepared in distilled water. The increase in corrosion current density caused by cations and anions corresponds to the following order (greater to lesser influence): $Na^+ > NH_4^+ > Mg^{2+}$. Magnesium ions form finer, more compact and less porous corrosion products than the other salts.

There are numerous studies on the influence of potentially aggressive ions on the corrosion of galvanized steel. The results obtained show that the critical concentrations for each ion are often determined by the pH value of the solution. However, the effect that the joint presence of two or more potentially aggressive ions generates has not yet been determined.

In the present work, we studied the effect of sulfate ions on galvanized steel in alkaline solution and the synergistic effect of sulfate ions with various cations and anions found in seawater or marine environments (Ca^{2+}, NH_4^+, HCO_3^-, Mg^{2+} and Cl^-), or in the atmosphere, and their influence on the corrosion of the galvanized steel wires.

2. Experimental Work

2.1. Solutions

Concentrations of different ions were used, based on those obtained in real solutions produced by the action of rainwater together with degradation processes of living beings' waste. Table 1 indicates the ionic composition of six synthetic solutions prepared from the following salts: Na_2SO_4, $Ca(OH)_2$, NH_4COOH, $NaHCO_3$, $Mg(COOH)_2$ and $NaCl$. Solution pH was set at 12.7 by adjusting with NaOH 2M. The main component was the sulfate ion, followed by magnesium, chloride and ammonium. Ions were introduced incrementally in order to determine which ones were responsible for corrosion initiation.

Table 1. Composition of synthetic solutions prepared for corrosion tests.

Solution	SO_4^{2-} [mol/L] 0.04	Ca^{2+} [mol/L] $4 \cdot 10^{-4}$	NH_4^+ [mol/L] $5 \cdot 10^{-3}$	HCO_3^- [mol/L] $4 \cdot 10^{-4}$	Mg^{2+} [mol/L] $9.6 \cdot 10^{-3}$	Cl^- [mol/L] $7.6 \cdot 10^{-3}$
1	X					
2	X	X				
3	X	X	X			
4	X	X	X	X		
5	X	X	X	X	X	
6	X	X	X	X	X	X

2.2. Corrosion Electrochemical Cells

Electrochemical cells were made in polypropylene bottles using a three-electrode system (Figure 1). A Ag/AgCl electrode was used as a reference electrode. Stainless steel mesh was used as a counter electrode and galvanized wire (nominal diameter of 0.519 cm) as a working electrode. Galvanized wires were cleaned with alcohol. Adhesive tape was used for limiting an exposed attack area of 4891 cm^2. To avoid carbonation, the solution was covered with a liquid paraffin layer.

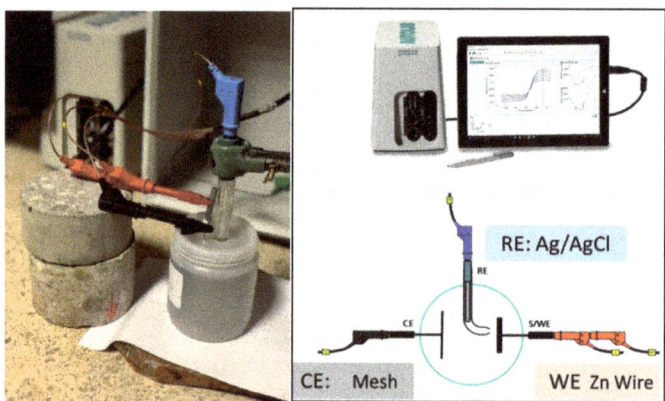

Figure 1. Corrosion cell and Autolab PGSTAT 204 potentiostat/galvanostat assembly together with experimental connections.

2.3. Techniques

2.3.1. Electrochemical Tests

Electrochemical tests were carried out with an Autolab PGSTAT 204 potentiostat/galvanostat from MetrohmAutolab BV®. NOVA 2.4.1 software with FRA 32 impedance module (Figure 1) was used. Linear polarization resistance (LPR) and electrochemical impedance spectroscopy (EIS) were also used. The LPR method was used to determine the instantaneous corrosion rate [22,23]. Measurement was carried out by applying a polarization scan from −20 mV to 20 mV around the open circuit potential (OCP) at a sweep rate of 0.1667 mV/s. Ohmic drop (R_Ω) obtained by an electrochemical impedance technique is then subtracted from this resistance. Therefore, charge transfer resistance between zinc surface and solution (R_p) (Equation (3)) is calculated as:

$$R_p = R_{p\ (LPR)} - R_{\Omega\ (EIS)} \tag{3}$$

Corrosion current density (I_{corr}) was obtained from the polarization resistance (R_p) calculated as the slope of the polarization resistance curve around the corrosion potential according to the Stern and Geary relationship (Equation (4)) [24] with parameter B = 13 [8,25–27] and the procedure proposed in UNE 112072 standard [22].

$$I_{corr} = B \cdot \frac{1}{R_p \cdot A} \tag{4}$$

Impedance measurements (EIS) were carried out by potentiostatic control in a frequency range between 10 mHz and 100 Khz, taking 10 points per decade. Amplitude of the input AC voltage signal was ±10 Mv (rms). This technique consists of taking measurements by applying a small signal of alternating current and constant voltage to a working electrode, making frequency sweeps of the applied signal [25,26,28].

2.3.2. Electron Microscopy SEM/EDS

At the end of the tests, the steel was extracted from the solution and dried in an oven at 40 °C for a week. Surface morphology was then observed using electron microscopy SEM/EDS. The attacked zone of the galvanized steel was observed by scanning electron microscope. A JEOL 6400 JSM microscope with EDS analysis was used with a resolution of 133 eV.

2.3.3. Optical Microscopy

In addition, an OLYMPUS SZX7 optical microscope with an OLYMPUS SC50 camera was used to characterize and observe the surface of wires after exposure to the corresponding solutions.

3. Results and Discussion
3.1. LPR Results

Figure 2 shows polarization resistance curves resulting from LPR measurements at the end of the test. Curves show the cathodic and anodic branches and corrosion potential. All zinc wires showed their potentials with an intermediate corrosion probability ($E_{corr} > -332$ mV) except for Solution 4, which produced somewhat higher values.

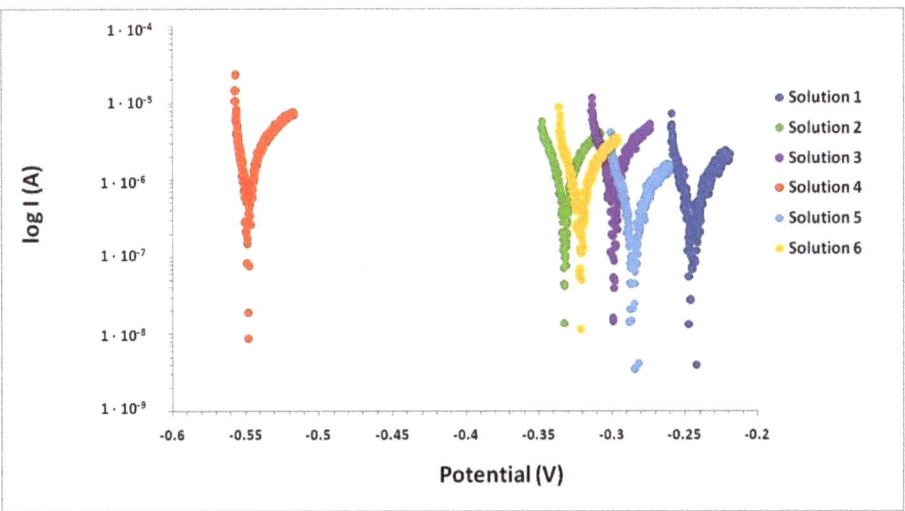

Figure 2. Polarization resistance curves of zinc wires in solution cells after 35 days of manufacture.

Table 2 shows values obtained for E_{corr}, R_p and I_{corr} of all synthetic solutions made during the entire test period. Wires evolve from high corrosion risk to less electronegative potentials. Figure 3 shows evolution in time of I_{corr} and E_{corr} of the wires.

Table 2. Electrochemical parameters obtained after 35 days of manufacture.

Day	Solution 1			Solution 2			Solution 3		
	E_{corr} (mV)	R_p (Ω)	I_{corr} (µA/cm^2)	E_{corr} (mV)	R_p (Ω)	I_{corr} (µA/cm^2)	E_{corr} (mV)	R_p (Ω)	I_{corr} (µA/cm^2)
0	−1403	27.33	97.24	−1373	56.23	47.26	−1396	28.75	92.45
1	−1396	31.45	84.51	−1386	42.26	62.89	−1389	28.30	93.92
8	−1379	20.57	129.20	−1375	55.35	63.35	−1375	21.17	125.54
28	−261	7290.27	0.36	−1351	31.50	84.37	-	-	-
35	−246	9383.37	0.28	−332	5039.88	0.53	−300	4306.20	0.62

Day	Solution 4			Solution 5			Solution 6		
	E_{corr} (mV)	R_p (Ω)	I_{corr} (µA/cm^2)	E_{corr} (mV)	R_p (Ω)	I_{corr} (µA/cm^2)	E_{corr} (mV)	R_p (Ω)	I_{corr} (µA/cm^2)
0	−1387	36.67	72.48	−1384	26.07	101.94	−1393	22.43	118.51
1	−1382	25.64	103.66	−1385	25.42	104.54	−1389	28.15	94.40
8	−1367	32.10	82.79	−1379	26.97	98.54	−1378	32.95	80.66
28	−593	2595.41	1.02	-	-	-	-	-	-
35	−548	3660.90	0.73	−287	14,025.21	0.19	−322	6303.55	0.42

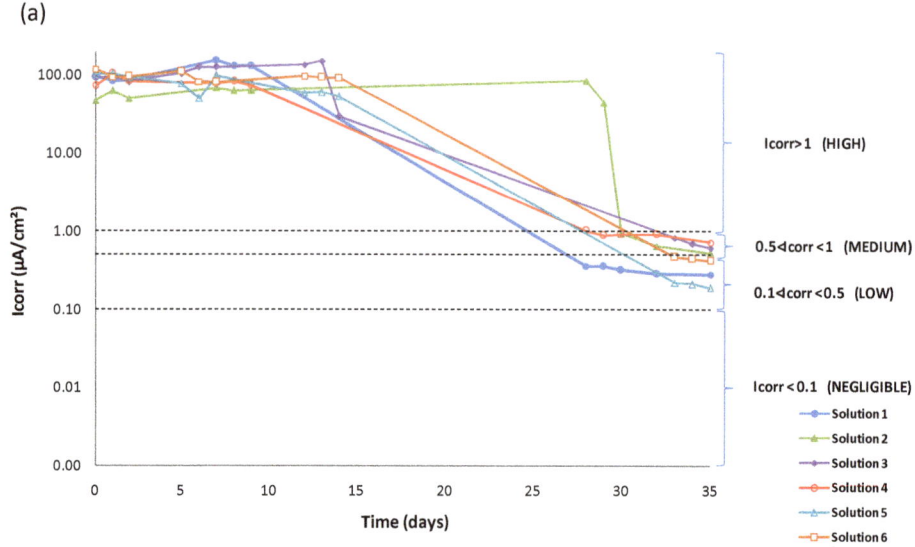

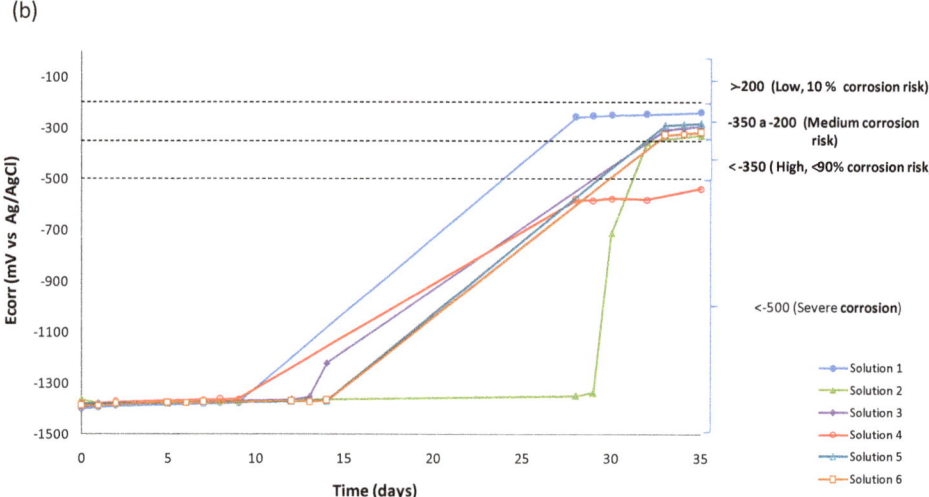

Figure 3. Evolution in time of (**a**) I_{corr} and (**b**) E_{corr} of the wires.

Starting point results were very electronegative for all solutions (−1.4 V) and showed corrosion current densities around 100 µA/cm². After 35 days of testing, the potential increased to less electronegative values, higher than −0.3 V (except for Solution 4). Corrosion current density decreased to 0.5 µA/cm². These values of E_{corr} were also recorded in other works. In particular, at a pH value of 13, Zn is actively dissolved, giving rise to a soluble phase of $Zn(OH)_4^{2-}$ in the potential range between −1.35 to −1.45 VSCE [29]. The corrosion current density decreases until they reach values between 0.1 and 1 µA/cm².

The solution containing only sulfate ions SO_4^{2-} (Solution 1) begins with active I_{corr} (97.24 µA/cm²) and a fairly electronegative E_{corr} (−1.4 V). Initial values recorded for E_{corr} and I_{corr} indicate that the galvanized layer upon contact with the alkaline medium dissolves anodically, with consequent evolution of hydrogen on the galvanized surface [6,7].

After ten days of testing, these electrochemical parameters changed. The wire became covered with a passive layer and the hydrogen evolution process slowed down. I_{corr} values decreased and reached 0.28 µA/cm^2 by the end of the test. This value is within the limits representing a low corrosion state ($0.1 > I_{corr} < 0.5$ µA/cm^2), as has also been shown in other works [10,12,25] that a passive layer is formed over time. This layer is capable of reducing the initial corrosion current density. This passive layer could be $Zn_4(SO_4)(OH)_6 \cdot 3H_2O$ [1] in the absence of Ca^{2+} ions.

According to Acha [10] and Liu [11], 0.04 M sulfate ions in saturated $Ca(OH)_2$ with a pH value of 12.4 should behave as depassivating ions. An increase of 0.3 units of pH in the presence of sulfates is not enough to passivate the steel. Vigneshwaran et al. [30] considered fixed amounts of sulfate ions of 2000 and 20,000 ppm (0.02 and 0.2 M respectively) to study carbon steel corrosion at pH values of 12.6 and 13.3, respectively. While steel corrodes at a pH of 12.6, at a pH of 13.3 the passive layer is not destabilized. Variation of just one-tenth in the pH value (12.7 vs. 12.6) changes the sulfate ion aggressiveness. With a sulfate ion, the steel is passivated. The corrosion ability of the sulfate ion depends on the pH and the type of steel.

When Ca^{2+} ions (Solution 2) were added to sulfate solution, the initial I_{corr} value of 47.26 µA/cm^2 was half of the I_{corr} value from Solution 1 and was the lowest value of all solutions. This initial corrosion decreased over time. However, it did so more slowly than in the rest of the solutions. The wire did not reach even a moderate corrosion until the 30th day of testing. The Ca^{2+} concentration was 100 times lower than that of the sulfate solution. The initial concentration of dissolved sulfate decreased by half when precipitating as calcium sulfate. The passive layer on the surface of the steel in the presence of calcium could be different from the previous solution and represent a slower development. Some authors [7,8] identified a passive layer of calcium hydroxyzincate $Ca(Zn(OH)_3)_2$. They indicated a stability limit for this layer at a pH value of 13.3. Above this pH, the larger size of the crystals does not allow them to cover the entire surface of the steel. The I_{corr} value at the end of the test (0.53 µA/cm^2) was within the limits that represent moderate corrosion ($0.5 > I_{corr} < 1$ µA/cm^2). At the studied pH, the presence of Ca^{2+} coating turned out to be less protective.

The protection of the wire in the presence of Ca^{2+} ions was modified when NH_4^+ ammonium ions were added to the solution (Solution 3). In the beginning, the I_{corr} (92.45 µA/cm^2) was at least twice that of Solution 2 (47.26 µA/cm^2) and was very similar to the sulfate solution. H Pan et al. [31] reported than the corrosion of NH_4^+ could be attributed to dissolution of the MgO inner layer and the $Mg(OH)_2$ outer passive layer. A similar process could take place in the case of ZnO and of $Zn(OH)_2$, increasing the initial corrosion rate. A higher ionic charge increased the solubility of calcium sulfate. The wire began to be covered by a passive layer after fifteen days of testing. The I_{corr} obtained after 35 days reached 0.62 µA/cm^2, within the limits that represent a state of moderate corrosion ($0.5 > I_{corr} < 1$ µA/cm^2).

When bicarbonate was added (Solution 4), an initial I_{corr} value of 72.48 µA/cm^2 was recorded. This was lower than the previous solution but higher than Solution 2. The solubility of zinc carbonate ($1.4 \cdot 10^{-11}$) is lower than that of calcium sulphate ($\approx 9.1 \cdot 10^{-6}$). Therefore, Zn ions could react with bicarbonate ions according to Equation (5) to produce corresponding carbonates with CO_2 released [13], thereby removing sulfate ions from the solution.

$$Zn^{2+} + 2HCO_3^- \leftrightarrow ZnCO_3 + H_2O + CO_2 \qquad (5)$$

At the end of the test, the I_{corr} (0.73 µA/cm^2) was the highest of the six solutions and it was within the limits that represent moderate corrosion with low tendency values ($0.5 > I_{corr} < 1$ µA/cm^2). The corrosion potential shifted to more negative values (−537 mV), which were the most electronegative of the six solutions. A passive film could be formed by hexagonal crystals of $ZnCO_3$ and monoclinic crystals of hydrozincite $Zn_5(CO_3)_2(OH)_6$. This passive layer would be the least protective.

Subsequently, initial corrosion current density adding Mg^{2+} (Solution 5) was 102 µA/cm^2. This was higher than previous solutions. After 15 days, the passive layer began to form. I_{corr} values after 35 days (0.19 µA/cm^2) were within the limits of low corrosion state with a negligible tendency (Icorr < 0.1 µA/cm^2). This corrosion current density was the lowest of all solutions, which means that any passive layer formed under these conditions would be the most protective. These results agree with those found by Neupane et al. [21]. They compared the dissolution effects of Na_2SO_4, NH_4SO_4 and $MgSO_4$ on galvanized steel. They found that the corrosion rate in the presence of NH_4^+ ions is higher than with Mg^{2+} ions. However, if the pH of the solution is not buffered, the addition of magnesium and ammonium sulfate, accompanied by a decrease in the initial pH of the solution, causes an increase in the corrosion rate, as shown in the study by Xu [20].

Lastly, once the behavior of the wire in chloride-free media was known, its behavior in a medium contaminated with Cl^- (Solution 6) was studied. The initial value of I_{corr} was 119 µA/cm^2, the highest of the six solutions, despite the fact that the concentration of this ion was lower than the sulfate ion solution. I_{corr} values at 35 days (0.42 µA/cm^2) were higher than the corrosion of the sulfate solution. This was within the limits of low corrosion state with a negligible tendency. A passive layer of simonkolleite $Zn_5Cl_2(OH)_8 \cdot H_2O$ would thus be less protective than zinc hydroxysulfate $Zn_4(SO_4)(OH)_6 \cdot 3H_2O$.

3.2. EIS Results

As noted above, impedance measurement is useful to complete the R_p calculation in the linear polarization resistance (LPR) method. In addition, it enables the determination of resistance of the different parts that make up a system.

One of the key aspects of this technique as a tool to research the electrical and electrochemical properties of systems is the direct relationship between the real behavior of a system and that of a circuit made up of a set discrete component of electrical components, called an equivalent circuit. The most accurate circuit will be the one with the fewest possible time constants, which would provide a clear physical meaning [32]. Nyquist and Bode diagrams are obtained with their respective adjustments through the equivalent circuit to determine the values of each parameter that make up the system.

Figure 4 shows the Nyquist and Bode diagram obtained with the adjustment through the equivalent circuit for Solution 1. The equivalent circuit used in this study consisted of two constants connected in series with the resistance of the electrolyte (Figure 5). Elements of the circuit had the following physical meanings: R_s was related to the resistance of the electrolyte (solution). Upon the addition of bicarbonate ions (Solution 4), the R_s increased due to the greater presence of solid carbonate species in the solution; the first time constants, R_c and C_c, were attributed to the resistance of the passive film on the steel surface [33] and their capacitance. The second time constants, R_{ct} and C_{ct}, were related to the charge transfer resistance or mass transfer resistance. The latter was comparable to that obtained by the linear polarization resistance method (R_p). A constant phase element (Q) was used instead of a pure capacitance in the adjustment because of the heterogeneity of the layer on the surface of the wire [32,34]. From the Nyquist diagram, values can be seen corresponding to the high-frequency zone ($1 \cdot 10^5$ Hz), given by the diameter of the first semicircle. It corresponds to R_c and C_c. R_{ct} and C_{ct} correspond to the diameter of the second semicircle in the low-frequency zone (0.01 Hz).

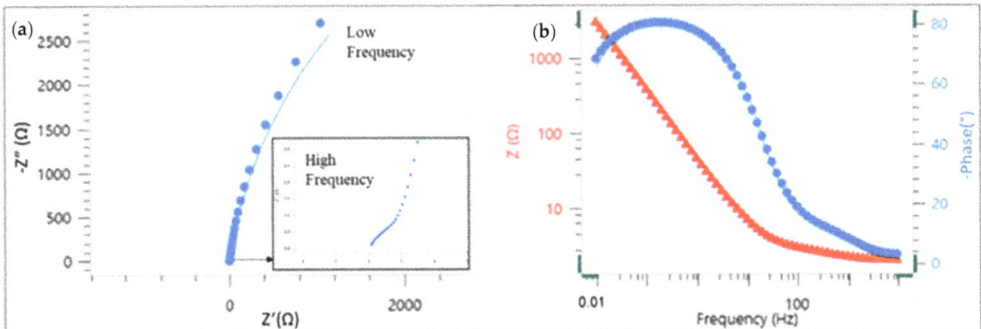

Figure 4. (**a**) EIS results with adjustment through equivalent circuit. (**b**) Nyquist and Bode diagram for Solution 1 at day 35.

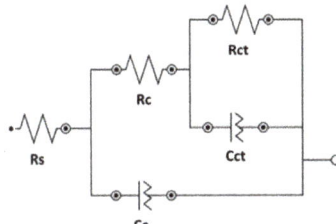

Figure 5. Equivalent circuit.

Figures 6 and 7 show the Nyquist diagrams with equivalent circuit adjustments resulting from study cases after 35 days. Parameters obtained from adjustments may be observed in Table 3. From the Nyquist diagrams, it is possible to confirm the beginning of the formation of a passive layer on the surface of the wire. It is also possible to observe the diameter of the second semicircle increasing.

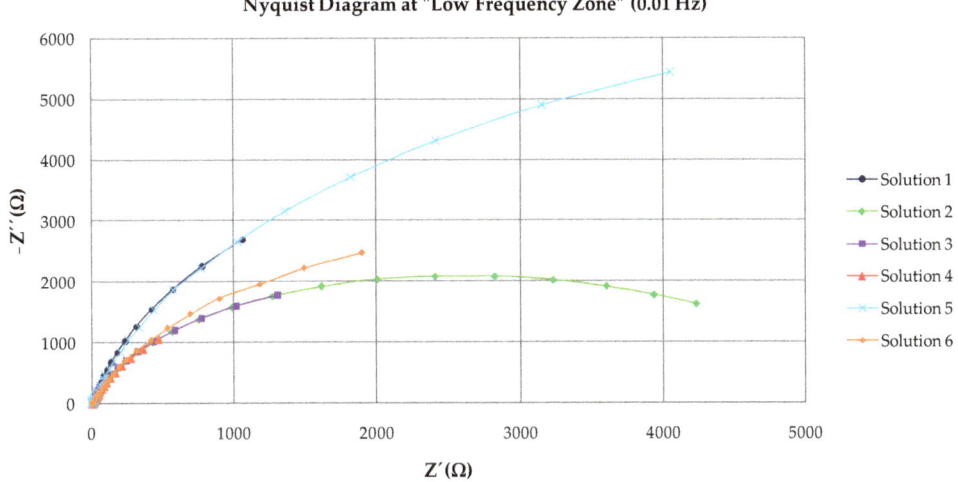

Figure 6. Nyquist diagram in "Low-Frequency Zone" (0.01 Hz).

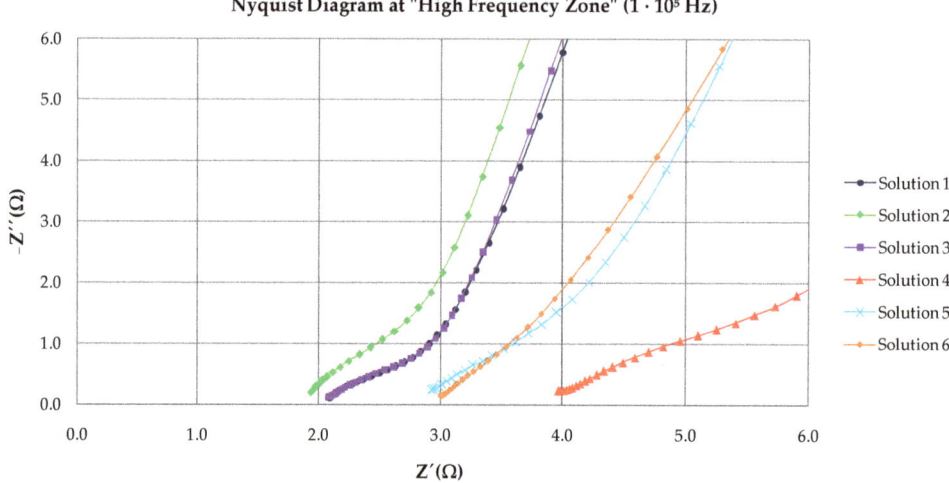

Figure 7. Nyquist diagram in "High-Frequency Zone" ($1 \cdot 10^5$ Hz).

Table 3. Parameters obtained through Equivalent Circuit of the Nyquist Diagram.

Solution	Rs ($\Omega \cdot cm^2$)	Rc ($\Omega \cdot cm^2$)	CPE $Y_0(\Omega^{-1} \cdot cm^{-2} \cdot s^n)$	Rct ($k\Omega \cdot cm^2$)	CPE $Y_0(\Omega^{-1} \cdot cm^{-2} \cdot s^n)$	χ^2
1	2.11	1.55	$1.92 \cdot 10^{-3}$	10,169	$6.36 \cdot 10^{-4}$	0.036
2	1.84	2.18	$9.20 \cdot 10^{-5}$	5346	$8.59 \cdot 10^{-5}$	0.024
3	2.13	2.95	$4.99 \cdot 10^{-4}$	4777	$3.99 \cdot 10^{-4}$	0.034
4	3.87	4.48	$3.86 \cdot 10^{-4}$	4665	$1.35 \cdot 10^{-3}$	0.036
5	2.93	2.87	$9.20 \cdot 10^{-5}$	14,923	$1.92 \cdot 10^{-4}$	0.019
6	3.03	4.83	$3.21 \cdot 10^{-4}$	7240	$2.51 \cdot 10^{-4}$	0.016

The equivalent circuit used seems to be the correct one because it has the fewest possible time constants with clear physical meaning. The deviation (χ^2) in all cases is less than 0.03. Charge-transfer resistance on the wire surface (R_{ct}) is then compared (Table 4) with the obtained R_p through linear polarization resistance to validate the results obtained through EIS. Equivalence is maintained in all cases except for Solution 4. The percentage difference between both methods is less than 14%. Although the difference with Solution 4 is 24%, both magnitudes are equal, which represents a state of medium corrosion in both cases ($0.5 > I_{corr} < 1 \mu A/cm^2$). Differences between outcomes are valid and are mainly attributed to the fact that LPR uses direct current while the EIS technique uses alternating current (sinusoidal disturbance of electric potential) of variable frequency to the studied material). The order of passive layer stability is confirmed by the two electrochemical techniques (Solution 5 < Solution 1 < Solution 6 < Solution 2 > Solution 3 > Solution 4).

Table 4. R_p obtained by LPR and EIS.

Solution	R_p (LPR) (Ω)	R_{ct} (EIS) (Ω)	Difference (%)
1	9383	10,169	8
2	5040	5346	6
3	4306	4777	10
4	3661	4665	24
5	14,025	14,923	6
6	6304	7240	14

3.3. Morphology of Steel Surface

Figure 8 shows images of different steel surfaces obtained by optical microscopy and SEM after electrochemical analysis in synthetic solutions. No traces of iron oxides are observed on the surface of the wires. The Solution 1 wire is homogeneously coated with prismatic crystals, possibly of $Zn_4(SO_4)(OH)_6 \cdot 3H_2O$. The Solution 3 wire shows a distributed oxide layer, leaving large voids on the surface. The size and coating of the crystals in the rest of the wires varies. This can be attributed to insoluble crystalline products such as $Zn_5(CO_3)_2(OH)_6$, $ZnCO_3$, $Zn_5Cl_2(OH)_8 \cdot H_2O$, ZnO, or others with the ability to passivate galvanized steel to a greater or lesser extent [15]. Some areas of exposure are darker in color due to increased detachment of the zinc layer and oxidation of the steel.

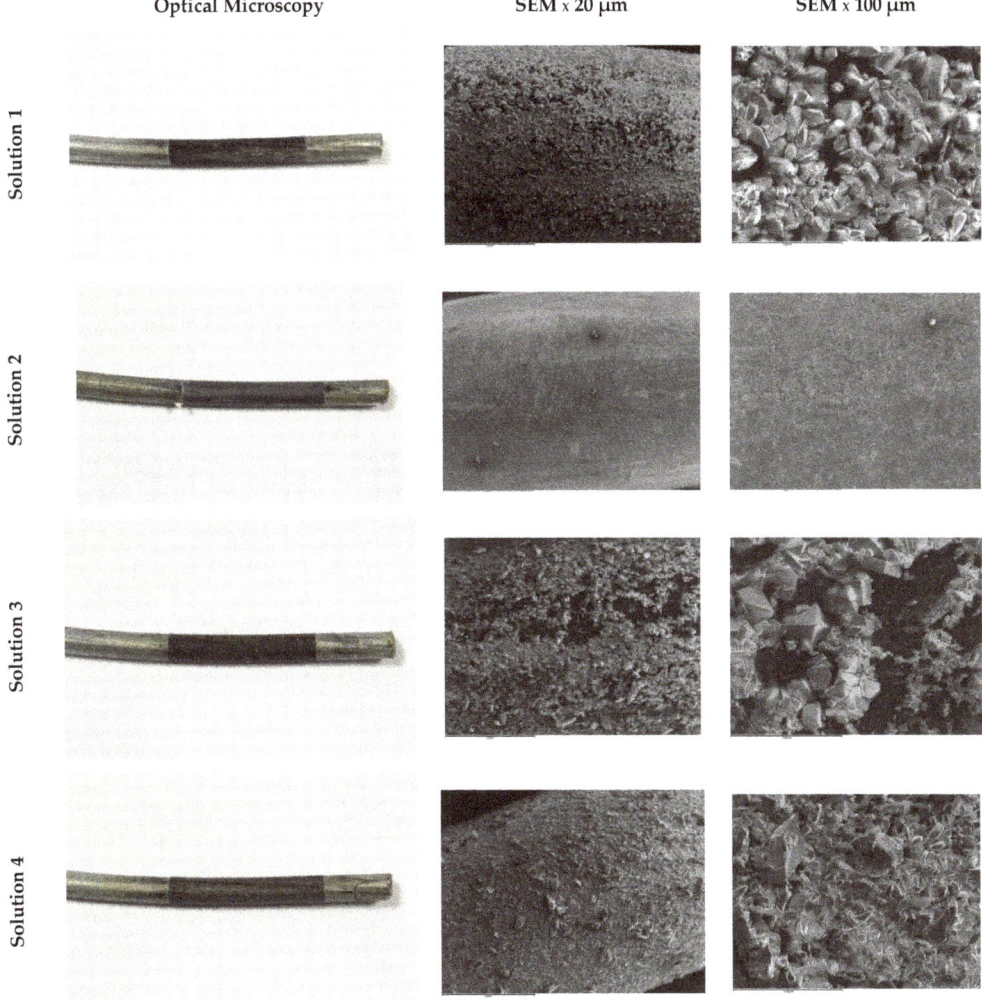

Figure 8. *Cont.*

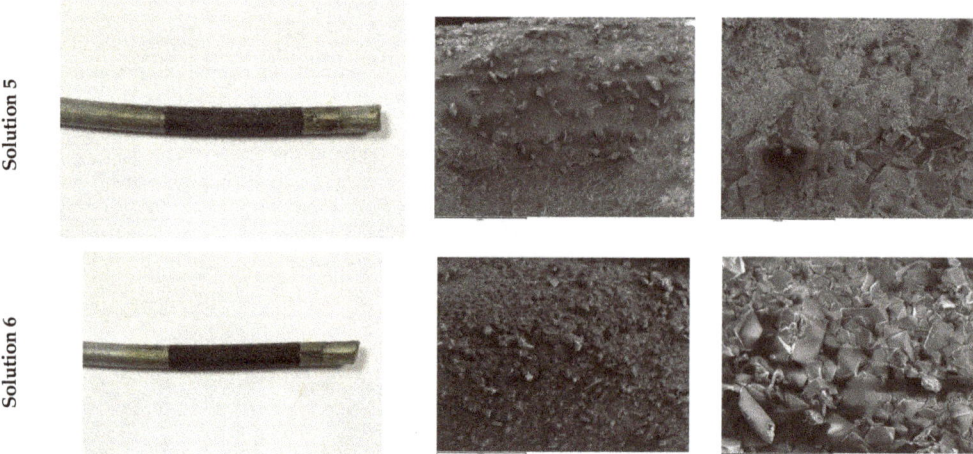

Figure 8. Optical microscopy and SEM results of the steel surfaces.

Only one isolated white crystal of calcium hydroxyzincate $Ca(Zn(OH)_3)_2$ was found on the surface of the wire exposed to Solution 2. Due to the high pH of the solution, the crystal becomes larger and does not have the capacity to homogeneously coat the surface of the wire (Figure 9) [14]. By observing the SEM appearance of this crystal, it was possible to identify a totally different morphology from those observed on the surface of the other wires. Small crystallized threads can be observed.

Figure 9. Optical microscopy and SEM crystal results $Ca(Zn(OH)_3)_2$ Solution 2.

4. Conclusions

Galvanized steel in contact with a strongly alkaline solution (pH = 12.7) and in the presence of sulfates dissolves anodically with corrosion potentials around -1.4 V and corrosion densities around 100 $\mu A/cm^2$. Over time, the surface of the steel is covered with a protective layer, and the corrosion potential increases until values of -0.25 V are attained, with a corresponding decrease in corrosion current density to 0.3 $\mu A/cm^2$, but greater than 0.1 $\mu A/cm^2$ in all cases. Consequently, the presence of sulfate ions enables the depassivating of galvanized steel at highly alkaline levels of pH.

The presence of other anions and cations together with the sulfate ions keeps the corrosion process active. The nature of the passive layer depends on the ions present. Cations and anions studied here contribute to the increase in the corrosion current density of the sulfate ions. The magnitude of this increase follows the following orders: $NH_4^+ > Ca^{2+} > Mg^{2+}$ and $HCO_3^- > Cl^- > SO_4^{2-}$.

At a pH of 12.7, the hydroxyzincate crystal formed on the surface of the steel immersed in the solution of sulfate and calcium ions is large and occurs in isolation without the ability to cover the entire surface of the steel.

Author Contributions: Conceptualization, A.B. and A.M.; methodology, A.B., C.A., A.M. and J.C.G.; software, A.B.; validation, A.B., C.A. and A.M.; formal analysis, A.B. and C.A.; investigation, A.B. and C.A.; resources, A.B. and C.A.; data curation, C.A.; writing—original draft preparation, A.B. and C.A.; writing—review and editing, A.B., C.A., A.M. and J.C.G.; visualization, A.M.; supervision, A.M. and J.C.G.; project administration, A.M. and J.C.G. All authors have read and agreed to the published version of the manuscript.

Funding: This research was funded by the Ministry of Science and Innovation of Spain by means of the Research Fund Projects RTI2018-100962-B-100 and PID2019-108978RB-C31R.

Institutional Review Board Statement: Not applicable.

Informed Consent Statement: Not applicable.

Data Availability Statement: Not applicable.

Conflicts of Interest: The authors declare no conflict of interest.

References

1. Powers, R.W.; Breiter, M.W. The anodic dissolution and passivation of zinc in concentrated potassium hydroxide solutions. *J. Electrochem. Soc.* **1969**, *116*, 719. [CrossRef]
2. Thirsk, H.R.; Armstrong, R.D.; Bell, M.F. *Electrochemistry*; The Royal Society of Chemistry: London, UK, 1974; Volume 4, ISBN 978-0-85186-037-4.
3. Muralidharan, V.S.; Rajagopalan, K.S. Kinetics and mechanism of passivation of zinc in dilute sodium hydroxide solutions. *Br. Corros. J.* **1979**, *14*, 231–234. [CrossRef]
4. Riecke, E. Investigations on the influence of zinc on the corrosion behavior of high strength steels. *Mater. Corros.* **1979**, *30*, 619–631. [CrossRef]
5. Parsons, R. Atlas of electrochemical equilibria in aqueous solutions. *J. Electroanal. Chem. Interfacial Electrochem.* **1967**, *13*, 471. [CrossRef]
6. Recio, F.; Alonso, C.; Gaillet, L.; Sánchez Moreno, M. Hydrogen embrittlement risk of high strength galvanized steel in contact with alkaline media. *Corros. Sci.* **2011**, *53*, 2853–2860. [CrossRef]
7. Macías, A.; Andrade, C. Corrosion rate of galvanized steel immersed in saturated solutions of Ca (OH)$_2$ in the PH range 12–13.8. *Br. Corros. J.* **1983**, *18*, 82–87. [CrossRef]
8. Blanco, M.T.; Andrade, C.; Macías, A. Estudio por SEM de los productos de corrosión de armaduras galvanizadas sumergidas en disoluciones de PH comprendido entre 12,6 y 13,6. *Mater. Constr.* **1984**, *34*, 55–66. [CrossRef]
9. Macías, A.; Andrade, C. Corrosion of galvanized steel in dilute Ca (OH)$_2$ solutions (PH 11.1–12.6). *Br. Corros. J.* **1987**, *22*, 162–171. [CrossRef]
10. Acha Hurtado, M. Corrosión Bajo Tensión de Alambres de Acero Pretensado en Medios Neutros con HCO$_3^-$ y alcalinos con SO$_4^{2-}$. Ph.D. Thesis, Universidad Complutense de Madrid, Madrid, Spain, October 1993; p. 321.
11. Liu, G.; Zhang, Y.; Ni, Z.; Huang, R. Corrosion behavior of steel submitted to chloride and sulphate ions in simulated concrete pore solution. *Constr. Build. Mater.* **2016**, *115*, 1–5. [CrossRef]
12. Bertolini, L.; Carsana, M.; High, P.H. Corrosion of Prestressing Steel in Segregated Grout. In Proceedings of the Modelling of Corroding Concrete Structures, Madrid, Spain, 22–23 November 2010; Andrade, C., Mancini, G., Eds.; Springer: Dordrecht, The Netherlands, 2011; pp. 147–158.
13. Roventi, G.; Bellezze, T.; Barbaresi, E.; Fratesi, R. Effect of carbonation process on the passivating products of zinc in Ca(OH)$_2$ Saturated Solution. *Mater. Corros.* **2013**, *64*, 1007–1014. [CrossRef]
14. Farina, S.B.; Duffó, G.S. Corrosion of zinc in simulated carbonated concrete pore solutions. *Electrochim. Acta* **2007**, *52*, 5131–5139. [CrossRef]
15. Zhu, F.; Persson, D.; Thierry, D.; Taxén, C. Formation of corrosion products on open and confined zinc surfaces exposed to periodic wet/dry conditions. *Corrosion* **2000**, *56*, 1256–1265. [CrossRef]
16. Ramirez, E.; González, J.A.; Bautista, A. The protective efficiency of galvanizing against corrosion of steel in mortar and in Ca(OH)$_2$ saturated solutions containing chlorides. *Cem. Concr. Res.* **1996**, *26*, 1525–1536. [CrossRef]
17. Yadav, A.; Nishikata, A.; Tsuru, T. Degradation mechanism of galvanized steel in wet–dry cyclic environment containing chloride ions. *Corros. Sci.* **2004**, *46*, 361–376. [CrossRef]
18. Liu, S.; Sun, H.; Sun, L.; Fan, H. Effects of PH and Cl$^-$ concentration on corrosion behavior of the galvanized steel in simulated rust layer solution. *Corros. Sci.* **2012**, *65*, 520–527. [CrossRef]
19. Persson, D.; Thierry, D.; Karlsson, O. Corrosion and corrosion products of hot dipped galvanized steel during long term atmospheric exposure at different sites world-wide. *Corros. Sci.* **2017**, *126*, 152–165. [CrossRef]

20. Xu, P.; Jiang, L.; Guo, M.-Z.; Zha, J.; Chen, L.; Chen, C.; Xu, N. Influence of sulfate salt type on passive film of steel in simulated concrete pore solution. *Constr. Build. Mater.* **2019**, *223*, 352–359. [CrossRef]
21. Neupane, S.; Hastuty, S.; Yadav, N.; Singh, N.; Gupta, D.K.; Yadav, B.; Singh, S.; Karki, N.; Kumari Das, A.; Subedi, V.; et al. Effects of NH_4^+, Na^+, and Mg^{2+} ions on the corrosion behavior of galvanized steel in wet–dry cyclic conditions. *Mater. Corros.* **2021**, *72*, 1388–1395. [CrossRef]
22. Metálicos, A. *UNE 112072*; Determinación de la Velocidad de Corrosión de Armaduras en Laboratorio Mediante Medida de la Resistencia a la Polarización; Asociación Española de Normalización: Madrid, Spain, 2011.
23. Jones, D.A. *Principles and Prevention of Corrosion*; Prentice-Hall, Inc.: Hoboken, NJ, USA, 1996; ISBN -0-13-359993-0.
24. Stern, M.; Geary, A.L. Electrochemical Polarization. *J. Electrochem. Soc.* **1957**, *104*, 56–63. [CrossRef]
25. Andrade, C.; Alonso, C. Corrosion rate monitoring in the laboratory and on-site. *Constr. Build. Mater.* **1996**, *10*, 315–328. [CrossRef]
26. Andrade, C.; Castelo, V.; Alonso, C.; González, J.A. The determination of the corrosion rate of steel embedded in concrete by the polarization resistance and AC impedance methods. *Corros. Eff. Stray Curr. Tech. Eval. Corros. Rebars Concr.* **1986**. [CrossRef]
27. González, J.A.; Albéniz, J.; Feliu, S. Valores de la constante *B* del método de resistencia de polarización para veinte sistemas metal-medio diferentes. *Rev. Metal.* **1996**, *32*, 10–17. [CrossRef]
28. Hladky, K.; Callow, L.M.; Dawson, J.L. Corrosion rates from impedance measurements: An introduction. *Br. Corros. J.* **1980**, *15*, 20–25. [CrossRef]
29. Bonk, S.; Wicinski, M.; Hassel, A.W.; Stratmann, M. Electrochemical characterizations of precipitates formed on zinc in alkaline sulphate solution with increasing PH values. *Electrochem. Commun.* **2004**, *6*, 800–804. [CrossRef]
30. Krishna Vigneshwaran, K.K.; Permeh, S.; Echeverría, M.; Lau, K.; Lasa, I. Corrosion of post-tensioned tendons with deficient grout, part 1: Electrochemical behavior of steel in alkaline sulfate solutions. *Corrosion* **2017**, *74*, 362–371. [CrossRef]
31. Pan, H.; Wang, L.; Lin, Y.; Ge, F.; Zhao, K.; Wang, X.; Cui, Z. Mechanistic study of ammonium-induced corrosion of AZ31 magnesium alloy in sulfate solution. *J. Mater. Sci. Technol.* **2020**, *54*, 1–13. [CrossRef]
32. Hu, J.; Koleva, D.; Petrov, P.; Breugel, K. Polymeric vesicles for corrosion control in reinforced mortar: Electrochemical behavior, steel surface analysis and bulk matrix properties. *Corros. Sci.* **2012**, *65*, 414–430. [CrossRef]
33. Ye, C.Q.; Hu, R.G.; Dong, S.G.; Zang, X.J.; Ho, R.Q.; Du, R.G.; Lin, C.J.; Pan, J.S. EIS analysis on chloride-induced corrosion behavior of reinforcement steel in simulated carbonated concrete pore solutions. *J. Electrochem. Chem.* **2013**, *688*, 275–288. [CrossRef]
34. Garces, P.; Andrade, M.; Saez, A.; Alonso, C. Corrosion of reinforcing steel in neutral and acid solutions simulating the electrolytic environments in the micropores of concrete in the propagation period. *Corros. Sci.* **2005**, *47*, 289–306. [CrossRef]

Article

Corrosion Damage to Joints of Lattice Towers Designed from Weathering Steels

Vít Křivý [1], Zdeněk Vašek [2], Miroslav Vacek [1,*] and Lucie Mynarzová [1]

[1] Department of Building Structures, Faculty of Civil Engineering, VSB—Technical University of Ostrava, 708 00 Ostrava, Czech Republic; vit.krivy@vsb.cz (V.K.); lucie.mynarzova@vsb.cz (L.M.)
[2] Liberty Ostrava a.s., 719 00 Ostrava, Czech Republic; zdenek.vasek@libertysteelgroup.com
* Correspondence: miroslav.vacek@vsb.cz; Tel.: +420-739-849-112

Abstract: The article dealt with the load-bearing capacity and durability of power line lattice towers designed from weathering steel. Attention was paid in particular to the bolted lap joints. The article evaluates the static and corrosion performance of bolted lap joints in long-term operating towers, and also presents and evaluates design measures that can be applied in the design of new lattice towers, or in the reconstruction of already operating structures. Power line lattice towers are the most extensive realization of weathering steel in the Czech Republic. On the basis of the inspections carried out to evaluate the working life of the transmission towers in operation, it can be stated that a sufficiently protective layer of corrosion products generally developed on the bearing elements of the transmission towers. However, the development of crevice corrosion at the bolted joints of the leg members is a significant problem. In this paper, the corrosion damage of bolted joints was evaluated considering two basic aspects: (1) the influence of crevice corrosion on the bearing capacity of the bolted joint was evaluated, using experimental testing and based on analytical and numerical calculations; (2) appropriate design measures applicable to the rehabilitation of developed crevice corrosion of in-service structures, or the elimination of crevice corrosion in newly designed lattice towers, was evaluated. Calculation analyses and destructive tests of bolted joints show that the development of corrosion products in the crevice does not have a significant effect on the bearing capacity of the joint, provided that there is no significant corrosion weakening of the structural elements, and bolts of class 8.8 or 10.9 are used. The results of the long-term experimental programme, and the experience from the rehabilitations carried out, show that, thanks to appropriate structural measures, specified in detail in the paper, the long-term reliable behaviour of the lattice towers structures is ensured.

Keywords: steel structures; crevice corrosion; lattice towers; bolted lap joints; weathering steel; experimental tests; numerical modelling

1. Introduction

Weathering steels have been used in the Czech Republic for steel structures since the mid-1970s. Bridges and electric power transmission systems are among the typical objects constructed from it [1]. Under appropriate atmospheric and design conditions, weathering steels form a layer of protective oxides on their surface that significantly slows the corrosion rate [2,3]. Weathering steels are used for structures with a design working life of up to 100 years, without additional corrosion protection measures [4]. These steels are mainly known by the trade name Corten [5], of which Atmofix steels are a similar variant [6]. For structures not affected by chloride deposition in coastal zones, two grades of weathering steels are practically applied—S355J2WP (Corten A) and S355J2W (Corten B).

With the correct application of weathering steels, it is possible to apply and utilise a number of significant advantages in the process of realization and long-term operation of structures made of these materials, in comparison with structures made of other grades of structural steels that need protection against corrosion in the long-term, using traditional corrosion protection systems. These potential advantages are summarised and expressed in the following basic terms:

- When manufacturing and assembling structures made of weathering steel, it is usually possible to reduce the scope of some of the main production operations and, as a result, partially reduce the price of the structure realization. In particular, costs can be saved on the corrosion protection coating systems. On the other hand, the final price is partly increased by the increased costs of the initial rolled material and welding consumables, or possibly by the costs associated with the necessary corrosion allowances. The resulting total cost for a newly manufactured and assembled weathering steel structure is typically 2 to 10% less than a similar structure designed from other structural steels protected by a corrosion protection coating system [7].
- The possible savings for the implementation of corrosion protection coating treatments, according to the previous paragraph, are directly followed by the possibility of achieving significant savings in the time required for the realization of the construction. This advantage is significant and important, especially for investors and contractors of large structures, such as motorway bridges [8].
- The essential, and most important, advantage of using weathering steels is a significant reduction in the amount of labour, time, and cost required to ensure the inspection and necessary maintenance of long-term structures [9]. For weathering steel structures, the range and cost of these important activities is relatively small compared to the range and cost of a complete corrosion protection restoration of a comparable structure made of other structural steels [10]. In the case of weathering steel structures, inspection and maintenance are mainly the observation of the conditions required for the protective function of the patina, in particular the necessary surface cleanliness of the material in all elements of the structure and, if needed, also carrying out the necessary local repairs, and providing corrosion protection in corrosion-damaged details of the structure where a sufficiently protective and stable patina has not formed [11].
- Removing the residues of old corrosion protection, and applying a new coating system when revitalising old structures protected by coating systems or metallization, are operations that can be harmful to the health of workers and the surrounding environment. Ensuring the environmentally sound execution of a repair or resurfacing is technically challenging, and increases the cost of execution. The elimination of these operations due to the use of weathering steels, therefore, represents an undeniable advantage for health and environmental protection [12].
- The natural appearance and dark brown or dark purple colour of the stable patina may be advantageous in relation to the colour of the surrounding environment or vegetation [13]. The ability to naturally modify partial irregularities or damage created in the texture or colour of the patina is also advantageous.

Steel lattice towers are typically used as bearing structures in electric power transmission systems. These structures represent the most extensive realization of weathering steel in the Czech Republic. Between 1974 and 1992, approximately 4000 transmission towers, and 130 substations, of 110 kV, 220 kV, and 400 kV were built. Power line lattice towers are manufactured in various shapes, always in the form of four-sided lattice structures, and braced in the individual faces by bracing members made of simple angles. The leg members [14] of the angles are lapped along the height by bolted lap joints, and the bracing angles are connected to the leg members by a single bolt [15].

The corrosion behaviour of the weathering steels of transmission towers has been systematically evaluated since 1975 [16–18]. In general, the protective corrosion products develop favourably on the bearing elements of the transmission towers, even on surfaces not exposed to direct weathering.

However, for the use of weathering steels, the same details were originally applied as for zinc-galvanized or paint-protected transmission towers. No specific requirements were given for the operational control and maintenance of the transmission towers. Therefore, inadequately designed details lead to the unfavourable development of corrosion products [19,20]. These are mainly the following details of the lattice towers bearing structure:

- Anchoring the structure to concrete foundations;
- Bolted joints of leg members.

Corrosion weakening of the bearing elements in the area of transition to the concrete foundation is a typical failure of all structures, not only of weathering steel transmission towers. There are known cases where the critical weakening of the structure in the area of transition to the concrete foundation was one of the main causes of the collapse of power line towers.

In the case of the bolted joint of the leg members, crevice corrosion [21,22] occurs in the crevice, between the splices and the angles to be connected [23], as illustrated in Figure 1. The time of surface wetting is prolonged in the crevice, impurities accumulate, and aeration and concentration differences are created, all of which favour the development of the corrosion process. The conditions necessary for the formation of a protective patina do not occur in the crevice. The accumulated corrosion products deform the splices in the connection.

Figure 1. Crevice corrosion in the bolted joint of the leg member.

Concern about the possible effects of adverse corrosion development on the bolted joints of leg members led transmission system operators in the Czech Republic to abandon the concept of using weathering steels in the construction of new power lines. In order to operate existing transmission powers responsibly, it is important to have verified technical data on the effect of crevice corrosion on the bearing capacity of the joints. In this paper, the results of loading tests of the leg members joints of two 110 kV power line transmission towers, taken from collapsed transmission towers, are presented. The main objective of the experimental programme, and subsequent computational analyses, was to verify the effect of crevice corrosion on the mechanical resistance of the transmission tower. The calculation analyses of bolted joints are performed using analytical equations, according to current standards [24,25]. For detailed numerical analyses of bolted joints using the finite element method, the recommendations given in technical publications [26–28] can be used.

High-voltage power lines with weathering steel transmission towers are still in operation. The transmission system operators carried out extensive rehabilitation work on most of the transmission towers, including the rehabilitation of the bolted joints in the lattice structure of the transmission towers. One of the objectives of the authors of the article was to provide concentrated information on the real technical condition of the operating transmission towers. The specific findings and recommendations are based primarily on the authors' long-term experience with the design, realization, and reliability assessment of these structures. The article deliberately provided mainly practically applicable information and recommendations for repair and maintenance.

In order to take advantage of the economic and environmental benefits of using weathering steels in the construction of lattice towers in the future, it is necessary to obtain demonstrable data confirming the functionality of the recommended design solutions for the bolted joints of leg members. For this purpose, a programme of experimental lattice towers was prepared, on which various bolted joints designs were tested in real conditions. After 9 years of the experimental programme, sufficiently reliable data are now available to confirm the suitability of the various design measures.

The issue of the corrosion behaviour of structures designed from weathering steels is very extensive, and this area received attention from researchers all over the world. Most of the scientific literature is devoted, in detail, to the study of the evolution of corrosion layers under different environmental conditions [29,30], or using accelerated tests [31,32]. The development of appropriate prediction models is important for the design of building structures [33]. This article, however, aimed at a somewhat different goal. The focus was mainly on the evaluation of the real influence of the development of corrosion products on the load-bearing capacity and durability of lattice truss towers with bolted joints, and on the design of appropriate structural measures. The presented results can be used by the scientific community, but also by experts from the construction industry, who are responsible for the design or operation of steel structures of power line transmission systems.

2. Materials and Methods

2.1. Loading Tests of Bolted Joints

The material for the test specimens was taken from the leg members of the collapsed power line transmission towers of 2 110 kV power lines *Neznášov—Týniště nad Orlicí* (overhead lines V1195 and V1196 in the Czech Republic). The transmission towers were made in the 1980s, using the weathering steel Atmofix made by Vítkovice Ironworks, Czech Republic. Eight test specimens were subsequently prepared from the collected leg members.

For the destructive tensile test of the joints, two pieces of the A-type joint, and six pieces of the K-type joint, were prepared. In all cases, L90/6 equal-leg angles were connected using P8 splices. The splices were connected to the profiles with 2×6 M20 bolts, of the class 5.6. The geometry of the test specimens is presented in Figures 2–4.

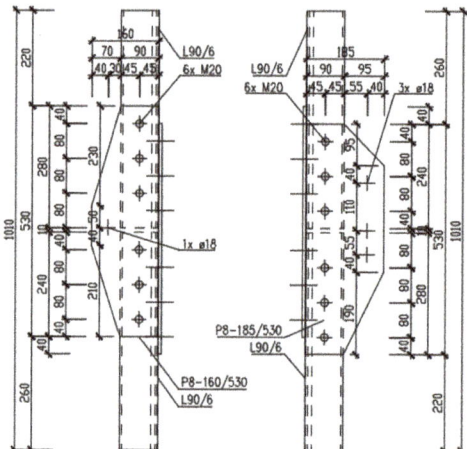

Figure 2. Detail of the A-type joint.

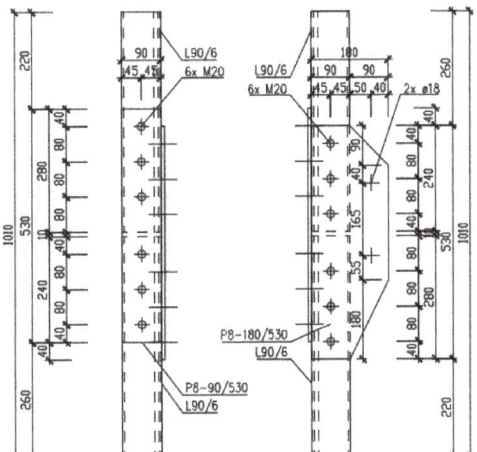

Figure 3. Detail of the K-type joint.

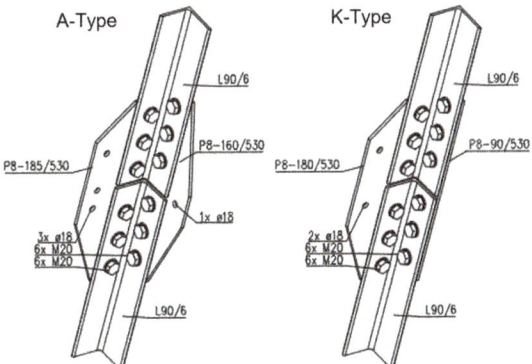

Figure 4. Details of the joint: an isometric view of the joints.

The chemical composition of the steel was determined using an optical emission spectrometer. Chemical analysis was carried out on two selected angles (test specimen A1 and K1), and one splice (test specimen K1). The chemical analysis confirms that the specimens taken from the collapsed lattice towers correspond to standard weathering steels, in accordance with the requirements of the standards [34,35]. The results are presented in Table 1. The mechanical properties of the steel were tested in the accredited laboratory of the Brno University of Technology [36]. The tensile testing, according to EN ISO 6891-1 [37], was carried out on eight flat test pieces. For the tensile testing, five pieces of leg members and three pieces of splices were made. The results are given in Table 2. The results of the chemical analysis and mechanical properties are compared with the original Czech national standard CSN 41 5217 [34], and the currently valid European standard EN 10025-5 [35]. The material meets the chemical and mechanical requirements for Atmofix steel (S355J2W).

Table 1. Loading tests of bolted joints: chemical composition of the test pieces.

Standard Test Piece	C	Mn	Si	P	S	Cu	Ni	Cr
EN 10025-5	max 0.19	0.45 1.60	max 0.55	max 0.35	max 0.030	0.20 0.60	max 0.70	0.35 0.85
CSN 41 5217	max 0.14	0.25 1.10	0.22 0.85	0.06 0.18	max 0.040	0.25 0.60	0.25 0.65	0.45 1.35
Leg member No. 1 (L90x6)	0.103	0.48	0.481	0.103	0.012	0.397	0.486	0.849
Leg member No. 2 (L90x6)	0.107	0.56	0.499	0.100	0.018	0.481	0.408	0.702
Splice No. 1 (P8)	0.111	0.59	0.378	0.091	0.009	0.367	0.378	0.810

Table 2. Loading tests of bolted joints: mechanical properties of test pieces.

Standard Test Piece	Yield Strength R_{eH} (MPa)	Tensile Strength R_m (MPa)	Extension A_5 (%)
EN 10025-5	355	470–630	22
CSN 41 5217	355	490–630	22
Leg member No. 1 (L90x6)	420	504	29.2
Leg member No. 2 (L90x6)	402	529	29.9
Leg member No. 3 (L90x6)	379	500	28.8
Leg member No. 4 (L90x6)	375	524	24.3
Leg member No. 5 (L90x6)	376	517	25.0
Splice No. 1 (P8)	355	505	26.5
Splice No. 2 (P8)	385	515	34.0
Splice No. 3 (P8)	359	497	26.9
minimum measured value	355	497	24.3
average of measured values	381.4	511.4	28.1
maximum measured value	420	529	34.0

During the tensile testing of the bolted joints, the relation between the tensile force F (kN) and the total deformation u (mm) of the test specimens was registered. At the same time, the gap opening (i.e., the change in the longitudinal distance between the ends of the connected angles) between the connected members was also measured. The uniform length of the test specimens between fixed ends is L_0 = 1010 mm, see Figure 5. The tensile testing was carried out until the specimen failed. Verification of the bearing capacity of the joint was performed using analytical relationships, in accordance with the valid European standards EN 1993-1-1 [24] and EN 1993-1-8 [25]. Numerical analysis was also carried out on the selected joint using ANSYS software. The joint was modelled using 3D solid elements and assuming the application of deformation load. The material properties of the steel and bolts were introduced using the Ramberg–Osgood stress–strain curve [38].

The microstructure of corrosion products in the crevice was analysed using a scanning electron microscope SEM and a EDAX analyser. The results are presented in Tables 3 and 4.

Table 3. Composition of the surface corrosion layer in the crevice (wt. %).

Fe	Cu	Ni	Mn	Cr	P	S	C
36.53	0.04	0.08	0.47	0.05	0.42	1.61	16.40
O	Na	Mg	Al	Si	Cl	K	Ca
38.60	0.83	0.80	1.20	2.13	0.26	0.39	0.19

Figure 5. Test piece arrangement in the tensile testing machine.

Table 4. Composition of the surface corrosion layer in the crevice (atom. %).

Fe	Cu	Ni	Mn	Cr	P	S	C
13.86	0.01	0.03	0.18	0.02	0.29	1.06	28.94
O	Na	Mg	Al	Si	Cl	K	Ca
51.12	0.77	0.70	0.94	1.61	0.16	0.21	0.10

2.2. Assessment of the Technical Condition of the Transmission Towers in Operation

The transmission towers that are part of very high-voltage power lines in the Czech Republic (overhead lines *V434 Slavětice—Čebín, V437/V438 Slavětice—Dürnrohr*) were selected for evaluation. The overhead lines were built in the 1980s, and have been in operation for approximately 40 years. In July 2021, an inspection of the steel bearing structure was carried out at 5 representative transmission towers, with a focus on the evaluation of the development of corrosion products in the area of the bolted joints of leg members.

The transmission towers are located in an agricultural area (corrosivity category C2). The structural design of the transmission towers can be seen from the photographs presented in Figure 6.

The transmission towers were visually inspected to evaluate the development of corrosion products, and the effectiveness of the implemented rehabilitation measures. The thickness of the corrosion products was measured on typical elements of the transmission towers structure, using the magnetic induction method with the Positector 6000 instrument. A total of 30 measurements were made on each of the evaluated surfaces. The actual thicknesses of the structural elements were continuously measured using a Positector UTG ME ultrasonic thickness gauge.

2.3. Experimental Lattice Towers

Experimental verification of the bolted joints in steel lattice structures designed from weathering steel was carried out in the premises of steelmaker Liberty Ostrava Inc. For this purpose, three experimental towers were built in 2012, on which the long-term monitoring of the development of corrosion products in the field of bolted joints with different structural design was carried out, as illustrated in Figure 7. This was a long-term experiment, the aim of which was to find a variant of the bolted joint of the leg members that minimized the formation of crevice corrosion in the long term, and at the same time be simple, inexpensive, and not complicate the assembly and maintenance of the structure.

Figure 6. The structural design of the evaluated transmission towers (Locality 1: Tasovice; Locality 2: Dyjákovičky; Locality 3: Ivančice; Locality 4: Rosice; Locality 5: Veverské Knínice).

Figure 7. View of the experimental towers' assembly at the time of installation.

During the experiment, the effectiveness of individual structural measures against the formation of crevice corrosion was continuously monitored. The design of the towers differed only in the joints of the leg members. In particular, the influence of the following factors was examined:

1. The effect of spacings between centres of fasteners and of end, and edge, distances from the centre of a fastener in the connection. Connections with the minimum allowable spacing and end/edge distances were expected to be less susceptible to crevice corrosion. Therefore, for tower 1, all splices were designed with smaller spacings and end/edge distances, compared to towers 2 and 3 (for tower 1, the minimum permissible values were chosen in accordance with EN 1993-1-8).
2. The effect of the thickness of the splice. It was assumed that joints with thicker splices are less susceptible to crevice corrosion. On towers 1 and 2, all splices were designed

with a nominal thickness of $t = 10$ mm, and on tower 3 the splices were designed with a nominal thickness of $t = 5$ mm.

3. The effect of bolt preloading. Joints with preloaded bolts were expected to be less susceptible to crevice corrosion. Half of the bolted joints on each tower included preloaded bolts (uncoated M20 bolts, strength class 8.8, bolt shank coated with Teflon Vaseline M8062, tightening moment 400 Nm).
4. The effect of treatment of the contact surface of the splices and leg members. It was assumed that connections with treated contact surfaces are less susceptible to crevice corrosion. Four types of contact surfaces were tested—contact surface of splices without treatment, contact surface of splices with paint (FEST-B S2141 paint), contact surface of splices with silicone sealant (neutral silicone OXIM; the method of application of silicone sealant is documented in Figure 8), and the contact surface of splices with Vaseline (Teflon Vaseline M8062; the method of application of Teflon Vaseline is illustrated in Figure 9). There are always two joints on each lattice tower with the appropriate contact surface treatment—one joint with non-preloaded bolts, the other joint with preloaded bolts.

Figure 8. Application of silicone sealant ((**left**) application of sealant to the splices; (**right**) sealant smoothed around the perimeter of the splice).

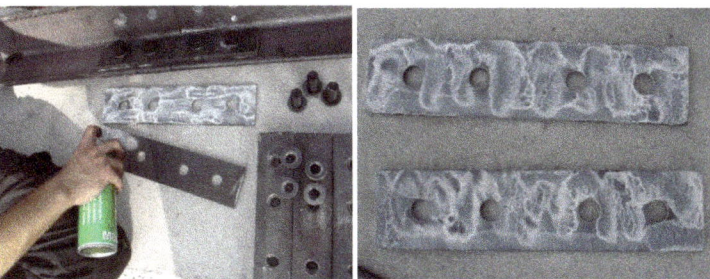

Figure 9. Application of Teflon Vaseline to the splices.

All towers have a rectangular ground plan of 1200 × 1500 mm, and the height of the towers is 2000 mm. Each tower has two vertical faces, and two sloping faces inclined from the vertical plane at an angle of 10°. The leg members are made of equal-leg angles L120 × 10, and the diagonals are made of equal-leg angles L50 × 5. Steel grade S355J2W was used. The angles used were produced on the rolling mills HCC and SJV of Liberty Ostrava Inc. The chemical composition of the leg members is listed in Table 5. The values of carbon equivalent, $CEV = 0.42\%$, and atmospheric corrosion index, $I = 6.3\%$, were calculated from the chemical composition [39,40].

Table 5. Chemical composition of the leg members.

C	Mn	Si	P	S	Cu	Ni	Cr	Al	V	N
0.11	1.00	0.21	0.11	0.007	0.33	0.17	0.49	0.030	0.04	0.011

The mechanical properties of the hot-rolled products are summarized in Table 6. The yield strength R_{eH} exceeds the characteristic value of 355 MPa by 20%, and the tensile strength is in the required range of 470–630 MPa. The A5 elongation value is 35%, above the required limit of 22%. Although the melt was rolled in S355J2W grade, the impact strength KV with 2 mm V-notch at 0 °C, −20 °C, and −50 °C was also determined. According to the results obtained, very good notch toughness of the tested steel is evident, even at extreme negative temperatures.

Table 6. Leg members: the mechanical properties of hot-rolled products.

Yield Strength R_{eH}	Tensile Strength R_m	Extension A_5	Impact Strength KV 7.5 mm (J)		
(MPa)	(MPa)	(%)	$T = 0\ °C$	$T = -20\ °C$	$T = -50\ °C$
427	541	29.7	189	157	71

The diagonals of the towers were rolled on the rolling mill SJV of Liberty Ostrava Inc. The results of mechanical properties are listed in Table 7. The lower values of impact strength, compared to the L120 × 10 equal-leg angles, are due to the smaller non-standard tested thickness. However, the minimum value of 13.5 (J) is still met.

Table 7. Diagonals: the mechanical properties of hot-rolled products.

Yield Strength R_{eH}	Tensile Strength R_m	Extension A_5	Impact Strength KV 7.5 mm (J)		
(MPa)	(MPa)	(%)	$T = 0\ °C$	$T = -20\ °C$	$T = -50\ °C$
419	553	32.9	79	50	-

The splices were also rolled on the rolling mill SJV as P80 × 10 and P60 × 10 flat pieces, and L80 × 5 angle. The mechanical properties of the splices are provided in Table 8.

Table 8. Splices: the mechanical properties of hot-rolled products.

Member	Yield Strength R_{eH}	Tensile Strength R_m	Extension A_5	Impact Strength KV 7.5 mm (J)		
	(MPa)	(MPa)	(%)	$T = 0\ °C$	$T = -20\ °C$	$T = -50\ °C$
P80 × 10	409	544	29.7	194	180	-
P60 × 10	395	574	30.3	184	166	-
L80 × 5	438	558	28.9	83	65	-

According to the analyses of chemical composition and mechanical properties, good compliance with the requirements of the EN 10025-5 standard for grade S355J2W can be stated. The experimental towers were built from material that meets the conditions of weathering steels of the required strength grade. All joints were designed with bolts. There are two bolted lap joints on each leg member. The towers are located in the premises of Liberty Ostrava Inc. (the area corresponds to the corrosivity category C3).

The surface of the angles and splices was left as rolled, including the mill scales. Only the coated splices were cleaned down to clean metal.

3. Results

3.1. Loading Tests of Bolted Joints

3.1.1. Calculation of the Resistance of the Bolted Connection

To compare with the results of the loading tests, the following section calculates the resistance of the bolted connection, according to the European standards EN 1993-1-1 and EN 1993-1-8.

The calculation is carried out for a K-type joint. Specimens for the loading tests are taken from transmission lattice towers made of Atmofix steel, which correspond to the current structural steel grade S355J2WP (yield strength f_y = 355 MPa, ultimate strength f_u = 510–680 MPa). The leg members of the L90 × 6 equal-leg angles are connected. In the connection, double-sided P8 × 90 splices are designed. The M20 non-preloaded bolts, of strength class 5.6, are used in the connection. The hole diameter for the bolts is d_0 = 22 mm. The bolt spacing and the end and edge distances are shown in Figure 3. The resistance of the following components of the connection is determined in sequence:

- Angle subjected to tension (calculation of $N_{pl,Rd}$ and $N_{u,Rd}$ for the angle);
- Contact splices subjected to tension (calculation of $N_{pl,Rd}$ and $N_{u,Rd}$ for the splices);
- Block tearing of a bolt group out of the angle (calculation of $V_{eff,2,Rd}$);
- Bolts subjected to shear (calculation of $F_{v,Rd}$ and $F_{b,Rd}$).

The design plastic resistance of the gross cross-section of the angle is determined, according to EN 1993-1-1, cl. 6.2.3(2a):

$$N_{pl,Rd} = \frac{A\,f_y}{\gamma_{M0}} = \frac{1050 \cdot 355}{1.0} \cdot 10^{-3} = 372.8 \text{ kN} \tag{1}$$

The design ultimate resistance of the net angle cross-section at holes for fasteners is determined, according to EN 1993-1-1, cl. 6.2.3(2b):

$$N_{u,Rd} = \frac{0.9\,A_{net}\,f_u}{\gamma_{M2}} = \frac{0.9 \cdot 816 \cdot 510}{1.25} \cdot 10^{-3} = 299.6 \text{ kN} \tag{2}$$

The design plastic resistance of the gross cross-section of the splices is determined, according to EN 1993-1-1, cl. 6.2.3(2a):

$$N_{pl,Rd} = \frac{A\,f_y}{\gamma_{M0}} = \frac{1440 \cdot 355}{1.0} \cdot 10^{-3} = 511.2 \text{ kN} \tag{3}$$

The design ultimate resistance of the net splice cross-section at holes for fasteners is determined, according to EN 1993-1-1, cl. 6.2.3(2b):

$$N_{u,Rd} = \frac{0.9\,A_{net}\,f_u}{\gamma_{M2}} = \frac{0.9 \cdot 1088 \cdot 510}{1.25} \cdot 10^{-3} = 399.5 \text{ kN} \tag{4}$$

The design block shear tearing resistance for a bolt group out of the angle is determined, according to EN 1993-1-8, cl. 3.10.2:

$$V_{eff,2,Rd} = 2\left(\frac{0.5 f_u A_{nt}}{\gamma_{M2}} + \frac{\frac{f_y}{\sqrt{3}} A_{nv}}{\gamma_{M0}}\right) = 2\left(\frac{0.5 \cdot 510 \cdot 204}{1.25} + \frac{\frac{355}{\sqrt{3}} \cdot 870}{1.0}\right) \cdot 10^{-3} = 439.9 \text{ kN} \tag{5}$$

The design shear resistance (shear plane passes through the unthreaded portion of the bolt) is determined, according to EN 1993-1-8, cl. 3.4.1, 3.6.1, and 3.7:

$$F_{v,Rd} = n \frac{\alpha_v f_{ub} A}{\gamma_{M2}} = 6 \cdot \frac{0.6 \cdot 500 \cdot 314}{1.25} \cdot 10^{-3} = 452.2 \text{ kN} \tag{6}$$

The design bearing resistance is determined, according to EN 1993-1-8, cl. 3.4.1, 3.6.1, and 3.7:

$$F_{b,Rd} = n\frac{k_1 \alpha f_u dt}{\gamma_{M2}} = 6 \cdot \frac{2.5 \cdot 0.606 \cdot 510 \cdot 20 \cdot 6}{1.25} \cdot 10^{-3} = 445.0 \text{ kN} \qquad (7)$$

On the basis of a comparison of the resistances corresponding to the possible failure modes of the bolted joint of the leg member, it can be stated that the design ultimate tension resistance of the net angle cross-section at holes for fasteners determines the resistance of the joint $N_{u,Rd} = 299.6$ kN.

3.1.2. Analytical Calculation of the Tensile Force in Bolts Due to Pressure of Corrosion Products in the Crevice

Bulky corrosion products develop in an imperfectly sealed crevice of a bolted joint. The corrosion products formed push the connected parts away from each other, so that the crevice opens up. The separation of the connected members is prevented by the bolts, where tensile forces are generated. The experiments carried out show that at the bolt location the gap opening is zero, and that with increasing distance from the bolts, the gap opening of the connection gradually increases. A permanent deformation of the connected members occurs.

A conservative estimation of the tensile force acting on the bolt $F_{t,Ed}$ can be determined on the basis of the assumption of plastic loading of the splice. For the calculation of the plastic bending resistance of the splice, the bolt spacing $L = p = 80$ mm, the splice width $b = 90$ mm, and the flange thickness $t = 8$ mm are considered. A simple beam model with uniform loading is assumed, and the tensile force in the bolt is equal to the reaction:

$$M_{pl,Rd} = \frac{W_{pl} f_y}{\gamma_{M0}} = M_{Ed} = \frac{1}{8} q_z L^2 \qquad (8)$$

$$F_{t,Ed} = q_z L = \frac{2 bt^2 f_y}{\gamma_{M0} L} = \frac{2 \cdot 90 \cdot 8^2 \cdot 355}{1.0 \cdot 80} \cdot 10^{-3} = 51.1 \text{ kN} \qquad (9)$$

The bolts in the connection are assessed for combined shear and tension, according to EN 1993-1-8, cl. 3.6.1. The shear force into the bolt is determined from the resistance of the least bearing member of the connection, i.e., the design ultimate resistance of the net angle cross-section at holes for fasteners:

$$F_{t,Rd} = n \frac{k_2 f_{ub} A_s}{\gamma_{M2}} = \frac{0.9 \cdot 500 \cdot 245}{1.25} \cdot 10^{-3} = 88.2 \text{ kN} \qquad (10)$$

$$\frac{F_{v,Ed}}{F_{t,Rd}} + \frac{F_{t,Ed}}{1.4 F_{t,Rd}} = \frac{49.9}{75.4} + \frac{51.1}{1.4 \cdot 88.2} = 0.662 + 0.414 = 1.076 > 1.0 \qquad (11)$$

The above assessment concludes that the combined action of extreme load effects, due to the axial stresses in the leg member, and the maximum tensile effects in the bolts, caused by the development of bulky corrosion products in the crevice, leads to bolts failure. This finding is valid for bolts of lower strength classes (4.6, 5.6) that were previously used in the design of transmission towers. Nowadays, bolts with strengths classes of 8.8 or 10.9 are commonly used, which minimizes the possibility of bolt failure in the connection under evaluation.

3.1.3. Results of Loading Tests

The loading tests are carried out until the failure of the test specimens, which occurs either by rupture of the angle of leg member, or by shear of the bolts. Summary results for both 'A' and 'K' test specimens are shown in Table 9. The ultimate load at failure of the test specimen is in the interval 441 to 484 kN, the average value being 466.9 kN. The load at the beginning of the plastic deformation of the test specimen in all cases is close to 350 kN (visible reduction of the axial stiffness of the specimens, as shown Figure 10). Significant differences are observed in the deformation values of the individual test specimens. In the opinion of the authors of the paper, these differences are mainly due to the different values of slip in the individual joints, and probably also the different levels of friction in the joints. The failure of the test specimen occurs by rupture of net cross-sectional area at holes for fasteners, as illustrated Figure 11. In one case, there is a shear failure of the bolts.

Table 9. Summary results of tensile tests.

Test Specimen	F_{max} [1] (kN)	u_{max} [2] (mm)	F_{el} [3] (kN)	u_{el} [4] (mm)	Failure Mode
A1	462	18.3		6.4	
A2	452	25.5		10.5	rupture of the angle
A3	468	22.4		11.2	
K1	484	26.6	~350	11.2	
K2	475	15.0		4.6	failure of the bolts
K3	472	25.1		12.4	
K4	481	27.2		10.4	rupture of the angle
K5	441	16.6		5.8	
minimum	441.0	15.0		4.6	
average	466.9	22.1	N/A	9.1	N/A
maximum	484.0	27.2		12.4	

[1] F_{max} ... the ultimate load at failure of the test specimen. [2] u_{max} ... total deformation of the test specimen under load F_{max}. [3] F_{el} ... the elastic capacity of the test specimen. [4] u_{el} ... total deformation of the test specimen under load F_{el}.

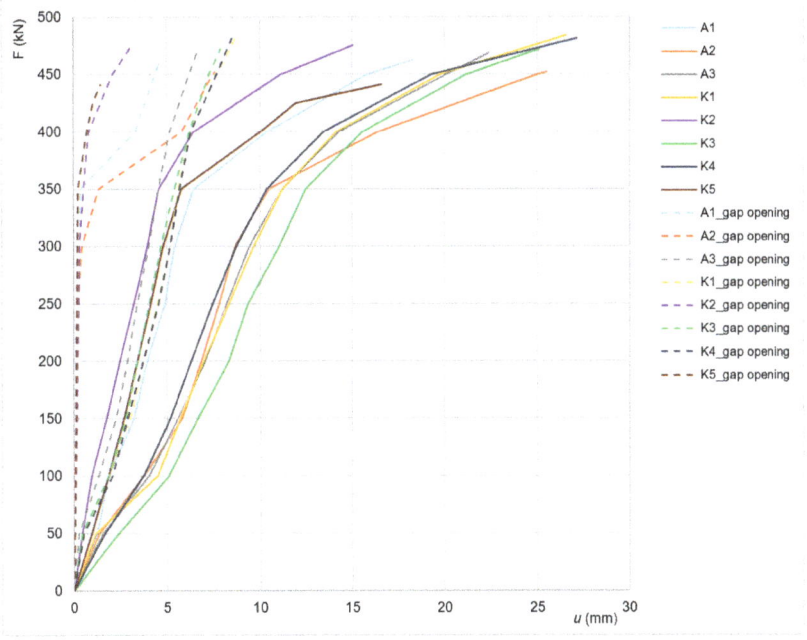

Figure 10. Tensile test results: the solid line shows the dependence of the tensile force F (kN) and the total deformation u (mm); the dashed line shows the dependence between the tensile force F (kN) and the gap opening between the connected angles.

Figure 11. Failure of the test specimens by rupture of the angles (test specimens K3 and K5).

3.1.4. Results of Numerical Models

Numerical analysis is performed for the K-type joint in ANSYS software. Both cases of possible failure are considered, i.e., without the influence of the crevice corrosion, and then with the influence of the crevice corrosion on the joint taken into account. The models are loaded with deformation loads. In the lower part of the model, a support with zero displacements is placed on the L90 × 6 profile, and a linear displacement is applied to the upper cross-sectional area, in a direction parallel to the specimen under load. The magnitude of the displacement corresponds to the measured real magnitude (i.e., the average of the measured u_{max} displacement values given in Table 9; i.e., 22.1 mm). The model mesh has a size of 2 mm (the major part is created using tetrahedron and hexahedron type of mesh; the minor part is created using the triangular prism and pyramid type of the mesh). Linear finite elements are used in the numerical model. The number of nodes in the model is 293,652. For contact between bodies, the *ANSYS contact tool* is used, with friction (a coefficient of friction 0.15).

For the K-type joint modelled without the effect of joint corrosion, the maximum von Mises equivalent stress of 497.7 MPa is obtained by numerical analysis. Figures 12 and 13 demonstrate the distribution of the von Mises equivalent stress, and the von Mises equivalent strain, and, therefore, the assumed location of the real failure is confirmed by destructive tests. The reaction force of the model is 474.0 kN, and the average value measured during destructive testing is 466.9 kN.

For a K-type joint modelled considering the effect of crevice corrosion, the von Mises equivalent stress of 499.9 MPa is obtained by numerical analysis. The consideration of the effect of crevice corrosion is based on a specific case of deformation of the splice of the K2 tested joint, where the measured maximum size of the crevice with corrosion products is equal to 10 mm. The model prescribes a 10 mm deformation of the vertical lines of the splices away from the L profile. The total elongation of the assembly is set, according to the K2 failure test, to 15 mm. The reaction force of the model is 486.7 kN, and the real measured force in the destructive test with bolt failure is 475 kN. Figures 14 and 15 below show, on the left, the results corresponding to the application of deformations from the crevice corrosion, and on the right, the results after the application of the subsequent load by the prescribed displacement.

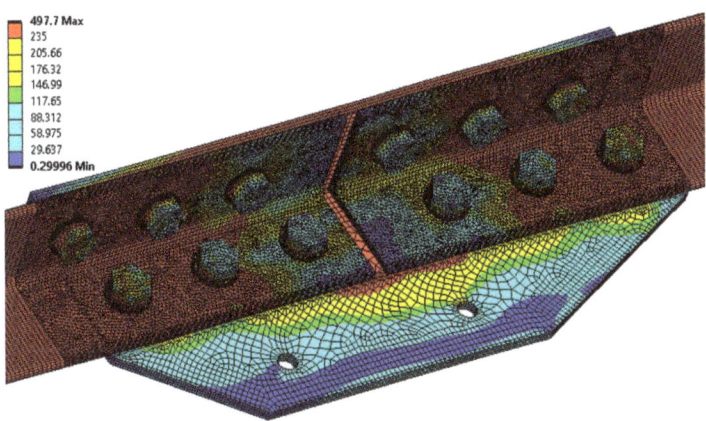

Figure 12. The von Mises equivalent stress (MPa) in the K-type joint (model without the effect of the crevice corrosion).

Figure 13. The von Mises equivalent strain in the selected part of the K-type joint (model without the effect of the crevice corrosion).

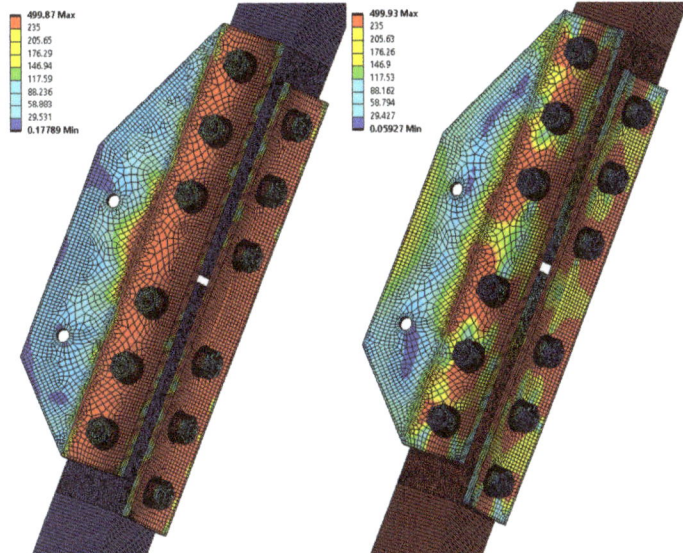

Figure 14. The von Mises equivalent stress (MPa) in the K2 joint (model considering the effect of the crevice corrosion); left: results corresponding to the application of deformations from crevice corrosion; right: results after subsequent loading with the prescribed displacement.

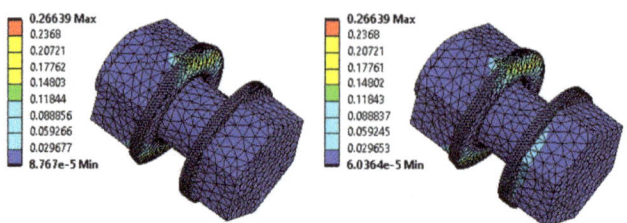

Figure 15. The von Mises equivalent strain in the K2 joint bolts (model considering the effect of the crevice corrosion); left: results corresponding to the application of deformations from crevice corrosion; right: results after subsequent loading with the prescribed displacement.

These results show that the bolts are significantly stressed from the beginning of the development of corrosion products in the crevice. With subsequent static tensile loading, the stresses in the bolts increase further. The numerical analysis is in good agreement with the results of the analytical calculation presented in Section 3.1.2.

3.2. Assessment of the Technical Condition of the Transmission Towers in Operation

3.2.1. Development of Corrosion Products on Leg Members and Diagonals

A protective adhesive compact layer of corrosion products, typical of directly wetted surfaces, develops on the surface of all the evaluated transmission towers. The average thickness of corrosion products is a suitable qualitative indicator of the favourable development of corrosion products on structures designed with weathering steels. The results of the thickness of corrosion products measurements (see Table 10) confirm the findings of the visual inspection. The thicknesses of the corrosion layers are similar on all the assessed transmission towers. The average corrosion thicknesses range from 136.3 μm to 197.7 μm, with a mean value of 161.0 μm. Thus, the criterion reported in the literature, [41,42] defining a maximum average thickness of corrosion products of 400 μm for sufficiently protective patinas, is reliably fulfilled. The value of the coefficient of variation ranges from 0.17 to 0.29, with a mean value of 0.22.

Table 10. Measured thicknesses of corrosion products.

Locality	Member	n	Mean (μm)	Std (μm)	Min (μm)	Max (μm)
Locality 1 GPS: 48.8226420N, 16.1303896E	leg member	30	161.0	37.8	110	226
	leg member	30	151.1	30.5	88	196
	diagonal	30	138.3	39.8	72	254
	diagonal	30	139.5	29.0	74	194
Locality 2 GPS: 48.7637931N, 16.0989825E	leg member	30	197.7	50.8	102	316
	leg member	30	157.1	42.5	90	270
	leg member	30	167.3	47.2	94	298
Locality 3 GPS: 49.1171492N, 16.3673119E	leg member	30	157.1	26.4	100	190
	diagonal	30	178.2	42.0	108	286
	diagonal	30	136.3	32.0	80	192
	diagonal	30	194.0	42.7	130	292

Table 10. Cont.

Locality	Member	n	Mean (μm)	Std (μm)	Min (μm)	Max (μm)
Locality 4 GPS: 49.1847531N, 16.4080061E	leg member	30	165.1	32.1	114	226
	leg member	30	194.0	40.8	116	282
	diagonal	30	150.6	29.9	104	218
	diagonal	30	146.9	27.5	102	192
	diagonal	30	148.9	26.7	100	206
Locality 5 GPS: 49.2328666N, 16.4083097E	leg member	30	170.5	34.7	110	244
	leg member	30	161.8	35.4	112	254
	diagonal	30	154.7	27.1	118	230
	diagonal	30	149.9	35.2	102	238

3.2.2. The Bolted Joints of the Leg Members

For all lattice towers situated at sites locality 1 to locality 5, as presented in Figure 6, the development of corrosion products in the crevice between the angle and the splices is identified. The bolted joints rehabilitation had not yet been performed on the transmission tower placed at locality 5. In the area of the bolted joint of the leg member, a typical development of corrosion products in the crevice between the splices and the leg member is observed, associated with the development of permanent plastic deformation of the splices, as illustrated in Figure 16. The thickness of corrosion products in the crevice reaches up to 10 mm.

Figure 16. Crevice corrosion at the bolted joint of leg members (Locality 5).

For the other transmission towers located at sites locality 1 to locality 4, the rehabilitation of the bolted joints was already carried out. The rehabilitation of the transmission towers was undertaken approximately 10 years ago. The repair of the transmission towers is carried out by workers wearing full body harnesses, and who are authorised to work at heights. In these difficult working conditions, it is very complicated to comply flawlessly with all the requirements recommended for the rehabilitation of the bolted joints of the leg members. For the transmission towers evaluated, failures of the additionally applied coating system are more frequently identified, especially on surfaces that the primer was applied to without prior proper surface cleaning (refer to Figure 17). However, from the point of view of the long-term reliable functioning of the structure, these failures of the coating system do not represent a significant problem, as the original protective layer of corrosion products remain under the peeling coating.

Figure 17. Failure of a paint system applied to a poorly prepared surface (the original protective layer of corrosion products remains under the peeling paint).

The protection of the crevice by the sealant is still functional, and no significant failures are identified on the transmission towers evaluated (see Figure 18). Thus, the bolted joint of the leg members is protected from further development of crevice corrosion, even after about 10 years following the application of the sealant.

Figure 18. Functional crevice protection with sealant (selected examples).

3.3. Experimental Lattice Towers

3.3.1. Development of Corrosion Products on Leg Members and Diagonals

The development of corrosion products on leg members and diagonals is monitored at regular intervals. The increase in the average corrosion thicknesses is noted in Figure 19. It is expected that the average thickness of the corrosion products will continue to increase gradually in the coming years (for the long-standing transmission towers in operation, an average thickness of corrosion product of 161 μm is found, see Chapter 3.2.1). During the manufacture of the towers, the surface of the steel elements is deliberately left as rolled, including the mill scales. The thin surface mill scale has mostly fallen off after 9 years of exposure. The rolled thicker scales still remain on some surfaces, but their surface is gradually diminishing.

Statically significant corrosion weakening is not yet observed; the differences in the thickness of the flanges at the time of installation of the lattice towers and after 9 years of exposure are minimal. As the experimental lattice tower project is designed to be long-term (assumed to be at least 25 years), detailed analyses of corrosion products sampled from the crevice of the de-installed bolted joints are not yet available.

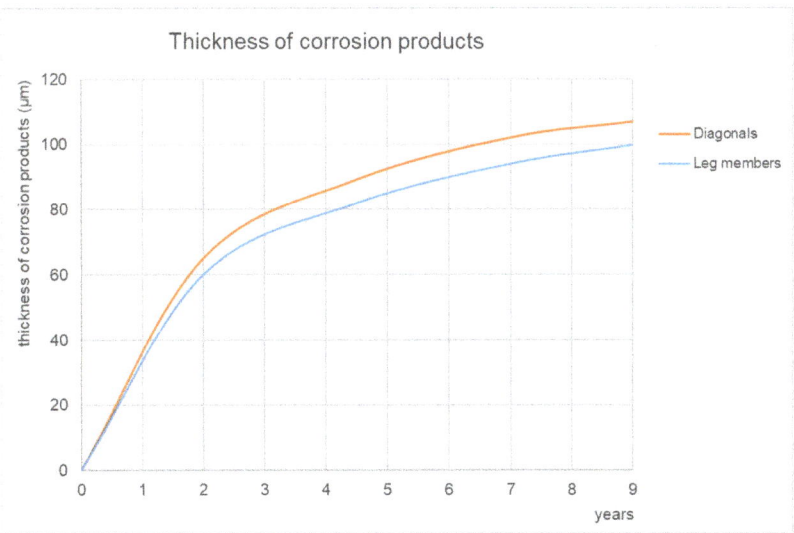

Figure 19. Experimental towers: development of thickness of corrosion products over time.

3.3.2. The Bolted Joints of the Leg Members

No phenomena related to the formation of crevice corrosion are observed in tower 1, which is designed with 10 mm thick splices, and with minimum bolt spacing and end/edge distances. No deterioration of the sealant is observed in the connections protected by sealant, as shown in Figure 20. For the other connection types, no mechanical damage to the splices is observed so far, although a lighter strip of corrosion products is visible on the top edge of the connection. No differences are observed between joints with preloaded and non-preloaded bolts.

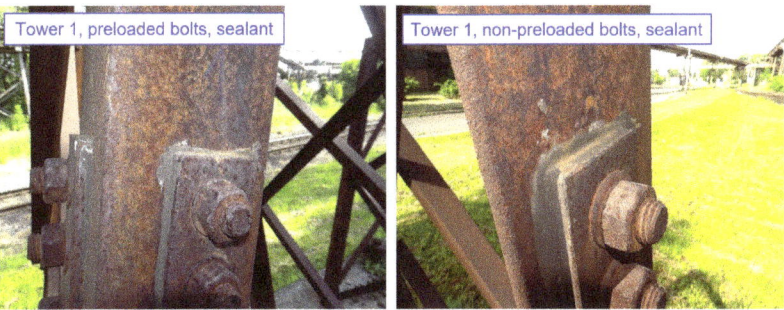

Figure 20. Tower 1: Selected examples of joints (functional sealant along the edge of the splice).

For tower 2, which is designed with 10 mm thick splices, and with normal bolt spacings and end/edge distances, phenomena related to the formation of corrosion products in the crevice between the bolt and the leg member are already observable. For bolted joints without treatment, the initial development of corrosion products in the crevice are observed (the thickness of the corrosion products at the top edge of the splice is still very small: up to about 0.5 mm, as demonstrated in Figure 21). For the other connections, the development of crevice corrosion is not yet identified, although a lighter strip of corrosion products is visible at the top edge, especially for the joints with coatings. All joints protected with sealant are functional. No significant differences between preloaded and non-preloaded bolts are observed. There is a visual influence on the development of corrosion products around the joints protected with Vaseline (this is only a visual failure, with a slowing of the development of corrosion products).

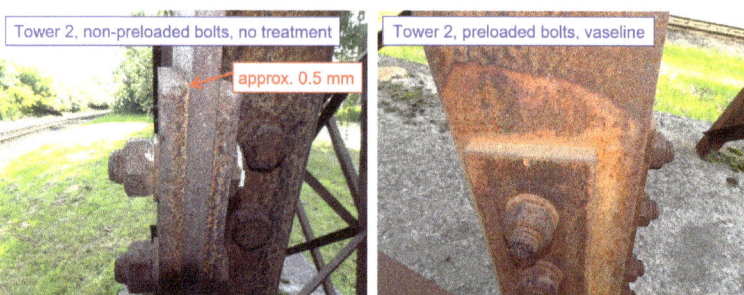

Figure 21. Tower 2: Selected examples of joints ((**left**): initial development of corrosion products in the crevice; (**right**): visual influence of the development of corrosion products in the area around the joints protected with Vaseline).

In tower 3, which is designed with 5 mm thick splices, and with normal bolt spacings and end/edge distances, development of crevice corrosion occurs at some bolted joints, as illustrated in Figure 22. For untreated bolted joints, corrosion products with a thickness of up to 3 mm are found at the top edge of the splice (the thickness of corrosion products is slightly less for preloaded bolted joints, approx. 2 mm).

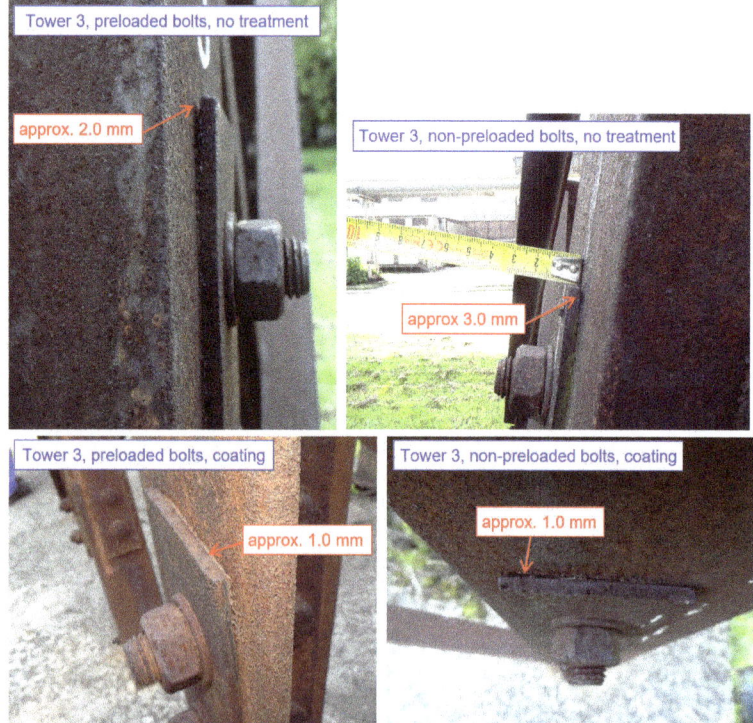

Figure 22. Tower 3: Selected examples of joints (the development of corrosion products in the crevice of joints without treatment, and in joints with coating on the inner surface of the splices).

Corrosion products with a thickness of 1 mm are also found on joints (both non-preloaded and preloaded bolted joints) protected by coating system. The joints protected with sealant are functional, as indicated in Figure 23.

Figure 23. Tower 3: Selected examples of joints (functional sealant along the edge of the splice).

4. Discussion

The environmental and economic benefits of using weathering steels for power line constructions can only be realised if the necessary technical data on the long-term behaviour of the structural elements, and their connections, are available. Therefore, in this paper, attention was paid to the issue of long-term reliable functioning of bolted joints of leg members, in relation to the possible development of corrosion products in the crevice between angles and splices. This paper examined the bolted joints of leg members from two basic aspects:

- Verification of the static resistance of bolted joints where crevice corrosion has already developed;
- Design and long-term verification of structural measures that eliminate the occurrence or further development of crevice corrosion in the long term.

4.1. Verification of the Static Resistance of Bolted Joints

The tension resistance of the bolted joint of the leg members was verified experimentally, on undamaged specimens taken from the collapsed transmission towers (the cause of the collapse of the towers was corrosion damage in the area of anchoring the leg members to the foundation). Analytical and numerical calculations of the resistance of the selected connection were performed for comparison. The performed calculations are in good agreement with the results of the loading tests, where in most cases the failure of the test specimen occurred due to tensile damage of the net angle cross-section at holes for fasteners. In only one case was the shear of the bolts responsible for the failure of the specimen (in this case, the increased tensile stress on the bolts from the pressure of bulky corrosion products in the crevice may have contributed to the failure mode). For all tested specimens, the value of the elastic capacity of the joint $F_{el} \cong 350$ kN is higher than the resistance calculated, according to the applicable standards $F_{T,Rd} = N_{u,Rd} = 299.6$ kN.

On the dismantled joints used for testing, it is observed that the thickness of the corrosion layer is greatest around the perimeter of the splice, decreasing towards the bolts. There is no significant corrosion weakening of the bolts. Thus, permanent deformation of the splices does not have a significant effect on the shear resistance of the bolts. This observation is consistent with the results of earlier studies carried out on bolted lap joints made of weathering steel [17,43].

The calculations presented in Section 3.1.2, as well as the numerical analyses presented in Section 3.1.4, indicate a possible negative effect related to the increase in tensile stresses on the bolts, caused by the development of corrosion products in the crevice. This process represents the direct effect of corrosion development in the crevice on the static load capacity of the structural member. When the extreme load effects from the axial stresses in the leg member are combined with the maximum tensile effects in the bolts, caused by the development of bulky corrosion products in the crevice, failure of the bolts may occur, provided that the bolts are designed using lower grades steels. This conclusion follows from the quantification of Equations (8)–(11).

The calculation analyses given in Sections 3.1.1, 3.1.2 and 3.1.4 do not include possible corrosion weakening of the cross-sections of the individual components of the joint. It is assumed that possible corrosion weakening may have the following consequences for the resistance of the connection:

- The shear and bearing resistance of the bolts will be minimally affected, because the bolt shank is located inside the connection, and the corrosion weakening of the bolt head and nut does not have a significant effect on the way the leg members are stressed.
- The greatest corrosion damage to the splices is expected in the gross cross-section between the bolts. However, the splices satisfy the resistance criterions with a margin in this cross-section, because they are not weakened by the bolt hole. In the net cross-section at the bolt, the effect of corrosion is less than in the cross-section between the bolts, due to the bolt tightening the crevice. The results of residual thickness measurements in the joint are found in [17], and the least corrosion weakening was found in the circumference of bolt holes.
- The corrosion weakening of the leg members designed from angles is similar to that of the contact splices. As a result of the smaller cross-sectional area, and the smaller thickness of the leg member, this effect may be more pronounced than for splices, and should always be assessed according to the specific condition of the towers. Therefore, the angle of the leg member is likely to remain the weakest component of the joint when considering corrosion weakening.

The results of the experimental and computational analyses of the bolted joints of the leg members correspond to the experience with the operation of the transmission towers. The cause of accidents of lattice towers designed from weathering steels in the Czech Republic is usually the loss of stability of the compressed elements of the lattice structure subjected to increased bending stresses (e.g., from the effects of wind loading), or significant corrosion damage in the area of the anchorage of the transmission tower to the concrete foundation. Adverse microclimatic conditions can occur in the anchorage area: the structure is affected by surrounding vegetation; structural elements are often permanently covered with soil; and frequent degradation of the foundation structures also contributes to increased wetting. In the case of transmission towers designed from weathering steel, a continuous cycle of wetting and drying of the surface is not ensured in locally affected areas and, therefore, a protective layer of corrosion products does not develop on the structural elements.

This results in corrosion weakening the bearing elements in the anchorage area. Selected typical examples of corrosion damage in the anchorage area of transmission towers are provided in Figure 24.

4.2. Design Solutions for New Transmission Towers

The experimental towers located on the premises of Liberty Ostrava Inc. are planned as a long-term corrosion test, which focuses primarily on the verification of design measures to eliminate the development of corrosion products in the crevice at the bolted joints of the leg members. Based on the results obtained after 9 years of exposure, the functionality of the individual design measures can already be reasonably evaluated.

The basic design measure is the design of sufficiently rigid splices, together with the use of minimum recommended bolt spacing and end/edge distances. Both of the above principles are applied to tower 1, where corrosion products at the crevice in any of the bolted joints evaluated have not yet been observed to develop. The effectiveness of the proposed measure is based on a basic static assumption. By increasing the thickness of the splices, and reducing the spacing and end/edge distances, a component with higher bending stiffness is installed in the bolted joint, which is able to resist transverse pressures from corrosion products arising in the crevice.

Figure 24. Selected examples of corrosion damage in the anchorage area of transmission towers.

The application of silicone sealant seems to be an appropriate measure. The principle of the method is to seal the edge between the splice and the angle against the direct effects of the atmosphere. The application of silicone sealant on the splice takes a maximum of 30 s, and the sealant is cheap. After placing the splice into the structure, and the subsequent tightening of the bolts, the sealant is pushed beyond the edges of the splice for both non-preloaded and preloaded joints, and the sealant is easily smoothed around the perimeter of the splice. After 9 years of exposure, all the sealant-coated joints are functional, and the sealant intact, with no signs of cracking or other degradation. The suitability of this design measure was also verified for operational transmission towers, where crevice sealing was one of the partial steps in the rehabilitation of bolted joints affected by the development of crevice corrosion (see Section 3.2.2). The use of sealants for the renovation of bolted joints is also recommended in the guidelines for the use of steels with increased resistance to atmospheric corrosion [42].

Based on the results obtained after 9 years of exposure of the towers, it is not possible to clearly evaluate the beneficial effect of bolt preloading. However, the results obtained from tower 3 suggest that bolt preloading in the connection reduces the risk of the development of corrosion products in the crevice. The beneficial effect of preloading is documented by mathematical numerical models [26,27]. The real effect of bolt preloading in weathering steel lattice towers is also planned to be monitored in detail in the remaining years of the experimental measurement.

On the other hand, measures based on coating the contact surfaces of the splices appear to be less effective. Although the development of bulky corrosion products in the crevice has not been observed in bolted joints with Teflon Vaseline-protected splices after 9 years of exposure, the adverse visual effect on the development of corrosion products around the connection significantly limits the applicability of this measure.

4.3. Recommended Method of Rehabilitation of Bolt Joints of the Leg Members

While a sufficiently protective layer of corrosion products usually develops on the common elements of the steel bearing structure of weathering steel lattice towers, locally unfavourable behaviour related to the development of crevice corrosion occurs in the bolted joints areas. Loading tests carried out on specimens taken from the collapsed transmission towers show that the static load capacity of the bolted joints is not significantly affected by adverse corrosion development in the crevice between the leg member and the splices (see Section 3.1). The results of load tests are very important for steel structure designers and transmission line administrators when evaluating the reliability of in-service steel lattice towers [44]. This finding is confirmed by experience with collapsed transmission towers, where the cause of failure is never the bolted joints of leg members, even though they show crevice corrosion. The collapse is usually caused by critical weakening of the cross-sections of the leg members at the anchorage to the foundations, or by overloading and buckling in strong winds. However, the significant plastic deformation of the splices, caused by the effects of crevice corrosion, is visually unfavourable, and raises concerns among the owners or administrators of the structure about the possible limitation of the bearing capacity or working life of the structure. Rehabilitation of these joints is required for this reason.

It is possible to rehabilitate the connection without dismantling and replacing the splices and bolts with new ones. In order to ensure the longest possible working life of the connection rehabilitated by repair without removing the splices and bolts, it is recommended to use to the following repair procedure [42]:

1. The removal of corrosion products. The crevice between the splice and the web is cleaned of corrosion products to a minimum depth equal to the thickness of the crevice. Cleaning of the external surfaces around the connection is carried out by manual cleaning, or by cleaning with small machinery. In accessible locations, cleaning by blasting, such as high-speed blasting with thermo-blast technology, is recommended to remove corrosion products instead of manual cleaning.
2. The application of the primer, with an overlap of up to 200 mm from the edge of the connection.
3. Sealing the joint. First, it is necessary to seal the inner space of the scratched crevice, then the space between the splice and the leg member is sealed and smoothed, so that the sealant is evenly distributed and does not overlap the splice.
4. Cover the sealant and the overlap from the edge of the connection with a coating (primer and topcoat).

Compatibility between the sealant and the coating system is necessary to achieve the expected life of the rehabilitation system. It is recommended to check the technical condition of the rehabilitation system at regular intervals (optimally, a period of 5 years).

5. Conclusions

The paper dealt with the bearing capacity and durability of steel lattice towers made of weathering steels. The main focus was the corrosion damage of the bolted joints of the leg members, and the effect on the load-bearing capacity and durability of the steel structure. Using experimental testing, and on the basis of analytical and numerical calculations, the effect of crevice corrosion on the bolted joint resistance was evaluated. Appropriate design measures, applicable for the rehabilitation of developed crevice corrosion of in-service structures, or the elimination of crevice corrosion in newly designed lattice towers, were evaluated. The main conclusions of the paper are as follows:

- The capacity of the bolted joint of the leg member is not significantly affected by the development of corrosion products in the crevice, provided that higher strength bolts (8.8 or 10.9) are used, and no significant corrosion weakening of the corner bolt occurs. This finding was verified both by tensile tests on specimens taken from actually operated transmission lattice towers, and by analytical and numerical calculations and static assessments.

- Using analytical or numerical models, it is possible to determine the static resistance of the joint affected by crevice corrosion with sufficient accuracy. Both calculation procedures given in Sections 3.1.1 and 3.1.4 can be used in the assessment of structures.
- The basic design measure is the design of sufficiently rigid splices. It is advisable to design the minimum recommended values for bolt spacing, and the end and edge distances. It is recommended to preload the bolts. This structural recommendation is supported by the results of long-term testing using experimental lattice towers.
- The application of silicone sealant seems to be the appropriate measure. The measure is effective in the long term, simple to apply, and also relatively cheap. The suitability of this measure is verified for new towers, and also for the rehabilitation of bolted joints in lattice structures in operation.

Author Contributions: Conceptualization, V.K.; methodology, V.K. and Z.V.; software, M.V.; validation, V.K.; formal analysis, V.K. and L.M.; investigation, V.K. and Z.V.; resources, V.K. and Z.V.; data curation, V.K.; writing—original draft preparation, V.K.; writing—review and editing, L.M.; visualization, V.K.; supervision, V.K.; project administration, V.K.; funding acquisition, V.K. All authors have read and agreed to the published version of the manuscript.

Funding: The evaluation of the influence of corrosion damage on the durability of the steel structure of the lattice towers was carried out within the project of the Grant Agency of the Czech Republic (project GA21-14886S Influence of material properties of high strength steels on durability of engineering structures and bridges). Financial support from VSB-Technical University of Ostrava, by means of the Czech Ministry of Education, Youth, and Sports, through the Institutional support for conceptual development of science, research, and innovations for the year 2021 (project No. SPP IP2071111) is also gratefully acknowledged.

Institutional Review Board Statement: Not applicable.

Informed Consent Statement: Not applicable.

Data Availability Statement: Data are contained within the article.

Conflicts of Interest: The authors declare no conflict of interest.

References

1. Lebet, J.P.; Lang, T.P. *Brücken Aus Wetterfestem Stahl*; ICOM Construction Métallique, EPFL: Lausanne, Switzerland, 2001.
2. Morcillo, M.; Díaz, I.; Chico, B.; Cano, H.; de la Fuente, D. Weathering steels: From empirical development to scientific design. A review. *Corros. Sci.* **2014**, *83*, 6–31. [CrossRef]
3. Morcillo, M.; Chico, B.; Díaz, I.; Cano, H.; de la Fuente, D. Atmospheric corrosion data of weathering steels. A review. *Corros. Sci.* **2013**, *77*, 6–24. [CrossRef]
4. Albrecht, P.; Naeemi, A. *Performance of Weathering Steel in Bridges*; NCHRP: Washington, DC, USA, 1984.
5. Bupesh Raja, V.K.; Palanikumar, K.; Renish, R.R.; Ganesh Babu, A.N.; Varma, J.; Gopal, P. Corrosion resistance of corten steel—A review. *Mater. Today Proc.* **2021**, *46*, 3572–3577. [CrossRef]
6. Kocich, J.; Ševčíková, J.; Tuleja, S. Effect of atmospheric corrosion on the mechanical properties of the weathering steel Atmofix 52A. *Corros. Sci.* **1993**, *35*, 719–725. [CrossRef]
7. Ungermann, D.; Hatke, P. *European Design Guide for the Use of Weathering Steel in Bridge Construction*, 2nd ed.; ECCS A3 Bridge Committee: Brussels, Belgium, 2021.
8. McConnell, J.; Shenton, H.W.; Mertz, D.R.; Kaur, D. Performance of Uncoated Weathering Steel Highway Bridges Throughout the United States. *Transp. Res. Rec.* **2014**, *2406*, 61–67. [CrossRef]
9. *CD 361 Weathering Steel for Highway Structures*; Highways England: London, UK, 2019.
10. Rigueiro, C.; Jehlička, P.; Ryjáček, P.; Wald, F. *Life Cycle Assessment of Steel Concrete Bridges I. General Issues and Examples*; CTU in Prague: Prague, Czech Republic, 2018. (In Czech)
11. Goto, H.; Moriya, S.; Naitoh, Y.; Yamamoto, M.; Fujishiro, M.; Saito, M. Examination of Repairing Methods of Weathering Steel by Painting. *Zair. Kankyo* **2010**, *59*, 10–17. [CrossRef]
12. Lambert, P. 6-Sustainability of metals and alloys in construction. In *Woodhead Publishing Series in Civil and Structural Engineering*; CRC Press Inc.: Boca Raton, FL, USA, 2016; pp. 105–128.
13. Targowski, W.; Kulowski, A. Influence of the Widespread Use of Corten Plate on the Acoustics of the European Solidarity Centre Building in Gdańsk. *Buildings* **2021**, *11*, 133. [CrossRef]
14. EN 1993-3-1; Eurocode 3: Design of Steel Structures—Part 3-1: Towers, Masts and Chimneys—Towers and Masts. CEN: Brussels, Belgium, 2006.

15. Shangwu, Z.; Delu, X.; Daijun, L.; Guang, L.; Jianwei, C.; Chao, F. A study on weathering steel bolts for transmission towers. *J. Constr. Steel Res.* **2020**, *174*, 106295.
16. Brockenbrough, R.L.; Schmitt, R.J. Considerations in the performance of bare high-strength low-alloy steel transmission towers. *IEEE Trans. Power Appar. Syst.* **1975**, *94*, 1–6.
17. Knotkova, D.; Vlckova, J. *Atmospheric Corrosion of Bolted Lap Joints Made of Weathering Steel*; ASTM Special Technical Publications: West Conshohocken, PA, USA, 1995; pp. 114–136.
18. Kreislova, K.; Geiplova, H.; Mindos, L.; Novakova, R. Corrosion Protection of Infrastructure of Power Industry. *Mater. Sci. Forum.* **2015**, *811*, 31–40. [CrossRef]
19. Krivy, V.; Urban, V.; Kreislova, K. Development and failures of corrosion layers on typical surfaces of weathering steel bridges. *Eng. Fail. Anal.* **2016**, *69*, 147–160. [CrossRef]
20. Krivy, V.; Kubzova, M.; Kreislova, K.; Urban, V. Characterization of corrosion products on weathering steel bridges influenced by chloride deposition. *Metals* **2017**, *7*, 336. [CrossRef]
21. Kelly, R.G.; Lee, J.S. Localized Corrosion: Crevice Corrosion. Encyclopedia of Interfacial Chemistry. In *Encyclopedia of Interfacial Chemistry*; Elsevier: Amsterdam, The Netherlands, 2018; pp. 291–301, ISBN 9780128098943.
22. Jones, D.A.; Wilde, B.E. Effect of Alternating Current on the Atmospheric Corrosion of Low-Alloy Weathering Steel in Bolted Lap Joints. *Corrosion* **1987**, *43*, 66–70. [CrossRef]
23. Leygraf, C.; Wallinder, I.O.; Tidblad, J.; Graedel, T. *Atmospheric Corrosion*, 2nd ed.; John Wiley & Sons, Inc.: Hoboken, NJ, USA, 2016.
24. *EN 1993-1-1*; Eurocode 3: Design of Steel Structures—Part 1-1: General Rules and Rules for Buildings. CEN: Brussels, Belgium, 2006.
25. *EN 1993-1-8*; Eurocode 3: Design of Steel Structures—Part 1-8: Design of Joints. CEN: Brussels, Belgium, 2006.
26. Ye, J.; Quan, G.; Yun, X.; Guo, X.; Chen, J. An improved and robust finite element model for simulation of thin-walled steel bolted connections. *Eng. Struct.* **2022**, *250*, 113368. [CrossRef]
27. Belardi, V.G.; Fanelli, P.; Vivio, F. Theoretical definition of a new custom finite element for structural modeling of composite bolted joints. *Compos. Struct.* **2021**, *258*, 113199. [CrossRef]
28. McCarthy, M.A.; McCarthy, C.T.; Lawlor, V.P.; Stanley, W.F. Three-dimensional finite element analysis of single-bolt, single-lap composite bolted joints: Part I—Model development and validation. *Compos. Struct.* **2005**, *71*, 140–158. [CrossRef]
29. Albrecht, P.; Hall, T.T. Atmospheric corrosion resistance of structural steels. *J. Mater. Civ. Eng.* **2003**, *15*, 2–24. [CrossRef]
30. Wang, Z.; Liu, J.; Wu, L.; Han, R.; Sun, Y. Study of the corrosion behavior of weathering steels in atmospheric environments. *Corros. Sci.* **2013**, *67*, 1–10. [CrossRef]
31. Raman, A.; Nasrazadani, S.; Sharma, L. Morphology of rust phases formed on weathering steels in various laboratory corrosion tests. *Metallography* **1989**, *22*, 79–96. [CrossRef]
32. Guo, X.; Zhu, J.; Kang, J.; Duan, M.; Wang, Y. Rust layer adhesion capability and corrosion behavior of weathering steel under tension during initial stages of simulated marine atmospheric corrosion. *Constr. Build. Mater.* **2020**, *234*, 117393. [CrossRef]
33. Strieška, M.; Koteš, P. Sensitivity of dose-response function for carbon steel under various conditions in Slovakia. *Transp. Res. Procedia* **2019**, *40*, 912–919. [CrossRef]
34. *ČSN 41 5217*; Steel Cr-Ni-Cu-P (National Standard). Office for Standards and Measurements: Prague, Czech Republic, 1986. (In Czech)
35. *EN 10025-5*; Hot Rolled Products of Structural Steels—Part 5: Technical Delivery Conditions for Structural Steels with Improved Atmospheric Corrosion Resistance. CEN: Brussels, Belgium, 2020.
36. Rozlívka, L.; Melcher, J. *Loading Tests of Bolted Joints of Lattice Transmission Towers Made of Atmofix Steel*; Research Report; Institute of Steel Structures, Ltd.: Frýdek Místek, Czech Republic; Brno University of Technology: Brno, Czech Republic, 2001. (In Czech)
37. *EN ISO 6892-1*; Metallic Materials—Tensile Testing—Part 1: Method of Test at Room Temperature. CEN: Brussels, Belgium, 2021.
38. Ramberg, W.; Osgood, W.R. *Description of Stress–Strain Curves by Three Parameters*; Technical Note No. 902; National Advisory Committee for Aeronautics: Washington, DC, USA, 1943.
39. *ASTM G101-04:2020*; Standard Guide for Estimating the Atmospheric Corrosion Resistance of Low-Alloy Steels. ASTM International: West Conshohocken, PA, USA, 2020.
40. Krivy, V.; Konecny, P. Real material properties of weathering steels used in bridge structures. *Procedia Eng.* **2013**, *57*, 624–633. [CrossRef]
41. Yamaguchi, E. Maintenance of weathering steel bridges. *Steel Constr. Today Tomorrow* **2015**, *45*, 12–15.
42. Krivy, V.; Kreislova, K.; Rozlivka, L.; Knotkova, D. *Guidelines for the Use of Steels with Improved Atmospheric Corrosion Resistance*; Czech Constructional Steelwork Association: Ostrava, Czech Republic, 2011. (In Czech)
43. Krejslova, K.; Knotkova, D. The Results of 45 Years of Atmospheric Corrosion Study in the Czech Republic. *Materials* **2017**, *10*, 394. [CrossRef] [PubMed]
44. Holický, M.; Viljoen, C.; Retlief, J.V. Assessment of existing structures. In *Advances in Engineering Materials, Structures and Systems: Innovations, Mechanics and Applications*; CRC Press: London, UK, 2019.

Article

Localized Corrosion Occurrence in Low-Carbon Steel Pipe Caused by Microstructural Inhomogeneity

Yun-Ho Lee [1,†], Geon-Il Kim [1,†], Kyung-Min Kim [1], Sang-Jin Ko [1], Woo-Cheol Kim [2] and Jung-Gu Kim [1,*]

1. School of Advanced Materials Science and Engineering, Sungkyunkwan University (SKKU), Suwon 16419, Korea; yunho0228@naver.com (Y.-H.L.); geonil1996@naver.com (G.-I.K.); kkm2628@gmail.com (K.-M.K.); tkdwls1315@naver.com (S.-J.K.)
2. Technical Efficiency Research Team, Korea District Heating Corporation, 92 Gigok-ro, Yongin 06340, Korea; kwc7777@kdhc.co.kr
* Correspondence: kimjg@skku.edu; Tel.: +82-31-290-7360
† These authors contributed equally to this work.

Abstract: In this study, the cause of failure of a low-carbon steel pipe meeting standard KS D 3562 (ASTM A135), in a district heating system was investigated. After 6 years of operation, the pipe failed prematurely due to pitting corrosion, which occurred both inside and outside of the pipe. Pitting corrosion occurred more prominently outside the pipe than inside, where water quality is controlled. The analysis indicated that the pipe failure occurred due to aluminum inclusions and the presence of a pearlite inhomogeneous phase fraction. Crevice corrosion occurred in the vicinity around the aluminum inclusions, causing localized corrosion. In the large pearlite fraction region, cementite in the pearlite acted as a cathode to promote dissolution of surrounding ferrite. Therefore, in the groundwater environment outside of the pipe, localized corrosion occurred due to crevice corrosion by aluminum inclusions, and localized corrosion was accelerated by the large fraction of pearlite around the aluminum inclusions, leading to pipe failure.

Keywords: failure analysis; low-carbon steel pipe; pitting corrosion; aluminum inclusions; pearlite inhomogeneity

1. Introduction

District heating (DH) systems produce heat that is used to provide steam and hot water to the residents of large cities [1,2]. These heating systems provide higher thermal efficiency and lower heating costs than small private boiler units [3]. The steam and hot water that are transported in DH systems expose pipes to corrosion risks. Failure of system pipes due to corrosion reduces thermal efficiency and negatively impacts the system's cost and reliability [1]. As such, failure analyses and pipe corrosion prevention are important for the maintenance of DH systems.

Low-carbon steel has been widely used as a pipe material in various industrial plants such as those that manage oil, gas, and water, as well as in DH systems [1,4–8]. Several studies have been conducted on the failure of low-carbon steel pipes. Kim et al. reported that crack propagation occurs due to stress concentrations and high hydrogen susceptibility in the weld zone (WZ) and heat-affected zone caused by poor welding, resulting in the failure of low-carbon steel pipe [1]. Lee et al. reported that the failure of low-carbon steel pipes are caused by stress corrosion cracking due to chloride presence and residual stress in the WZ [5]. Heyes et al. observed the fatigue cracking as a result of oxygen-induced pitting in the pipe [9]. Various other failure types and mechanisms can be found in available literatures [10].

In most studies, the failure of low-carbon steel pipe occurs in the form of stress corrosion cracking (SCC) or corrosion fatigue cracking (CFC) when stress is applied. However,

the failure shown in this study was due to pitting corrosion. Low-carbon steel is not passivated as stainless steel is, and corrosion generally occurs uniformly over the pipe. Pitting corrosion is rarely observed on the low-carbon steel pipe [11]. Therefore, it is necessary to investigate the cause of pitting corrosion of low-carbon steel pipe to prevent specific corrosion and sudden leakage.

This study analyzes the failure of a low-carbon steel pipe due to corrosion, which has not been reported in the actual use of low-carbon steel pipe. Compliance with material specification was evaluated using inductively coupled plasma atomic emission spectroscopy (ICP-AES) component analysis. The cause of the failure was then analyzed through visual inspection, optical microscopy (OM), metallographic examination, scanning electron microscopy (SEM) with energy dispersive spectroscopy (EDS), electron probe microanalyzer (EPMA) and atomic force microscopy (AFM). Furthermore, the electrochemical properties of the failed low-carbon steel were evaluated using the potentiodynamic polarization and galvanostatic polarization tests.

2. Materials and Methods

2.1. Description of the Pipeline

Before conducting the failure analysis of the low-carbon steel pipe, the specifications and environment were investigated. According to the user's description, the failed low-carbon steel pipe transported hot water through underground pipeline in DH systems. The low-carbon steel pipe satisfied standard KS D 3562 (ASTM A135, Electric-Resistance-Welded Steel Pipe Grade A) [1,12–14]. The outer diameter of the pipe was 457.2 mm with a wall thickness of 6.4 mm. As shown in Figure 1, the buried pipe is surrounded by polyurethane foam as a heat insulator and a high-density polyethylene (HDPE) pipe as an outer casing. Internal corrosion can occur inside of the pipe caused by the transported water in the DH system. External corrosion can occur on the exterior surface of the pipe due to degradation of the HDPE pipe and subsequent penetration of groundwater [15,16]. Tables 1 and 2 present the chemical composition of the DH water inside of the pipe and the synthetic groundwater outside of the pipe, respectively [15,17]. In DH systems, the design service life of a low-carbon steel pipe is 40 years; however, the pipe observed in this study failed after only 6 years of operation [1].

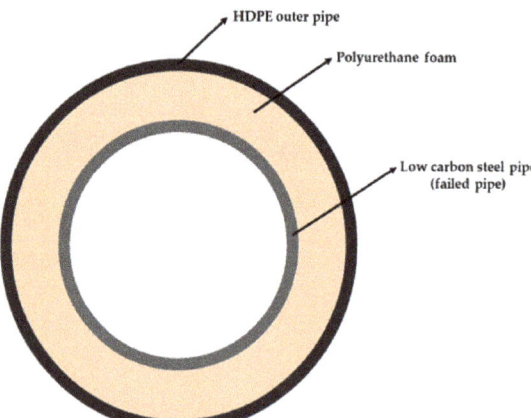

Figure 1. A schematic illustration of the heat transport pipe used in a DH system.

Table 1. Chemical composition of district heating water (ppm) used in a district heating system.

pH	NaCl	Mg(OH)$_2$	CaCO$_3$	NH$_4$OH
9.5	15.01	0.48	2.65	10.28

Table 2. Chemical composition of synthetic ground water (ppm).

pH	CaCl$_2$	MgSO$_4$·7H$_2$O	NaHCO$_3$	H$_2$SO$_4$	HNO$_3$
6.8	133.2	59.0	208.0	48.0	21.8

2.2. Metallurgical Analyses

In this study, it was investigated whether the material used for the pipe satisfies the KS D 3562 standard through chemical composition analysis. The shape of the failed pipe was visually investigated via OM. The metallographic examination was performed to confirm the microstructure uniformity of the pipe metal. To investigate the microstructure of the pipe metal, the specimen was polished using a 1-µm diamond suspension, and thereafter etched by a 2% Nital etching solution for 20 s [1,4]. The volume fractions of pearlite and ferrite phases in the microstructure according to the location of corrosion in the failed pipe were measured by OM using Image J software (version 1.8.0) [18]. For the OM image, the fraction of each phase was calculated by counting the number of pixels in ferrite and pearlite and dividing by the total number of pixels. For microanalysis of the microstructure, topography and surface potential were measured using AFM and kelvin probe force microscopy (KPFM), a mode of AFM. AFM measurements were performed using a commercial AFM system (NX10, Park Systems, Suwon, Korea). KPFM measurements were performed using a conductive Pt/Cr coated tip (Multi75E-G, BudgetSensors, Sofia, Bulgaria) in lift mode with a tip-to-sample distance of 20 nm, and an AC modulation voltage of 2V$_{rms}$ at 17 kHz. Measurements were performed at 10 µm × 10 µm and 2 µm × 2 µm, respectively. The fracture properties were analyzed using SEM/EDS (SEM-7800F Prime, JEOL Ltd., Tokyo, Japan) and EPMA (JXA-8530F, JEOL Ltd., Tokyo, Japan). Prior to SEM/EDS and EPMA analyses, specimens were pickled, polished with 1000-grit size silicon carbide paper, and then rinsed with deionized water and cleaned with ethanol.

2.3. Electrochemical Tests

Potentiodynamic polarization and galvanostatic polarization tests were performed using a VSP 300 (Bio-Logic SAS, Seyssinet-Pariset, France). To conduct these electrochemical tests, a three-electrode system comprising low-carbon steel pipe specimen taken from the failed pipe as the working electrode (WE), two pure graphite rods as the counter electrodes (CE), and a saturated calomel electrode (SCE) with a Luggin capillary as the reference electrode (RE) was used. To confirm the cause of the pitting corrosion, specimens were prepared for the pitting corrosion part (specimen A) and the uniform corrosion part (specimen B) of the failed pipe. For electrochemical tests, all specimens were polished with a 1000-grit silicon carbide paper, rinsed with ethanol, and dried with nitrogen gas. The area of all specimens was controlled to a size of 1 cm^2 using a sealant. A groundwater solution was used for the electrochemical tests (Table 2) because it was more corrosive to the pipe. The temperature was maintained at 60 °C to reflect the temperature of the distinct heating system [2,14,16]. Before conducting the electrochemical tests, the WE was immersed in the test solution for 6 h to obtain a stable open-circuit potential (OCP) as the corrosion potential (E_{corr}). The potentiodynamic polarization tests were conducted using a potential sweep of 0.01 mV/s from −0.25 V vs. E_{corr} to 1.6 V vs. E_{corr}. The galvanostatic polarization tests were performed at 5 mA/cm^2 to accelerate corrosion. The total amount of coulombic charge was 143.86 mAh, which is equivalent to 6 months of uniform corrosion in specimen B. To evaluate the same amount of corrosion, the same coulombic charge was applied to both specimens.

3. Results and Discussion

3.1. Chemical Composition Analysis of the Low-Carbon Steel Pipe

Table 3 shows the chemical composition analysis of the low-carbon steel pipe material, as well as the KS D 3562 standard. It was confirmed that the composition of the failed pipe material complied with the KS D 3562 standard. A key finding of note was the detection of a small amount of Al component in the failed pipe, where KS D 3562 does not dictate any Al component.

Table 3. Chemical compositions of the failed low-carbon steel pipe and KS D 3562 standard (wt. %).

Elements	C	Si	Mn	P	S	Al
Failed pipe	0.08	0.02	0.42	0.011	0006	0.04
KS D 3562	0.25	0.35	0.30–0.90	0.04	0.004	–

3.2. Visual and Macroscopic Inspections

Based on visual inspection of the failed pipe, leakage occurred in the form of pitting, and the severe pitting corrosion occurred near the part of the pipe where the leaking occurred (Figure 2). Figure 3 shows surface images of the pitting corrosion part where leakage occurred, as well as pitting corrosion near the leakage. Figure 4 shows cross-sectional images of the pitting corrosion part where the leakage occurred and the pitting corrosion near the leakage. As stated previously, pitting corrosion occurred on both the exterior and interior of the failed pipe wall. However, since low-carbon steel is a material that does not have passivation, pitting corrosion cannot occur due to the failure of passivation. In low-carbon steel, localized corrosion can be caused by crevice corrosion by specific inclusions, galvanic corrosion between dissimilar metals, and under deposit corrosion (UDC) by solid particles (sand, debris, and iron oxides) [14,19]. In other words, it is necessary to investigate the above possibilities openly.

Figure 2. Photographs of the failed pipe: (**a**) water leakage resulting from pitting in the failed pipe, (**b**) external surface of the failed pipe around the leakage area.

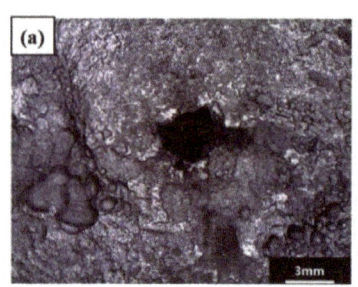
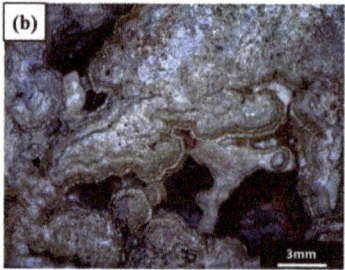
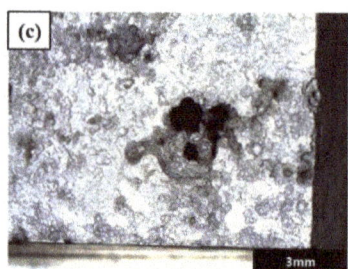

Figure 3. Surface image of the pitting part: (**a**) pitting corrosion area of the leakage (the outside of the pipe), (**b**) pitting corrosion area near the leakage (the outside of the pipe), and (**c**) pitting corrosion area near the leakage (the inside of the pipe).

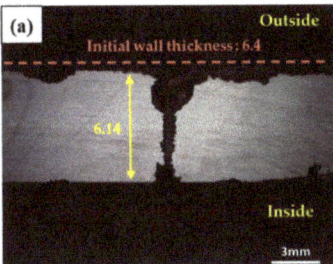 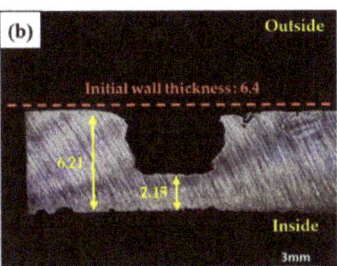 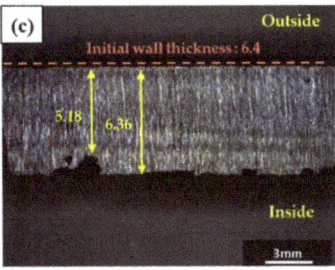

Figure 4. Cross-sectional images of the pitting part: (**a**) pitting corrosion where the leakage occurred, (**b**) pitting corrosion near the leakage, and (**c**) pitting corrosion near the leakage.

Moreover, the depth of the pitting corrosion on the exterior of the failed pipe was significantly deeper than the interior. It is considered to have been due to the difference between the environment of the inside and outside of the pipe (Tables 1 and 2). In addition, to reduce the risk of corrosion, the DH water that flowed through the pipe was maintained at a dissolved oxygen level below 200 ppm through periodic water quality management [15]. In low-carbon steel, localized corrosion hardly occurs when the concentration of dissolved oxygen is low [20]. In addition, in the case of the exterior of the pipe, localized corrosion may be further accelerated due to the non-uniform distribution of groundwater penetrating into the heat insulator. Therefore, it is considered that the corrosion perforation of the pipeline is induced by the external corrosion.

3.3. Metallographic Examinations and Atomic Force Microscopy

Figure 5 shows the microstructure of the pitting corrosion region (specimen A) and the uniform corrosion region (specimen B) of the failed pipe. The darker-colored section is pearlite, which is a layered structure composed of ferrite and cementite, and the brighter-colored section is ferrite [21]. Figure 5 and Table 4 show that there is a larger fraction of pearlite in specimen A than in specimen B. Pearlite is susceptible to corrosion as its constituent phases, ferrite and cementite, have dissimilar electrochemical potentials that cause microgalvanic corrosion when exposed to corrosive electrolytes [21–23].

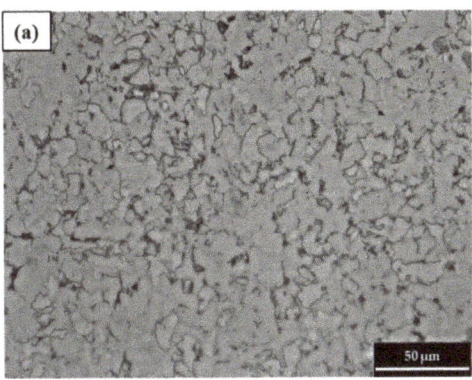 

Figure 5. The optical microstructures of specimens: (**a**) specimen A (pitting corrosion region), and (**b**) specimen B (uniform corrosion region).

Table 4. Volume fraction of the pearlite and ferrite phases according to the specimen A (pitting corrosion region) and specimen B (uniform corrosion region) in the failed pipe.

	Pearlite (%)	Ferrite (%)
Specimen A	13.68 ± 0.58	86.32 ± 0.58
Specimen B	5.57 ± 0.34	94.43 ± 0.34

Figure 6 shows the AFM and KPFM analyses of the pearlite section of the failed pipe. Figure 6a,b show the correlation between pearlite and the surrounding pro-eutectoid ferrite by measuring topology and surface potential within a size of 10 µm × 10 µm, respectively. In Figure 6a, pearlite has a lamella structure, and the surface potential of pearlite (A site) is approximately 44.4 mV higher than that of the surrounding pro-eutectoid ferrite (B site) in Figure 6b and Table 5. This indicates that the corrosion of pro-eutectoid ferrite is locally accelerated by microgalvanic corrosion between pearlite and the surrounding pro-eutectoid ferrite [19,24]. Figure 6c,d show the correlation between cementite and ferrite in the pearlite by measuring topology and surface potential within a size of 2 µm × 2 µm, respectively. Figure 6c shows the lamella structure of pearlite, indicating that the bright protrusions are cementite while the dark region is the ferrite [25]. This is because the corrosion of ferrite was more corroded than that of cementite during the etching process using 2% Nital etching solution. Figure 6d and Table 5 show that the surface potential of cementite in the pearlite (C site) is approximately 34.78 mV higher than that of ferrite in the pearlite (D site). Thus, the corrosion of ferrite is locally accelerated by microgalvanic corrosion between ferrite and cementite in the pearlite [19,24].

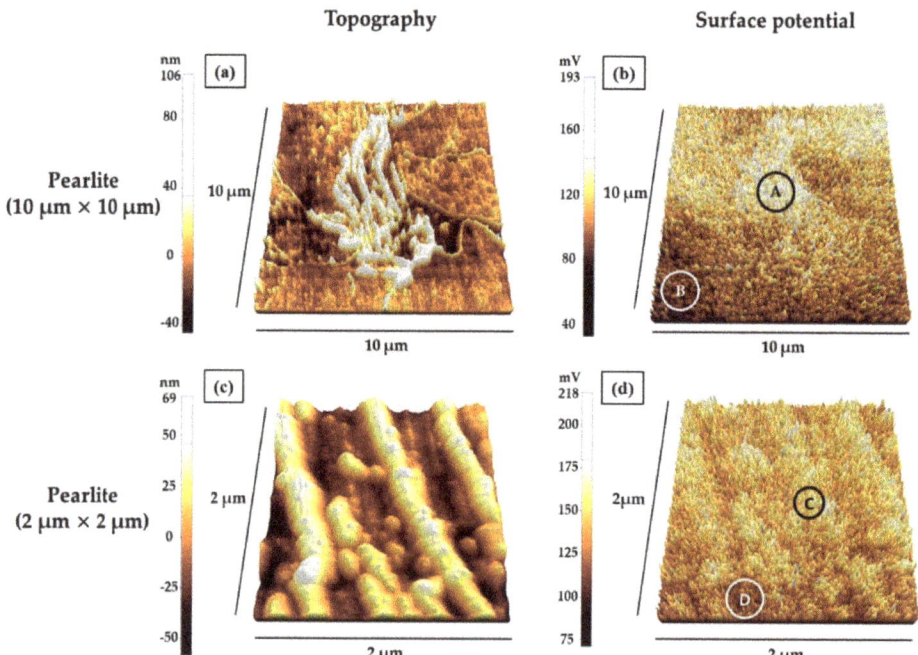

Figure 6. Topography and surface potential of pearlite (specimen A): (**a**) topography of pearlite, (**b**) surface potential of pearlite, (**c**) topography inside of pearlite, and (**d**) surface potential inside of pearlite.

Table 5. Surface potential of different phase area and their differences.

Phase	Position	Potential (mV)		Surface Potential Difference (mV)
		Mean	Dev	
Pearlite	A	124.99	15.85	44.40
Ferrite	B	80.59	13.97	
Cementite in pearlite	C	147.22	13.56	34.78
Ferrite in pearlite	D	112.44	13.26	

Therefore, the inhomogeneity of the microstructure forces a difference in corrosion rate. A pipe section exhibiting pitting corrosion with a lot of pearlite experiences severe corrosion, whereas a uniform corrosion section of the failed pipe with a low pearlite phase incurs only minor thickness reduction.

3.4. Microscopic Analyses

To identify the cause of severe pitting corrosion, microscopic analyses were performed on the pitting corrosion near the leakage area using SEM/EDS and EPMA (Figure 7). Figure 7a,b showed several inclusions around the pitting corrosion site. The EDS elemental mapping of the inclusions in Figure 7c reveals that the particle is primarily composed of Al. The size range of these Al inclusions is 10–20 μm. In the chemical composition analysis, 0.04% Al was contained in the failed pipe, as shown in Table 3.

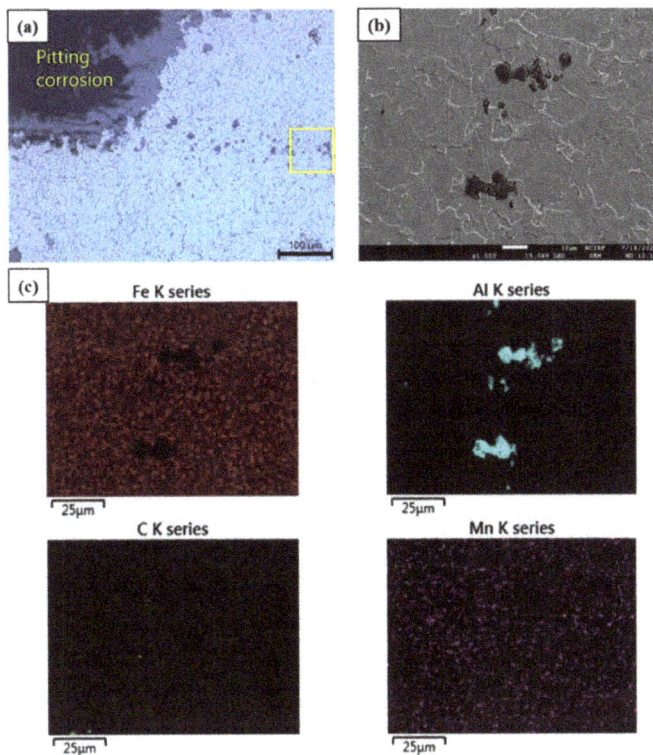

Figure 7. The OM, and SEM/EDS analyses of the pitting corrosion near the location of the leakage; (**a**) OM analysis, (**b**) SEM analysis (yellow box), and (**c**) EDS mapping analysis (yellow box).

Al inclusions were considered to have been the result of Al use as a deoxidizer in the steelmaking process [26]. Al is known to have uniform distribution and fine particle size (less than 2 μm) upon proper heat treatment; accordingly, inclusions typically have little influence on corrosion behavior [27]. However, the failed pipe appeared to have larger-sized Al inclusions as well as an uneven distribution compared with the results in the literature [27,28]. This may have been the result of improper or incomplete homogenization during the steelmaking process. Furthermore, during the pipe manufacturing process, microcrevices formed at matrix–inclusion interfaces due to dissimilarities in the strain values and thermal expansion coefficients, which may have caused crevice corrosion, thereby accelerating localized corrosion [19,29].

Figure 8 shows the EPMA analysis of a cross section where pitting corrosion occurred in the failed pipe. Carbon agglomerations were partially observed in the pitting corrosion part. It appeared that the pitting corrosion region had a larger pearlite fraction than the uniform corrosion region in the failed pipe. Accordingly, the carbon agglomerations occurred due to the high fractional presence of cementite in the pearlite. As shown in the AFM and KPFM analyses results (Figure 6 and Table 5), when there are many pearlite phases, the material is vulnerable to corrosion due to microgalvanic corrosion between pearlite and surrounding pro-eutectoid ferrite, ferrite, and cementite [19,24,30]. The ferrite is corroded locally by microgalvanic corrosion around cementite.

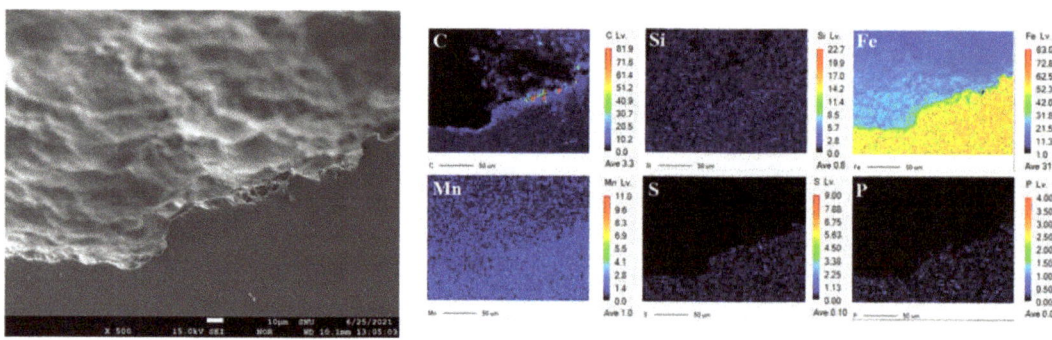

Figure 8. The EPMA analysis of a cross section where pitting corrosion occurred in the failed pipe.

3.5. Open-Circuit Potential Measurement

Figure 9 shows the open-circuit potential (OCP) of specimen A and specimen B in the groundwater solution. Specimen A had a higher E_{corr} than specimen B, and had a relatively large potential fluctuation of approximately 50 mV. Specimen A has a higher E_{corr} due to a larger fraction of pearlite, which has a relatively noble potential compared to ferrite. Pearlite has a higher E_{corr} than ferrite due to an increase in cathodic sites that cause oxygen reduction ($O_2 + 2H_2O + 4e = 4OH^-$) [31,32]. In addition, due to the higher corrosion activity caused by larger fraction of pearlite and presence of Al inclusions, OCP fluctuation is shown on specimen A [33,34].

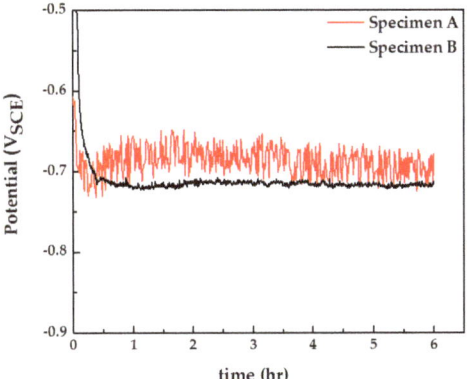

Figure 9. Open-circuit potential of specimen A (pitting corrosion region) and specimen B (uniform corrosion region) with immersion time in the groundwater at 60 °C.

3.6. Potentiodynamic Polarization Test

Figure 10 and Table 6 show the results of the potentiodynamic polarization test of specimen A and specimen B in the groundwater solution. The corrosion current density was controlled by the oxygen reduction reaction [31]. The corrosion current density was analyzed using the Tafel extrapolation method. Once the corrosion current density was determined, the corrosion rate can be calculated using the following equation [11]:

$$\text{Corrosion rate (mm/year)} = 0.00327 \frac{a \cdot i_{corr}}{n \cdot D} \quad (1)$$

where a is the atomic weight, i_{corr} is the corrosion current density, n is the number of equivalents exchanged, and D is the density of the low-carbon steel. Specimen A had a corrosion current density twice that of specimen B. The high corrosion rate in specimen A is due to the accelerated corrosion caused by Al inclusions and the larger pearlite phase fraction. In the vicinity of Al inclusions, localized corrosion occurs due to crevice corrosion, increasing the corrosion rate [19]. Larger pearlite phase fraction accelerates the corrosion rate via galvanic corrosion between pearlite and pro-eutectoid ferrite and between cementite and ferrite in pearlite [31]. In addition, when a larger fraction of pearlite exists around the Al inclusion, corrosion is accelerated by the larger fraction of pearlite, and aggressive ions, such as the Cl^- ion, are concentrated around the Al inclusion. This further accelerates crevice corrosion in Al inclusions. If there is no crevice around the Al inclusions, crevice corrosion does not occur. In addition, when pitting formed as crevice corrosion and galvanic corrosion progressed around the Al inclusions and the pearlite, the surface area became wider than the initial area due to morphological changes [29,35]. Equation (2) shows the anode current density according to the anodic overpotential in the activation polarization [11].

$$i_a = i_0 \exp\left(\frac{\alpha \cdot n \cdot F \cdot \eta_a}{2.3 \cdot R \cdot T}\right) \quad (2)$$

Table 6. The electrochemical parameters resulting from the polarization measurements of specimen A and specimen B in the groundwater at 60 °C.

	E_{corr} (mV$_{SCE}$)	i_{corr} (A/cm^2)	Corrosion Rate (mm/yr)
Specimen A	−621.04	6.47×10^{-5}	0.75
Specimen B	−704.63	3.33×10^{-5}	0.39

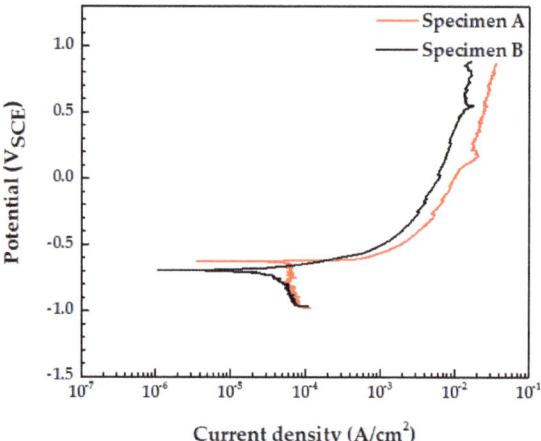

Figure 10. Potentiodynamic polarization curves of specimen A (pitting corrosion region) and specimen B (uniform corrosion region) in the groundwater at 60 °C.

In Equation (2), i_a is the current density by the anodic overpotential, i_0 is the exchange current density, α is the fraction of η_a taken by the ionization reaction, n is the number of equivalent exchanged, F is the Faraday's constant, η_a is the anodic overpotential, R is the gas constant, and T is the temperature. Equation (3) shows the correlation between area and current.

$$i = \frac{I}{A} \quad (3)$$

In Equation (3), i is the current density, A is the reaction area, and I is the current. When the same overpotential was applied, the generated current (I) increased in proportion to the increased area. However, to obtain the current density, the reaction area (A) is equally divided by 1 cm^2 for the generated current (I). Therefore, the change in roughness due to localized corrosion of specimen A causes higher current density on the potentiodynamic polarization curve.

3.7. Galvanostatic Polarization Test

The galvanostatic polarization test was performed to accelerate corrosion. The acceleration time for the galvanostatic polarization test was calculated using the Faraday's law as shown below [15]:

$$i_{real} \cdot t_{real} = \frac{m \cdot F \cdot n}{a} = i_{accelerated} \cdot t_{accelerated} \quad (4)$$

where m is the reacted mass (g), i is the current density (A/cm^2), t is the time (s), a is the atomic weight (g/mol), F is the Faraday's constant (96,500 C/mol), and n is the number of electrons exchanged. The same coulombic charge was applied to observe the corrosion behavior for the same amount of corrosion. The total coulombic charge was 143.86 mAh, and the applied current was 5 mA/cm^2. Figure 11 shows surface and cross-sectional images of specimen A and specimen B after galvanostatic polarization test. Pitting corrosion occurred in specimen A, and uniform corrosion occurred in specimen B. This indicates that pitting corrosion is related to the presence of Al inclusions and inhomogeneity of the pearlite.

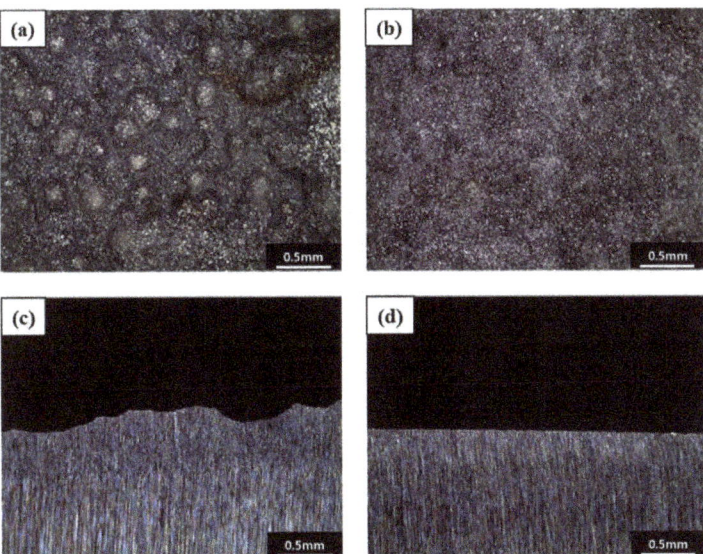

Figure 11. Surface and cross-sectional images of the specimens after galvanostatic polarization test: (**a**) surface image of specimen A (pitting corrosion region), (**b**) surface image of specimen B (uniform corrosion region), (**c**) cross-sectional image of the specimen A, and (**d**) cross-sectional image of the specimen B.

3.8. Mechanism

Figure 12 shows the failure mechanism of the failed low-carbon steel pipe due to the Al inclusions and a large amount of pearlite formed locally during the steelmaking process. Due to the coefficient of thermal expansion differences between Al and Fe, microcrevices form around Al inclusions during the pipe manufacturing process. As crevice corrosion is initiated, pH drops and Cl$^-$ ions are concentrated in the microcrevice to maintain charge neutrality [19]. The concentration of Cl$^-$ ions further accelerates crevice corrosion and corrosion products accumulate on this part of the pipe. Additionally, oxygen-concentration cells form, which accelerate localized corrosion, and eventually the Al inclusions fall off [19,29].

The large fraction of pearlite has a higher corrosion rate due to microgalvanic corrosion between the surrounding pro-eutectoid ferrite and pearlite, and between cementite and ferrite in pearlite [19,24]. Due to the relatively accelerated corrosion rate of the large fraction of pearlite near the Al inclusions, the concentration of Cl$^-$ ions into Al vicinity and the accumulation of corrosion products are accelerated. This promotes the formation of an oxygen-concentration cell in the vicinity of the Al inclusion.

However, the presence of a large fraction of pearlite alone cannot cause localized corrosion such as pitting. The cementite inside the pearlite locally accelerates the surrounding ferrite corrosion by microgalvanic corrosion, but over time, the rust is covered by the dissolution of ferrite, and the corrosion proceeds in the form of a uniform corrosion [19]. In other words, a large fraction of pearlite accelerates corrosion locally, but cannot lead to pitting corrosion. That is, the large fraction of pearlite accelerates crevice corrosion via Al inclusions and promotes the formation of oxygen-concentration cell, thereby accelerating localized corrosion in the vicinity of the Al inclusions.

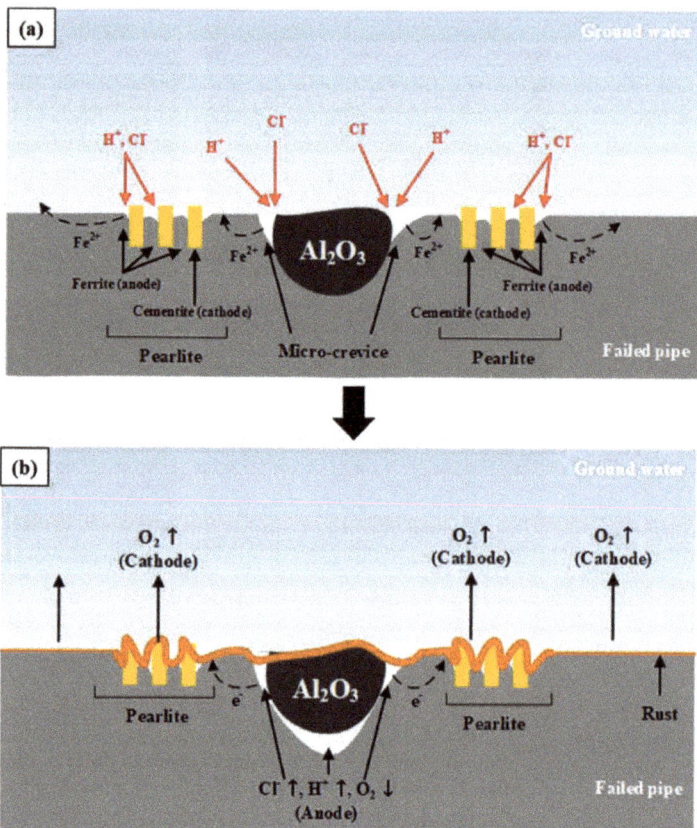

Figure 12. Failure mechanism of the failed low-carbon steel pipe based on aluminum inclusion and the larger phase fraction of the pearlite: (**a**) initial stage, and (**b**) later stage.

4. Conclusions

In this study, the failure analysis of a low-carbon steel pipe used in DH system was investigated using visual inspection, ICP-AES, OM, AFM, SEM/EDS, EPMA, and electrochemical tests. According to the results of the failure analysis, the following conclusions were drawn.

- Leakage occurred in the form of pitting corrosion, which was observed both inside and outside of the failed pipe. In particular, severe pitting corrosion occurred on the outside of the pipe, exposed to the soil environment. Al inclusions and a larger phase fraction of pearlite were observed near the leaking section. Crevice corrosion occurred in the microcrevice around the Al inclusions, and the large phase of pearlite around Al inclusions accelerated the localized corrosion in the microcrevice. Localized corrosion was accelerated near the Al inclusions and the large fraction of pearlite in the groundwater environment outside of the pipe, resulting in the pipe's failure.
- The corrosion rate of the specimen taken where the pitting corrosion was present in the failed pipe was approximately double that of the specimen taken from the uniform corrosion part of the failed pipe. Furthermore, the corrosion type was similar to that observed in the actual failed pipe. This confirms the pipe failure had been caused by Al inclusions and the inhomogeneity of the pearlite.

5. Recommendations

- It is recommended that the uniform distribution of fine-sized pearlite and Al inclusions be produced through proper liquid steel homogenization and heat treatment during the steelmaking process.
- It is recommended that a standard for the chemical composition of Al be established within the existing KS D 3562 standard.

Author Contributions: Conceptualization, Y.-H.L. and G.-I.K.; methodology, Y.-H.L. and G.-I.K.; validation, Y.-H.L., G.-I.K., K.-M.K., S.-J.K., W.-C.K. and J.-G.K.; formal analysis, Y.-H.L. and G.-I.K.; investigation, Y.-H.L. and G.-I.K.; resources, Y.-H.L. and G.-I.K.; data curation, Y.-H.L., G.-I.K., K.-M.K. and S.-J.K.; writing—original draft preparation, Y.-H.L. and G.-I.K.; writing—review and editing, Y.-H.L., G.-I.K., K.-M.K., S.-J.K., W.-C.K. and J.-G.K.; visualization, Y.-H.L. and G.-I.K.; supervision, J.-G.K.; project administration, J.-G.K.; funding acquisition, W.-C.K. and J.-G.K. All authors have read and agreed to the published version of the manuscript.

Funding: This research was supported by the Korea District Heating Corporation (No. 0000000014524).

Institutional Review Board Statement: Not applicable.

Informed Consent Statement: Not applicable.

Data Availability Statement: Not applicable.

Conflicts of Interest: The authors declare no conflict of interest.

References

1. Kim, Y.-S.; Kim, J.-G. Failure analysis of a thermally insulated pipeline in a district heating system. *Eng. Fail. Anal.* **2018**, *83*, 193–206. [CrossRef]
2. Song, S.-J.; Cho, S.; Kim, W.-C.; Kim, J.-G. Failure analysis of electric-heater tube for heat-storage tank. *Eng. Fail. Anal.* **2018**, *87*, 69–79. [CrossRef]
3. Benonysson, A.; Bøhm, B.; Ravn, H.F. Operational optimization in a district heating system. *Energy Convers. Manag.* **1995**, *36*, 297–314. [CrossRef]
4. Lee, J.; Han, S.; Kim, K.; Kim, H.; Lee, U. Failure analysis of carbon steel pipes used for underground condensate pipeline in the power station. *Eng. Fail. Anal.* **2013**, *34*, 300–307. [CrossRef]
5. Lee, D.Y.; Kim, W.C.; Kim, J.G. Effect of nitrite concentration on the corrosion behaviour of carbon steel pipelines in synthetic tap water. *Corros. Sci.* **2012**, *64*, 105–114. [CrossRef]
6. Tavares, S.; Pardal, J.; Mainier, F.; Da Igreja, H.; Barbosa, E.; Rodrigues, C.; Barbosa, C.; Pardal, J. Investigation of the failure in a pipe of produced water from an oil separator due to internal localized corrosion. *Eng. Fail. Anal.* **2016**, *61*, 100–107. [CrossRef]
7. Bolzon, G.; Rivolta, B.; Nykyforchyn, H.; Zvirko, O. Mechanical analysis at different scales of gas pipelines. *Eng. Fail. Anal.* **2018**, *90*, 434–439. [CrossRef]
8. Mohtadi-Bonab, M.; Eskandari, M. A focus on different factors affecting hydrogen induced cracking in oil and natural gas pipeline steel. *Eng. Fail. Anal.* **2017**, *79*, 351–360. [CrossRef]
9. Heyes, A. Oxygen pitting failure of a bagasse boiler tube. *Eng. Fail. Anal.* **2001**, *8*, 123–131. [CrossRef]
10. Duarte, C.A.; Espejo, E.; Martinez, J.C. Failure analysis of the wall tubes of a water-tube boiler. *Eng. Fail. Anal.* **2017**, *79*, 704–713. [CrossRef]
11. Jones, D.A. *Principles and Prevention of Corrosion*; Prentice Hall: Upper Saddle River, NJ, USA, 1996.
12. ASTM. *A135: Standard Specification for Electric-Resistance-Welded Steel Pipe*; ASTM: West Conshohocken, PA, USA, 2014.
13. KS D. *3562: Carbon Steel Pipes for Pressure Service*; KS D: Eumseong, Korea, 2021.
14. Kim, Y.-S.; Kim, J.-G. Corrosion behavior of pipeline carbon steel under different iron oxide deposits in the district heating system. *Metals* **2017**, *7*, 182. [CrossRef]
15. Hong, M.-S.; So, Y.-S.; Lim, J.-M.; Kim, J.-G. Evaluation of internal corrosion property in district heating pipeline using fracture mechanics and electrochemical acceleration kinetics. *J. Ind. Eng. Chem.* **2021**, *94*, 253–263. [CrossRef]
16. Choi, Y.-S.; Chung, M.-K.; Kim, J.-G. Effects of cyclic stress and insulation on the corrosion fatigue properties of thermally insulated pipeline. *Mater. Sci. Eng. A* **2004**, *384*, 47–56. [CrossRef]
17. Chung, N.-T.; Hong, M.-S.; Kim, J.-G. Optimizing the Required Cathodic Protection Current for Pre-Buried Pipelines Using Electrochemical Acceleration Methods. *Materials* **2021**, *14*, 579. [CrossRef] [PubMed]
18. Katiyar, P.K.; Sangal, S.; Mondal, K. Effect of various phase fraction of bainite, intercritical ferrite, retained austenite and pearlite on the corrosion behavior of multiphase steels. *Corros. Sci.* **2021**, *178*, 109043.
19. Liu, C.; Cheng, X.; Dai, Z.; Liu, R.; Li, Z.; Cui, L.; Chen, M.; Ke, L. Synergistic effect of Al_2O_3 inclusion and pearlite on the localized corrosion evolution process of carbon steel in marine environment. *Materials* **2018**, *11*, 2277. [CrossRef]

20. Xue, F.; Wei, X.; Dong, J.; Etim, I.-I.N.; Wang, C.; Ke, W. Effect of residual dissolved oxygen on the corrosion behavior of low carbon steel in 0.1 M NaHCO$_3$ solution. *J. Mater. Sci. Technol.* **2018**, *34*, 1349–1358. [CrossRef]
21. Liu, H.; Wei, J.; Dong, J.; Chen, Y.; Wu, Y.; Zhou, Y.; Babu, S.D.; Ke, W. Influence of cementite spheroidization on relieving the micro-galvanic effect of ferrite-pearlite steel in acidic chloride environment. *J. Mater. Sci. Technol.* **2021**, *61*, 234–246. [CrossRef]
22. Katiyar, P.K.; Misra, S.; Mondal, K. Comparative corrosion behavior of five microstructures (pearlite, bainite, spheroidized, martensite, and tempered martensite) made from a high carbon steel. *Metall. Mater. Trans. A* **2019**, *50*, 1489–1501. [CrossRef]
23. Hao, X.; Dong, J.; Etim, I.-I.N.; Wei, J.; Ke, W. Sustained effect of remaining cementite on the corrosion behavior of ferrite-pearlite steel under the simulated bottom plate environment of cargo oil tank. *Corros. Sci.* **2016**, *110*, 296–304. [CrossRef]
24. Sun, C.; Ko, S.-J.; Jung, S.; Wang, C.; Lee, D.; Kim, J.-G.; Kim, Y. Visualization of electrochemical behavior in carbon steel assisted by machine learning. *Appl. Surf. Sci.* **2021**, *563*, 150412. [CrossRef]
25. Katiyar, P.K.; Misra, S.; Mondal, K. Corrosion behavior of annealed steels with different carbon contents (0.002, 0.17, 0.43 and 0.7% C) in freely aerated 3.5% NaCl solution. *J. Mater. Eng. Perform.* **2019**, *28*, 4041–4052. [CrossRef]
26. Väinölä, R.; Holappa, L.; Karvonen, P. Modern steelmaking technology for special steels. *J. Mater. Process. Technol.* **1995**, *53*, 453–465. [CrossRef]
27. Wakoh, M.; Sano, N. Behavior of alumina inclusions just after deoxidation. *ISIJ Int.* **2007**, *47*, 627–632. [CrossRef]
28. Wang, Y.; Cheng, G.; Wu, W.; Li, Y. Role of inclusions in the pitting initiation of pipeline steel and the effect of electron irradiation in SEM. *Corros. Sci.* **2018**, *130*, 252–260. [CrossRef]
29. Villavicencio, J.; Ulloa, N.; Lozada, L.; Moreno, M.; Castro, L. The role of non-metallic Al$_2$O$_3$ inclusions, heat treatments and microstructure on the corrosion resistance of an API 5L X42 steel. *J. Mater. Res. Technol.* **2020**, *9*, 5894–5911. [CrossRef]
30. Wei, J.; Dong, J.; Zhou, Y.; He, X.; Wang, C.; Ke, W. Influence of the secondary phase on micro galvanic corrosion of low carbon bainitic steel in NaCl solution. *Mater. Charact.* **2018**, *139*, 401–410. [CrossRef]
31. Katiyar, P.K.; Misra, S.; Mondal, K. Effect of different cooling rates on the corrosion behavior of high-carbon pearlitic steel. *J. Mater. Eng. Perform.* **2018**, *27*, 1753–1762. [CrossRef]
32. Hao, X.; Dong, J.; Mu, X.; Wei, J.; Wang, C.; Ke, W. Influence of Sn and Mo on corrosion behavior of ferrite-pearlite steel in the simulated bottom plate environment of cargo oil tank. *J. Mater. Sci. Technol.* **2019**, *35*, 799–811. [CrossRef]
33. Kadowaki, M.; Muto, I.; Katayama, H.; Masuda, H.; Sugawara, Y.; Hara, N. Effectiveness of an intercritical heat-treatment on localized corrosion resistance at the microstructural boundaries of medium-carbon steels. *Corros. Sci.* **2019**, *154*, 159–177. [CrossRef]
34. Xia, D.-H.; Song, S.; Behnamian, Y.; Hu, W.; Cheng, Y.F.; Luo, J.-L.; Huet, F. electrochemical noise applied in corrosion science: Theoretical and mathematical models towards quantitative analysis. *J. Electrochem. Soc.* **2020**, *167*, 081507. [CrossRef]
35. Kim, S.K.; Park, I.J.; Lee, D.Y.; Kim, J.G. Influence of surface roughness on the electrochemical behavior of carbon steel. *J. Appl. Electrochem.* **2013**, *43*, 507–514. [CrossRef]

Article

Method for Mitigating Stray Current Corrosion in Buried Pipelines Using Calcareous Deposits

Sin-Jae Kang [1], Min-Sung Hong [2] and Jung-Gu Kim [1,*]

[1] School of Advanced Materials Science and Engineering, Sungkyunkwan University (SKKU), Suwon 16419, Korea; paul7751@hanmail.net
[2] Department of Nuclear Engineering, University of California at Berkeley, Berkeley, CA 94720, USA; mshong@berkeley.edu
* Correspondence: kimjg@skku.edu; Tel.: +82-31-290-7360

Citation: Kang, S.-J.; Hong, M.-S.; Kim, J.-G. Method for Mitigating Stray Current Corrosion in Buried Pipelines Using Calcareous Deposits. *Materials* **2021**, *14*, 7905. https://doi.org/10.3390/ma14247905

Academic Editor: Vít Křivý

Received: 10 November 2021
Accepted: 18 December 2021
Published: 20 December 2021

Publisher's Note: MDPI stays neutral with regard to jurisdictional claims in published maps and institutional affiliations.

Copyright: © 2021 by the authors. Licensee MDPI, Basel, Switzerland. This article is an open access article distributed under the terms and conditions of the Creative Commons Attribution (CC BY) license (https://creativecommons.org/licenses/by/4.0/).

Abstract: Stray current corrosion in buried pipelines can cause serious material damage in a short period of time. However, the available methods for mitigating stray current corrosion are still insufficient. In this study, as a countermeasure against stray current corrosion, calcareous depositions were applied to reduce the total amount of current flowing into pipelines and to prevent corrosion. This study examined the reduction of stray current corrosion via the formation of calcareous deposit layers, composed of Ca, Mg, and mixed Ca and Mg, at the current inflow area. To verify the deposited layers, scanning electron microscopy (SEM), energy dispersive X-ray spectroscopy (EDS), and X-ray diffraction (XRD) were performed. The electrochemical tests revealed that all three types of calcareous deposits were able to effectively act as current barriers, and that they decreased the inflow current at the cathodic site. Among the deposits, the $CaCO_3$ layer mitigated the stray current most effectively, as it was not affected by $Mg(OH)_2$, which interferes with the growth of $CaCO_3$. The calcium-based layer was very thick and dense, and it effectively blocked the inflowing stray current, compared with the other layers.

Keywords: stray current corrosion; pipeline; calcareous deposit; corrosion mitigation; cathodic protection

1. Introduction

Stray current corrosion, which is a drastic corrosion phenomenon due to external current sources, can cause serious damage to buried pipelines in a short period of time. With the increasing number of buildings, facilities, and subways that use high voltages in modern society, the amount of stray current is also increasing, inducing more stray current corrosion in buried pipelines [1–6]. In particular, stray current from subways, power towers, and high-voltage facilities flows into buried pipelines that have lower resistances than soil. The area into which the stray current flows is negatively charged, resulting in anticorrosion, while the area out of which it flows is positively charged, resulting in corrosion [1,7,8]. The inflow and outflow of the stray current occur at random areas throughout pipelines, making it difficult to detect, and difficult to prevent the associated corrosion [9–11]. In addition, most pipelines are usually installed in urban areas, where stray current can be introduced to the pipelines, causing both economic and human-related losses [12,13]. In the UK, GBP 500 million is spent annually on infrastructure restoration and repair due to stray current corrosion [14]. To solve this issue, drainage systems and electrical shields have been applied to buried pipelines to prevent corrosion [1,15–17]. However, they are expensive and cannot be applied to all pipelines. Since it is hard to predict where stray current corrosion will occur, it is difficult to ensure that all pipelines are protected.

To overcome these issues, we applied calcareous deposits for protection against stray current corrosion. Generally, calcareous deposits are a type of combined deposit based on calcium (Ca) and magnesium (Mg), and they are generated under cathodic protection in seawater. Generally, calcareous deposits act as electrical barriers, and they therefore have the potential to be an excellent solution with regard to mitigating stray current corrosion. The cathodic sites of pipelines, where the inflow current is introduced, have the same conditions when cathodically protected. Therefore, the supply of Ca and Mg can generate a calcareous deposit in a soil environment [18–22].

In this study, potentiostatic polarization tests were performed to form calcareous deposits with different compositions [23]. After forming the calcareous deposits, the surface morphology was analyzed using scanning electron microscopy (SEM), energy dispersive X-ray spectroscopy (EDS), and X-ray diffraction (XRD). Electrochemical impedance spectroscopy (EIS) experiments were also performed after each potentiostatic polarization test. Finally, potentiostatic acceleration tests were undertaken to verify the effects of the calcareous deposits on the stray current corrosion, and to determine the most effective composition for protection against it.

2. Materials and Methods

2.1. Specimen and Solution Preparation

As shown in Table 1, the specimens used in this study were made of SPW-400 (low-carbon steel), which is the most common material for pipelines.

Table 1. Chemical composition of SPW 400 (wt.%).

Fe	C	P	S	Si	Mn
Bal	0.130 max.	0.018 max.	0.070 max.	0.240 max.	0.560 max.

The SPW-400 was cut into square sections (10 mm × 10 mm × 2 mm), which were used as the working electrodes (WE). The specimens were ground with SiC paper (600-grit), after which they were cleaned with deionized water, and then dried with N_2 gas. Table 2 lists the chemical composition of the synthetic soil solution used in the experiments. For the formation of the calcareous deposits using potentiostatic polarization tests, Ca and Mg, which are the main components of these deposits, were added to the synthetic soil solution, separately and together. The elements were based on the following chemicals: $Mg(OH)_2$ (Mg: 1000 ppm), $CaCO_3$ (Ca: 1000 ppm), and Ca and Mg (500 ppm each).

Table 2. Chemical composition of synthetic soil solution (ppm).

$CaCl_2$	$MgSO_4\ 7H_2O$	$NaHCO_3$	H_2SO_4	HNO_3
133.2	59.0	208.0	48.0	21.8

2.2. Formation of Calcareous Deposits

A potentiostatic polarization test using a three-electrode system was performed to form the calcareous deposits. The specimens were connected to a WE, a carbon rod was used as the counter electrode (CE), and a saturated calomel electrode (SCE) was used as the reference electrode (RE). The area of each test specimen exposed to electrolytes was 1 cm². The open-circuit potential (OCP) was established within 30 min, after which the electrochemical tests were performed. Potentiostatic polarization tests were undertaken to form the calcareous depositions. The tests were performed at −1.0 V_{SCE} to put the specimens in the cathodic state. The current was inflowed for over 30 h at room temperature (25 °C), while the solution was rotated at 350 rpm.

2.3. Surface Analyses

The surface analysis of the experimental specimens was performed after the potentiostatic polarization tests. The morphology and the cross-sectional images of the calcareous deposits were observed using SEM and SEM/EDS (JSM-7900F, JEOL Ltd., Tokyo, Japan) to verify the type of calcareous deposit on the specimen. XRD (Dmax-2500V/PC, Rigaku,

Tokyo, Japan) measurements were also performed on the calcareous deposits to verify their types. The XRD analysis was conducted using Cu Kα radiation (λ = 1.54056 Å), in a 2θ range of 0–60°, at a scan rate of 0.02.

2.4. Electrochemical Test

EIS measurements were performed in a frequency range of 100 kHz–10 mHz, with a 10-mV amplitude. The impedance plots were interpreted on the basis of an equivalent circuit, using a fitting procedure performed by ZsimpWin software (ZsimpWin 3.20, Echem Software, Warminster, PA, USA). Stray current corrosion tests were performed in a stray current simulation cell, as shown in Figure 1. The 304 stainless steel rods used as the CE were enclosed with insulating tape to reduce the current dispersion, and the SCE was used as the RE. The specimens used for the inflow part of the current and those used for the outflow part of the current were electrically connected to each other. The tests were conducted in a synthetic soil solution, and 3.5 V_{SCE} was applied for 100 h at room temperature (25 °C). The specimens for the outflow part of the current were weighed and recorded before the potentiostatic acceleration tests. After the tests, the specimens were cleaned, rinsed, and reweighed. All electrochemical tests were performed using a VSP-300 model potentiostat (Biologic SAS, Seyssinet-Pariset, France).

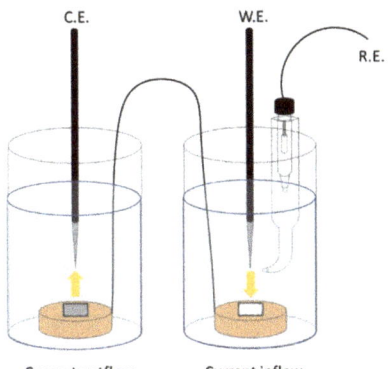

Figure 1. Schematic of stray current simulation cell.

3. Results

3.1. Formation of Calcareous Deposits

The potentiostatic tests were performed at −1.0 V_{SCE} for 30 h to form three kinds of calcareous deposits on the carbon steel. Figure 2 shows the potentiostatic test results over the 30-h period. The current density decreased with time for all solution types. This can be explained by the electrochemical reactions on the surfaces of the cathodic site. When the specimens were negatively charged in the solution containing both Ca and Mg, the dissolved oxygens were converted into OH^- ions, leading to an increase in the pH on the surface. Because of the increasing number of OH^- ions, Mg ions reacted with them, forming an $Mg(OH)_2$ deposition on the metal surface. In addition, the increase in OH^- ions affected the carbonate equilibrium at the metal surface. Thus, a $CaCO_3$ layer was deposited on the metal surface. These processes can be described by the following reactions [21–25]:

$$O_2 + 2H_2O + 4e^- \rightarrow 4OH^- \quad (1)$$

$$Mg^{2+} + 2OH^- \rightarrow Mg(OH)_2 \text{ (s)} \quad (2)$$

$$OH^- + HCO_3^- \rightarrow CO_3^{2-} + H_2O \quad (3)$$

$$Ca^{2+} + CO_3^{2-} \rightarrow CaCO_3 \text{ (s)} \quad (4)$$

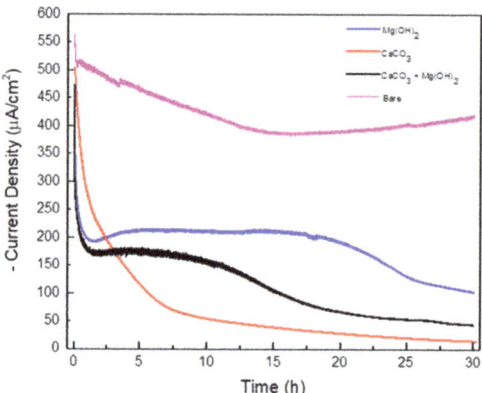

Figure 2. Current density vs. time curves during potentiostatic test (applied voltage: −1.0 V$_{SCE}$; solutions: synthetic soil solution with Mg(OH)$_2$ CaCO$_3$ and Mg(OH)$_2$, CaCO$_3$; testing time: 30 h).

These calcareous deposits decreased the O$_2$ diffusion to the metal surface as a physical and electrical coating layer, and hindered the oxygen reduction reaction [23,26]. Therefore, the current density was decreased because of the formation of calcareous deposits on the metal surface, as shown in Figure 2. However, in the case of the solution with only Mg, the Mg(OH)$_2$ layer was porous and gel-like rather than solid. It offered a relatively lower protective property at the surface of the metal compared with the other layers, and it did not significantly decrease the current inflow to the metal [27]. Therefore, the specimen that deposited only Mg(OH)$_2$ had the highest current density. In contrast, the CaCO$_3$ layer has the property of forming a solid and dense layer. Because the CaCO$_3$ layer with these properties grew without any interference, the current density decreased rapidly to the lowest current density value measured in this study [28]. The specimen that deposited both CaCO$_3$ and Mg(OH)$_2$ had a current density higher than that of the specimen with only CaCO$_3$, and a current density lower than that of the specimen with only Mg(OH)$_2$. This is because the Mg(OH)$_2$ hindered the growth of the CaCO$_3$, meaning that the formed CaCO$_3$ layer was thin and unstable [28,29].

3.2. Surface Analysis

The cross-sectional SEM images and EDS mapping results of the calcareous deposits after the 30-h potentiostatic polarization tests are shown in Figures 3–5. Figure 3 shows this information for the calcareous deposit based only on Mg, revealing a thin Mg(OH)$_2$ layer deposited on the carbon steel. Figure 4 shows this information for the deposit based only on Ca, revealing a relatively thicker CaCO$_3$ layer deposited on the carbon steel than the other deposit layers. Figure 5 shows the SEM image and EDS mapping results of the mixed CaCO$_3$ and Mg(OH)$_2$ deposit. This calcareous deposit layer included both Ca and Mg, and can therefore be regarded as a combined CaCO$_3$ and Mg(OH)$_2$ layer. In addition, it was confirmed that the Mg(OH)$_2$, which hindered the growth of the CaCO$_3$, resulted in a thinner CaCO$_3$ layer compared with the specimen containing only CaCO$_3$ [28,29].

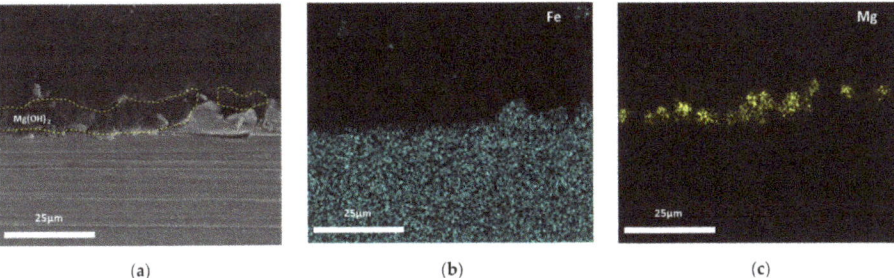

Figure 3. The cross-sectional SEM images and EDS mapping results of $Mg(OH)_2$: (**a**) cross-sectional image; (**b**) Fe (EDS mapping); and (**c**) Mg (EDS mapping).

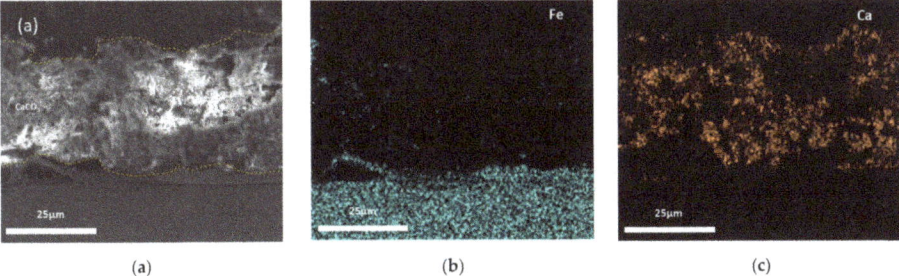

Figure 4. The cross-sectional SEM images and EDS mapping results of $CaCO_3$: (**a**) cross-sectional image; (**b**) Fe (EDS mapping); and (**c**) Ca (EDS mapping).

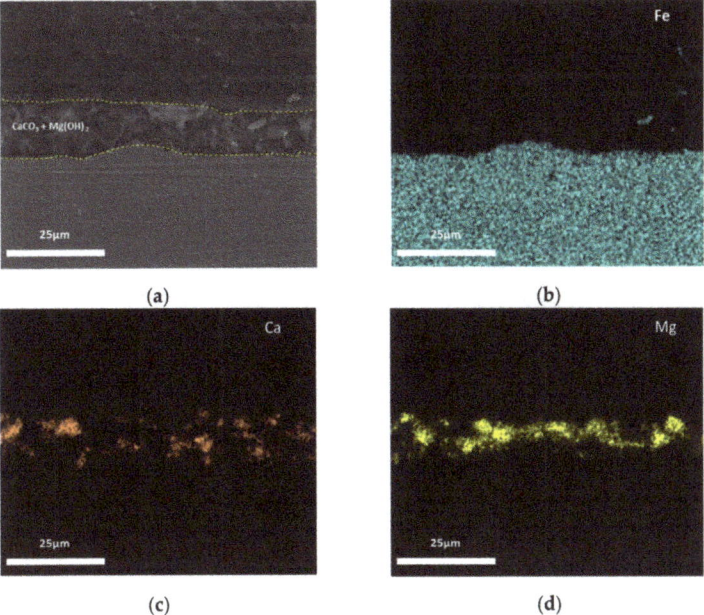

Figure 5. The cross-sectional SEM images and EDS mapping results of $CaCO_3 + Mg(OH)_2$: (**a**) cross-sectional image; (**b**) Fe (EDS mapping); (**c**) Ca (EDS mapping); and (**d**) Mg (EDS mapping).

Figure 6 shows the XRD patterns used to verify the three types of calcareous deposits on the surfaces of the specimens. It was confirmed that the calcareous deposit layer from the Mg(OH)$_2$-added soil solution was Mg(OH)$_2$. In addition, the calcareous deposit layer from the CaCO$_3$-added soil solution was CaCO$_3$. Finally, the calcareous deposit layer from the CaCO$_3$ and Mg(OH)$_2$-added soil solution consisted of CaCO$_3$ and Mg(OH)$_2$. When the calcareous deposition layer is formed in a solution to which CaCO$_3$ and Mg(OH)$_2$ are added, not only CaCO$_3$ and Mg(OH)$_2$ are formed, but (Ca,Mg)CO$_3$ (JCPDS 43-0697), which is similar to the peaks of CaCO$_3$ (JCPDS 05-0586), is also formed as the product. The width of the peaks of the calcareous deposit layer from the CaCO$_3$ and Mg(OH)$_2$-added soil solution are greater than that of the peaks in the other two patterns because of the overlapping XRD peaks of the (Ca,Mg)CO$_3$.

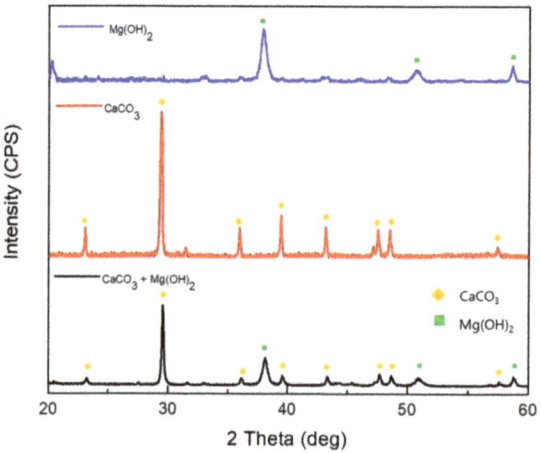

Figure 6. XRD results for the 3 types of calcareous deposits after 30 h of potentiostatic polarization.

3.3. Electrochemical Impedance Spectroscopy

After the 30-h potentiostatic polarization tests, EIS measurements were performed. The Nyquist plots of the data from the electrodes giving different types of calcareous deposits are shown in Figure 7a. The Nyquist plots consist of a capacitive semicircle at a high frequency. Figure 7b presents the equivalent electrical circuit for bare steel, where R_s is the solution resistance, R_{ct} is the charge transfer resistance, and CPE_1 is the double-layer capacitance formed by the electrical double layer that exists at the interface between the electrolyte and electrode [26,30,31]. Figure 7c shows the equivalent electrical circuit that describes the formation of porous calcareous deposits on the surface of the steel. In Figure 7c, R_s is the solution resistance; CPE_2 is the dielectric nature of the calcareous deposits, which is associated with the thickness of the calcareous deposit layer; R_{film} is the pore resistance; CPE_1 is the capacitance generated by the metal dissolution reaction and by the electric double layer at the solution/metal interface; and R_{ct} is the charge transfer resistance caused by the metal dissolution reaction [25]. Here, a CPE is used instead of a capacitor to compensate for the nonhomogeneity of the system frequency. The impedance of a CPE is described by the following equation:

$$Z_{CPE} = A^{-1}(j\omega)^{-n} \qquad (5)$$

where A^{-1} is the proportionality coefficient (with units, $\Omega^{-1}\, s^n\, cm^{-2}$); ω is the angular frequency (rad s^{-1}); $j^2 = -1$ is an imaginary number; and n is an empirical exponent ($0 \leq n \leq 1$) that measures the deviation from the ideal capacitive behavior [32–34].

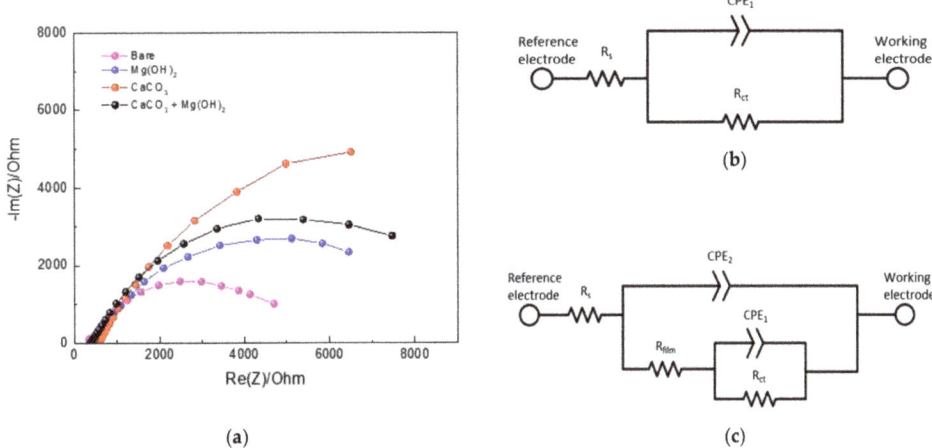

Figure 7. (a) Result of EIS test according to the types of calcareous deposits. (b) Equivalent circuit diagram of bare carbon steel. (c) Equivalent circuit diagram of carbon steel covered by calcareous deposits.

The results of the EIS fitting using the ZSimpWin software are shown in Table 3. The R_{film} values were the largest in the $CaCO_3$ layer, followed by the $CaCO_3$ and $Mg(OH)_2$ mixed layer, and then the $Mg(OH)_2$ layer. Similar to the results for the R_{film} value, $CaCO_3$ had the largest R_{ct} value, followed by the $CaCO_3$ and $Mg(OH)_2$ mixed layer, and then the $Mg(OH)_2$ layer. These results indicate that the $CaCO_3$ layer worked better as a protective layer than the others. At the same time, the CPE_1 values tend to be the opposite of R_{ct}. This is because the active area of metal dissolution decreases as the calcareous deposition becomes wider and thicker on the metal surface. The CPE_2 values of the $CaCO_3$ layer were higher than those of the other calcareous deposit layers, meaning that the $CaCO_3$ layer was the thickest. This is in agreement with the SEM image results. Figure 8 shows the total resistance values of the bare steel and the three types of calcareous deposits. All specimens with calcareous deposits had a higher total resistance than the bare specimen, demonstrating that the calcareous deposits provided protection to the bare specimens.

Table 3. The results of the EIS fitting using the circuit.

Type of Deposit	R_s ($\Omega \cdot cm^2$)	CPE_1 CPE	Y_0 (0 < n < 1)	R_{film} ($\Omega \cdot cm^2$)	CPE_2 CPE	Y_0 (0 < n < 1)	R_{ct} ($\Omega \cdot cm^2$)
Bare	393.7	1.912×10^{-4}	0.7527	-	-	-	4785
$Mg(OH)_2$	511.3	2.955×10^{-4}	0.8693	523.5	3.893×10^{-4}	0.7953	6100
$CaCO_3$	568.2	5.955×10^{-4}	0.7615	865.9	1.702×10^{-4}	0.7231	15,210
$CaCO_3 + Mg(OH)_2$	375.9	2.923×10^{-4}	0.7533	643.4	1.795×10^{-4}	0.7959	8872

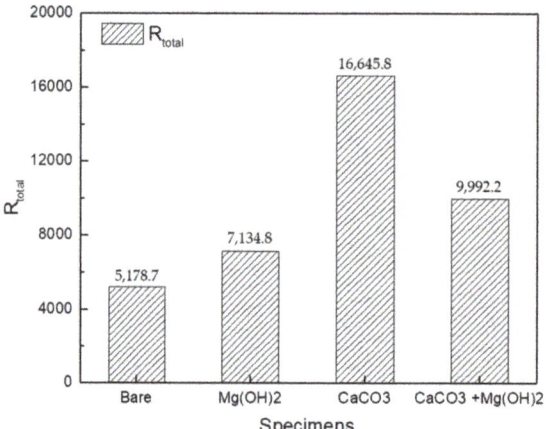

Figure 8. Comparison of total resistance on the bare and 3 types of calcareous deposited specimens.

3.4. Corrosion Acceleration Test

A potentiostatic test was performed at 3.5 V_{SCE} for 100 h to verify the stray current corrosion mitigation of the calcareous deposits [35]. Figure 9 shows the total electric charge at the current inflow area when 3.5 V_{SCE} was applied for 100 h, along with the mass loss of the specimen at the current outflow area. The total electrical charge value was obtained using the following equation [14,36]:

$$Q = \int_{ti}^{tf} I \, dt \qquad (6)$$

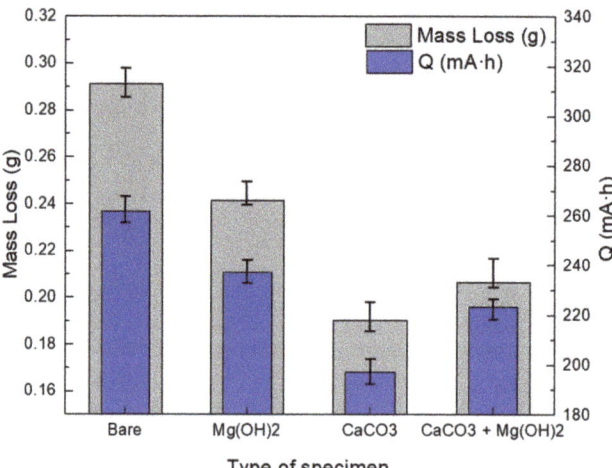

Figure 9. Mass losses of corrosion specimen and quantities of inflow current after potentiostatic acceleration test.

Since the calcareous deposits decreased the inflow current, the total electric charge values of all specimens with a calcareous deposit were lower than that of the bare specimen. During the corrosion acceleration tests, a crack occurred in the unstable $Mg(OH)_2$ layer. Therefore, the total electric charge of the specimen with the $Mg(OH)_2$ layer is higher than those of the other specimens with calcareous deposits. In addition, the total electric charge of the specimen with the CaCO$_3$ layer is lower than those of the other specimens, meaning that the CaCO$_3$ layer was the most protective against inflow current. Figure 10 shows that the specimen on which CaCO$_3$ is deposited receives the lowest inflow current and acts as a stable electrical barrier layer. This is because the CaCO$_3$ layer grows in a solid and stable form, compared to the Mg(OH)$_2$ deposition layer, and without the hindering of Mg(OH)2, it forms a thicker deposition layer than other layers [28,29]. Therefore, it acts as an electrical and physical barrier that blocks the external inflow current more efficiently than other layers. The total electric charge of the specimen with the CaCO$_3$ and Mg(OH)$_2$ mixed layer is higher than that of the specimen with the CaCO$_3$ layer, and lower than that of the specimen with the Mg(OH)$_2$ layer. Generally, more current flowing in means more current flowing out. Therefore, as current inflow increases, the areas where the current outflows become more susceptible to corrosion. As a result, among the specimens representing the current outflow area, the mass loss was the lowest in the specimen with the CaCO$_3$ layer deposited. The mass reduction then increased in the following order, with respect to the deposit layer composition: CaCO$_3$; the CaCO$_3$ and Mg(OH)$_2$ mixed layer; the Mg(OH)$_2$ layer; and the bare specimen.

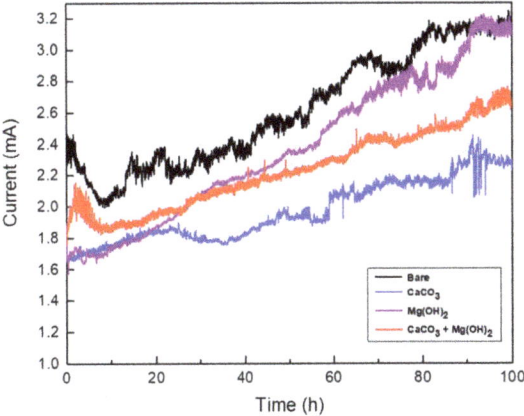

Figure 10. Current vs. time curves during potentiostatic test (applied voltage: 3.5 V$_{SCE}$; solutions: synthetic soil solution; testing time: 100 h).

4. Conclusions

This study evaluated the stray current corrosion mitigation of calcareous deposits on carbon steel in a synthetic soil solution using the electrochemical tests, SEM, EDS, and XRD. On the basis of the experiments, the following conclusions can be drawn:

- In the potentiostatic test, the current densities in all types of calcareous deposit layers decreased with the test time;
- The specimen with the CaCO$_3$ layer had the lowest current density. In the surface analysis, the specimen in the CaCO$_3$ solution has the thickest layer compared to the Mg(OH)$_2$ and mixed solutions;
- In the EIS test, the specimen immersed in the CaCO$_3$ solution had the highest R$_{film}$ and R$_{ct}$, indicating that the calcareous deposit of CaCO$_3$ is the most protective layer;
- The potentiostatic acceleration test demonstrated that the CaCO$_3$ layer had the lowest total electric charge among the specimens with calcareous deposits. In addition, the mass loss by the current outflow was the lowest in those with a CaCO$_3$ layer.

Consequently, stray current corrosion can be effectively mitigated if CaCO$_3$ powders are buried together with the pipelines and deposited when the soil solution and stray current are introduced to the pipelines.

Author Contributions: Conceptualization, S.-J.K.; methodology, S.-J.K.; validation, M.-S.H. and J.-G.K.; formal analysis, S.-J.K.; investigation, S.-J.K.; resources, M.-S.H.; data curation, S.-J.K. and M.-S.H.; writing—original draft preparation, S.-J.K.; writing—review and editing, S.-J.K. and M.-S.H.; visualization, S.-J.K.; supervision, J.-G.K.; project administration, J.-G.K.; funding acquisition, M.-S.H. and J.-G.K. All authors have read and agreed to the published version of the manuscript.

Funding: This work was supported by the National Research Foundation of Korea (NRF) grant, funded by the Korean Government (MEST) (No. NRF-2019R1A2B5B01070453).

Institutional Review Board Statement: Not applicable.

Informed Consent Statement: Not applicable.

Data Availability Statement: Not applicable.

Conflicts of Interest: The authors declare no conflict of interest.

References

1. Chen, Z.; Koleva, D.; van Breugel, K. A review on stray current-induced steel corrosion in infrastructure. *Corros. Rev.* **2017**, *35*, 397–423. [CrossRef]
2. Yoo, Y.H.; Nam, T.H.; Choi, Y.S.; Kim, J.G.; Chung, L. A galvanic sensor system for detecting the corrosion damage of the steel embedded in concrete structures: Laboratory tests to determine the cathodic protection and stray-current. *Met. Mater. Int.* **2011**, *17*, 623–629. [CrossRef]
3. Guo, Y.B.; Liu, C.; Wang, D.G.; Liu, S.H. Effects of alternating current interference on corrosion of X60 pipeline steel. *Pet. Sci.* **2015**, *12*, 316–324. [CrossRef]
4. Wang, X.; Wang, Z.; Chen, Y.; Song, X.; Yang, Y. Effect of a DC stray current on the corrosion of X80 pipeline steel and the cathodic disbondment behavior of the protective 3PE coating in 3.5% NaCl solution. *Coatings* **2019**, *9*, 29. [CrossRef]
5. Chao, Y.; Gan, C.; Zili, L.; Yalei, Z.; Chengbin, Z. Study the influence of DC stray current on the corrosion of X65 steel using electrochemical method. *Int. J. Electrochem. Sci.* **2015**, *10*, 10223–10231.
6. Riskin, J. *Electrocorrosion and Protection of Metals: General Approach with Particular Consideration to Electrochemical Plants*; Elsevier Science: Amsterdam, Netherlands, 2008.
7. Bertolini, L.; Carsana, M.; Pedeferri, P. Corrosion behaviour of steel in concrete in the presence of stray current. *Corros. Sci.* **2007**, *49*, 1056–1068. [CrossRef]
8. Chen, Z.; Qin, C.; Tang, J.; Zhou, Y. Experiment research of dynamic stray current interference on buried gas pipeline from urban rail transit. *J. Nat. Gas Sci. Eng.* **2013**, *15*, 76–81. [CrossRef]
9. Richard, W.S.; David, D.; Carl, E.L., Jr. *Stray Current Corrosion Due to Utility Cathodic Protection*; Structural Engineering and Engineering Materials SM Report No. 45; University of Kansas Center for Research, Inc.: Lawrence, KS, USA, 1997; Chapter 2; pp. 10–17.
10. Zhu, Q.; Cao, A.; Zaifend, W.; Song, J.; Shengli, C. Stray current corrosion in buried pipeline. *Anti-Corros. Methods Mater.* **2011**, *58*, 234–237. [CrossRef]
11. Lin, Y.; Li, K.; Su, M.; Meng, Y. Research on stray current distribution of Metro based on Numerical Simulation. In Proceedings of the 2018 IEEE International Symposium on Electromagnetic Compatibility and 2018 IEEE Asia-Pacific Symposium on Electromagnetic Compatibility (EMC/APEMC), Suntec City, Singapore, 14–18 May 2018.
12. Kim, Y.S.; Kim, J.G. Failure analysis of a thermally insulated pipeline in a district heating system. *Eng. Fail. Anal.* **2018**, *83*, 193–206. [CrossRef]
13. Hong, M.S.; So, Y.S.; Kim, J.G. Optimization of cathodic protection design for pre-insulated pipeline in district heating system using computational simulation. *Materials* **2019**, *12*, 1761. [CrossRef] [PubMed]
14. Tang, K. Stray current induced corrosion of steel fibre reinforced concrete. *Cem. Concr. Res.* **2017**, *100*, 445–456. [CrossRef]
15. Szeliga, M.J. "Stray Current Corrosion", in: Peabody's Control of Pipeline Corrosion, 2nd ed.; NACE International: Houston, TX, USA, 2001; p. 211.
16. Wang, C. Stray Current Distributing Model in the Subway System: A review and outlook. *Int. J. Electrochem. Sci.* **2018**, *13*, 1700–1727. [CrossRef]
17. Du, G.; Wang, J.; Jiang, X.; Zhang, D.; Yang, L.; Hu, Y. Evaluation of Rail Potential and Stray Current with Dynamic Traction Networks in Multitrain Subway Systems. *IEEE Trans. Transp. Electrif.* **2020**, *6*, 784–796. [CrossRef]
18. Ghanbari, E.; Lillard, R.S. The influence of $CaCO_3$ scale formation on AC corrosion rates of pipeline steel under cathodic protection. *Corrosion* **2017**, *74*, 551–565. [CrossRef]
19. Barchiche, C.; Deslouis, C.; Gil, O.; Refait, P.; Tribollet, B. Characterisation of calcareous deposits by electrochemical methods: Role of sulphates, calcium concentration and temperature. *Electrochim. Acta* **2004**, *49*, 2833–2839. [CrossRef]
20. Ce, N.; Paul, S. The effect of temperature and local pH on calcareous deposit formation in damaged thermal spray aluminum (TSA) coatings and its implication on corrosion mitigation of offshore steel structures. *Coatings* **2017**, *7*, 52. [CrossRef]

21. Devos, O.; Jakab, S.; Gabrielli, C.; Joiret, S.; Tribollet, B.; Picart, S. Nucleation-growth process of scale electrodeposition – influence of the magnesium ions. *J. Cryst. Growth* **2009**, *311*, 4334–4342. [CrossRef]
22. Zhang, L.; Shen, H.-J.; Sun, J.-Y.; Sun, Y.-N.; Fang, Y.-C.; Cao, W.-H.; Xing, Y.-Y.; Lu, M.-X. Effect of calcareous deposits on hydrogen permeation in X80 steel under cathodic protection. *Mater. Chem. Phys.* **2018**, *207*, 123–129. [CrossRef]
23. Hong, M.S.; Hwang, J.H.; Kim, J.H. Optimization of the cathodic protection design in consideration of the temperature variation for offshore structures. *Corrosion* **2018**, *74*, 123–133. [CrossRef]
24. Yan, J.-F.; Nguyen, T.V.; White, R.E.; Griffin, R.B. Mathematical modeling of the formation of calcareous deposits on cathodically protected steel in seawater. *J. Electrochem. Soc.* **1993**, *140*, 733–742. [CrossRef]
25. Deslouis, C.; Festy, D.; Gil, O.; Rius, G.; Touzain, S.; Tribollet, B. Characterization of calcareous deposits in artificial sea water by impedance techniques-I. Deposit of $CaCO_3$ without $Mg(OH)_2$. *Electrochim. Acta* **1998**, *43*, 1891–1901. [CrossRef]
26. Li, C.J.; Du, M. The growth mechanism of calcareous deposits under various hydrostatic pressures during the cathodic protection of carbon steel in seawater. *RSC Adv.* **2017**, *7*, 28819–28825. [CrossRef]
27. Deslouis, C.; Festy, D.; Gil, O.; Maillot, V.; Touzain, S.; Tribollet, B. Characterization of calcareous deposits in artificial sea water by impedances techniques: 2-deposit of $Mg(OH)_2$ without $CaCO_3$. *Electrochim. Acta* **2000**, *45*, 1837–1845. [CrossRef]
28. Carré, C.; Zanibellato, A.; Jeannin, M.; Sabot, R.; Gunkel-Grillon, P.; Serres, A. Electrochemical calcareous deposition in seawater. A review. *Environ. Chem. Lett.* **2020**, *18*, 1193–1208. [CrossRef]
29. Barchiche, C.; Deslouis, C.; Festy, D.; Gil, O.; Refait, P.; Touzain, S.; Tribollet, B. Characterization of calcareous deposits in artificial seawater by impedance techniques 3- Deposit of of $CaCO_3$ in the presence of $Mg(OH)_2$. *Electrochim. Acta* **2003**, *48*, 1645–1654. [CrossRef]
30. Kim, Y.-S.; Kim, S.-H.; Kim, J.-G. Effect of 1, 2, 3-benzotriazole on the corrosion properties of 316L stainless steel in synthetic tap water. *Met. Mater. Int.* **2015**, *21*, 1013–1022. [CrossRef]
31. An, J.-H.; Lee, J.; Kim, Y.-S.; Kim, W.-C.; Kim, J.-G. Effects of Post Weld Heat Treatment on Mechanical and Electrochemical Properties of Welded Carbon Steel Pipe. *Met. Mater. Int.* **2018**, *25*, 304–312. [CrossRef]
32. Zhang, P.Q.; Wu, J.X.; Zhang, Q.; Lu, X.Y.; Wang, K. Pitting mechanism for passive 304 stainless steel in sulphuric acid media containing chloride ions. *Corros. Sci.* **1993**, *34*, 1343–1354. [CrossRef]
33. Lopez, D.A.; Simison, S.N.; De Sanchez, S.R. The influence of steel microstructure on CO_2 corrosion. EIS studies on the inhibition efficiency of benzimidazole. *Electrochim. Acta* **2003**, *48*, 845–854. [CrossRef]
34. Lee, D.Y.; Kim, W.C.; Kim, J.G. Effect of nitrite concentration on the corrosion behaviour of carbon steel pipelines in synthetic tap water. *Corros. Sci.* **2012**, *64*, 105–114. [CrossRef]
35. Hong, M.S.; So, Y.S.; Lim, J.M.; Kim, J.G. Evaluation of internal corrosion property in district heating pipeline using fracture mechanics and electrochemical acceleration kinetics. *J. Ind. Eng. Chem.* **2021**, *94*, 253–263. [CrossRef]
36. Silva, J.C.D.E.; Panicali, A.R.; Barbosa, C.F.; Caetano, C.E.F.; Paulino, J.O.S. Electric charge flow in linear circuits. *Electr. Power Syst. Res.* **2019**, *170*, 57–63. [CrossRef]

Article

Evaluation of the Influence of the Combination of pH, Chloride, and Sulfate on the Corrosion Behavior of Pipeline Steel in Soil Using Response Surface Methodology

Nguyen Thuy Chung [1], Yoon-Sik So [1], Woo-Cheol Kim [2] and Jung-Gu Kim [1],*

[1] School of Advanced Materials Science & Engineering, Sungkyunkwan University, 300 Chunchun-Dong, Jangan-Gu, Suwon 440-746, Korea; chung.ngthuy@g.skku.edu (N.T.C.); soy2871@gmail.com (Y.-S.S.)
[2] Technical Efficiency Research Team, Korea District Heating Corporation, 92 Gigok-ro, Yongin 06340, Korea; Kwx7777@kdhc.co.kr
* Correspondence: kimjg@skku.edu

Citation: Chung, N.T.; So, Y.-S.; Kim, W.-C.; Kim, J.-G. Evaluation of the Influence of the Combination of pH, Chloride, and Sulfate on the Corrosion Behavior of Pipeline Steel in Soil Using Response Surface Methodology. *Materials* **2021**, *14*, 6596. https://doi.org/10.3390/ma14216596

Academic Editor: Vít Křivý

Received: 30 September 2021
Accepted: 29 October 2021
Published: 2 November 2021

Publisher's Note: MDPI stays neutral with regard to jurisdictional claims in published maps and institutional affiliations.

Copyright: © 2021 by the authors. Licensee MDPI, Basel, Switzerland. This article is an open access article distributed under the terms and conditions of the Creative Commons Attribution (CC BY) license (https://creativecommons.org/licenses/by/4.0/).

Abstract: External damage to buried pipelines is mainly caused by corrosive components in soil solution. The reality that numerous agents are present in the corrosive environment simultaneously makes it troublesome to study. To solve that issue, this study aims to determine the influence of the combination of pH, chloride, and sulfate by using a statistical method according to the design of experiment (DOE). Response surface methodology (RSM) using the Box–Behnken design (BBD) was selected and applied to the design matrix for those three factors. The input corrosion current density was evaluated by electrochemical tests under variable conditions given in the design matrix. The output of this method is an equation that calculates the corrosion current density as a function of pH, chloride, and sulfate concentration. The level of influence of each factor on the corrosion current density was investigated and response surface plots, contour plots of each factor were created in this study.

Keywords: response surface methodology; Box–Behnken design; modeling; soil environment; carbon steel corrosion

1. Introduction

Buried pipeline systems are a key part of global infrastructure. Any considerable disturbance to the performance of these systems frequently has negative consequences for regional businesses, economies, and citizen living circumstances [1]. Due to the complexity of the medium and surroundings such as soil components, corrosion is one of the leading reasons for pipeline degradation and failure [2]. Therefore, it is necessary to thoroughly understand the processes as well as a comprehension of the cause of external corrosion steel pipelines in soil [3].

The factors influencing soil corrosion include soil components, moisture content, temperature, resistivity, differential aeration, soil type, permeability, and the presence of sulfate-reducing bacteria [4–8]. Due to the complexities of its environment, soil corrosion is not easy to study. A large number of studies on soil corrosion are necessary for the future. This article focuses on the influence of the chemical component, and three corrosive agents selected are pH, chloride, sulfate concentration.

Several authors have previously studied the effects of pH, chloride, and sulfates in synthetic soil and particle soil [9–12]. However, to our knowledge, no studies have examined the combined influence of these three factors. To study the combined effects of these parameters, different combinations were investigated using statistically designed experiments. However, when using a statistical synthesis method, the number of experiments can be very large. Finding an optimal statistical method is essential. Recently, two widely applied methods are Artificial Neural Networks (ANNs) and response surface methodology (RSM) [13–15] and response surface methodology (RSM) has been selected and used herein.

The main objective of this study was to enhance a clear understanding of the effects of the soil environment, especially the influencing factors, on the corrosion of steel pipelines. RSM, based on the design of experiments (DOE), is a set of statistical and mathematical tools for experimental design and optimization of the effect of process variables [16–18]. RSM reduces the number of trials and recognizes the influence of component parameters on corrosion current density. To obtain the input statistics of corrosion current density from RSM, electrochemical tests were carried out to formulate an equation that calculates the corrosion current density using the parameters of pH, chloride concentration, and sulfate concentration. After that, the equation was confirmed to be reliable by analysis of variance (ANOVA), and the level of influence of each factor on corrosion current density was investigated by T-test. Finally, to determine the interactions between the corrosive agents, surface contour plots of each factor were created.

2. Investigation Scheme Design of Experiment

To achieve the desired aim, the investigation was planned in the sequence shown in Figure 1.

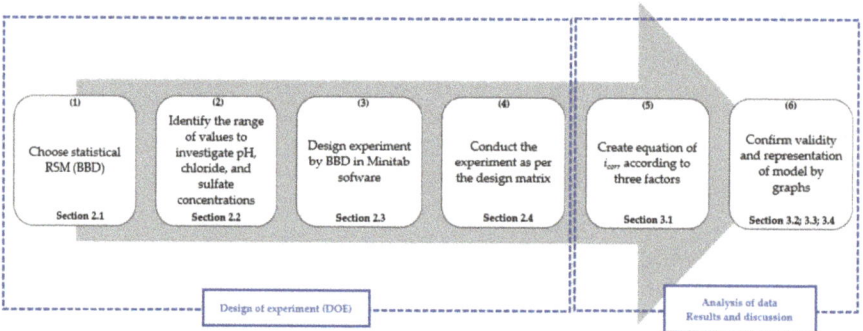

Figure 1. Six steps were implemented to investigate the influence of pH, chloride concentration, and sulfate concentration on the corrosion current density of carbon steel in a soil environment.

The above steps will be explained in detail in the following sections.

2.1. Response Surface Methodology and Box–Behnken Design

To investigate the influence of several factors on a response, a conventional statistical method can be very useful. However, this approach requires a large number of combinations of different parameter values to determine the effects of those combinations. For example, in this study using the three factors of pH, chloride, and sulfate to obtain the corrosion current density, the maximum, median, and minimum were determined for each factor; therefore, the number of experiments would be at least $3 \times 3 \times 3 = 27$ experiments and thus time-consuming. To simplify the procedure, RSM, a multivariate statistical tool that offers a modern approach to investigating the combined variables, was implemented because it has several advantages [19,20]. Firstly, RSM provides accuracy and process optimization with a good perspective for predictive model innovation; secondly, response surfaces are graphical representations used to illustrate the interactive effects of factors effects on response; and most importantly, a limited number of experiment runs are designed, in comparison to a conventional approach for the same number of estimated factors [20]. RSM is useful for analyzing the effects of parameters and their interactions with each other.

Estimation of coefficients, prediction of responses, and confirmation of the adequacy of the model are process optimization. The response is represented by Equation (1) [19],

$$Y = f(X_1, X_2, \ldots, X_n) \pm E, \qquad (1)$$

Here, Y is the response, f is the response function, $X_1 \ldots X_n$ are the variables of action called factors, n is the number of variables, and E is the experimental error. These mathematical models are typically polynomials with an unknown structure; however, to fully present linear interactions and quadratic effect, it is preferable to employ a second-order polynomial model [21]. Experiment design could have a significant effect on the accuracy of approximation as well as the cost of building the response surface [22]. In this study, X_1, X_2, and X_3 are the three experimental variables studied—that is, pH, chloride concentration, and sulfate concentration, respectively. The response (i_{corr}) was related to the selected variables by the second-order polynomial regression model given in Equation (2) [19],

$$Y = \beta_0 + \sum_{i=1}^{n} \beta_i X_i + \sum_{i=1}^{n} \beta_{ii} X_i^2 + \sum_{i=1}^{n-1} \sum_{j>i}^{n} \beta_{ij} X_i X_j + E, \qquad (2)$$

Here, β_0 represents the intercept or regression coefficient; β_i, β_{ii}, and β_{ij} represent the linear, quadratic, and interaction parameters, respectively; X_i and X_j are the coded values of the process variables.

Different designs can be used in statistical experiments; the variation between these designs is determined by the number of runs (experiments) and the experiment points chosen. Central composite design (CCD) and Box–Behnken design (BBD) are two extremely helpful and common for fitting a second-order model. Simple factorial or fractional designs are used to create both designs. Simple factorial or fractional factorial designs are used to create both designs. Both designs are built up from simple factorial or fractional factorial designs. BBD is a spherical, three-level fractional factorial design including a center point and middle point of edges of a circle circumscribed on a sphere while CCD is a fractional factorial design. The advantage of BBD is that it just requires a reasonable group of factors to determine the complex response function and voids experiments performed under extreme conditions, and BBD was selected and applied to analyze the influence of combination factors on the response [23]. Response surface methodology has recently been a widespread approach to optimize and analyze diverse products because of the availability of various computer software programs, such as Design Expert, Statistica, MATLAB, and Minitab, which make the application of RSM easier.

2.2. Identifying the Range of the Values to Investigate and Preparing Reagents

Before designing the experiment, the effective factors of a system and the range of these factors must be identified. In this study, three factors (pH, chloride, and sulfate) were used and the tested range of these components was determined based on a real soil environment in Korea (pH: 6.8; $CaCl_2$: 133.2 ppm; $MgSO_4.7H_2O$: 59.0 ppm; $NaHCO_3$: 208.0 ppm; H_2SO_4: 48 ppm; HNO_3: 21.8 ppm) [24]. Table 1 lists the initial concentration of the solution and range of the factors that influence corrosion.

Table 1. Initial concentrations and investigated value range of factors in synthetic soil solution [24].

Investigated Factor	pH	Chloride [Cl^-], ppm	Sulfate [SO_4^{2-}], ppm
Initial value	6.8	85.2	70.04
Investigated value range	4–8	85.2–1085.2	70.04–670.04

2.3. Design of Experiment and Design Matrix

Based on the testing range, the design of the experiment was carried out using the Minitab 19 software. For the RSM using Box–Behnken design, an experiment number of $N = k^2 + k + c_p$ was required, where k is the factor number and c_p is experiment runs at the center point. It can be viewed as a cube in Figure 2, which consists of a central point and middle points on the edges. For the statistical calculations, the variables were coded as the minimum, central, and maximum levels of each variable, designated as −1, 0, and 1, respectively, as shown in Table 2. Two ways to collect enough data is to use replicates and add just a few center points. Replicating could be time-consuming; therefore, this design includes 3 experiment runs at the center point. In this study, the number of factors, k, was 3 and c_p was 3 times the number of central points (red point in Figure 2); therefore, the total number of experiments was 15 at three levels, as shown in Table 3. This number was significantly reduced compared with the conventional methods.

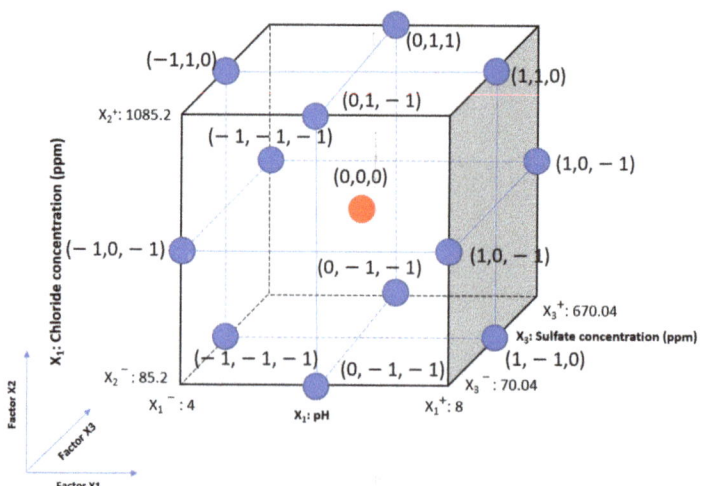

Figure 2. Cube plot design of experiments for corrosion current density according to pH, chloride, and sulfate concentrations by BBD.

Table 2. Actual investigated values of factors and their corresponding coded levels.

Variable	Code Values		
	−1 (Minimum)	0 (Medium)	1 (Maximum)
X_1, pH	4	6	8
X_2, Chloride (ppm)	85.20	585.20	1085.20
X_3, Sulfate (ppm)	70.04	370.04	670.04

The results were input into the Minitab 19 software for further analysis. On examining the fit summary, the quadratic model was found to be statistically significant for the response.

Table 3. Experimental design based on the Box–Behnken Design matrix for the three variables.

Standard Run	Coded Parameter			Real Parameter		
	X_1	X_2	X_3	pH	$[Cl^-]$, (ppm)	$[SO_4^{2-}]$, (ppm)
1	−1	−1	0	4	85.2	370.04
2	1	−1	0	8	85.2	370.04
3	−1	1	0	4	1085.2	370.04
4	1	1	0	8	1085.2	370.04
5	−1	0	−1	4	585.2	70.04
6	1	0	−1	8	585.2	70.04
7	−1	0	1	4	585.2	670.04
8	1	0	1	8	585.2	670.04
9	0	−1	−1	6	85.2	70.04
10	0	1	−1	6	1085.2	70.04
11	0	−1	1	6	85.2	670.04
12	0	1	1	6	1085.2	670.04
13	0	0	0	6	585.2	370.04
14	0	0	0	6	585.2	370.04
15	0	0	0	6	585.2	370.04

2.4. Conducting Electrochemical Tests

Next, electrochemical tests were carried out using a three-electrode system configuration based on the matrix values described previously. The experiment was performed using a corrosion cell in which carbon steel SPW400 (0.04% wt. S; 0.04% wt. P; 0.25% wt. C; Bal. Fe—Korean standard) with a 2 mm thickness (a common material for buried pipelines) was used as the working electrode along with a reference electrode (saturated calomel electrode; RE, Qrins), and a counter electrode (two pure graphites; CE, Qrins, Seoul, Korea) [25]. The specimen was polished with SiC paper from 200 to 600 grit sizes, the surface was coated with silicone paste, leaving an exposed area of 1 cm². NaOH and HNO₃ (Samchun, Pyeongtaek-si, Korea) were used to adjust the pH values, and NaCl and Na₂SO₄ (Samchun, Pyeongtaek-si, Korea) were used to adjust chloride and sulfate concentrations, respectively. An open-circuit potential (OCP) was established within 3 h before the electrochemical tests. Tafel plots were obtained by potentiodynamic polarization tests. To prevent mutual polarization, the potentiodynamic polarization tests were separated into anodic and cathodic polarization. The potential range of the cathodic polarization was from 0 V vs. OCP to −0.25 V vs. OCP, whereas that for the anodic polarization was from 0 vs. OCP to 0.5 V vs. OCP, both with scan rates of 0.166 mV/s. Each experiment was carried out with the maintained process variable conditions given in the design matrix.

3. Results and Discussion

3.1. Electrochemical Test, Second-Order Polynomial Equation, and Statistical Analysis

Figure 3 shows the potentiodynamic polarization curves used to obtain the corrosion current density for the RSM with BBD. The corrosion current density was calculated using the Tafel extrapolation method.

The results of the corrosion current density for the RSM experiment design are displayed in Table 4; the collected data results were inserted into the DOE. For convenient comparison, the order of corrosion current density increases from the minimum value to the maximum value. Based on the results shown in Table 4, the highest corrosion current density was found in experiment 12, which was a combination of pH 6 and the highest chloride and sulfate concentrations of 1085.20 and 670.04 ppm, respectively, in the synthetic soil solution. Meanwhile, the lowest corrosion rate was obtained in experiment number 9, with a pH of 6 and the lowest chloride and sulfate concentrations of 85.20 and 70.04 ppm, respectively. Therefore, the results show that chloride and sulfate concentrations seem to be a negative influence, meaning that increases in these two factors will increase corrosion current density.

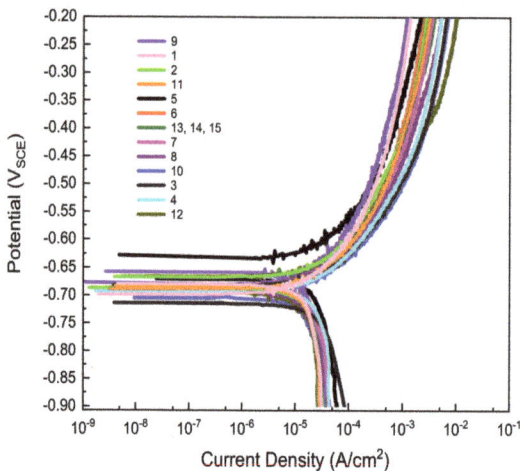

Figure 3. Potentiodynamic polarization curves of carbon steel with variations in pH, chloride, and sulfate concentrations based on an experimental design of RSM with BBD.

Table 4. Experimental observations and predicted data for response variables obtained from the BBD.

Experiment Run	Factor			Response i_{corr}, $\mu A/cm^2$	
	pH	Chloride, ppm	Sulfate, ppm	Experiment Observation	Predicted
9	6	85.2	70.04	4.2	4.22
1	4	85.2	370.04	4.9	4.88
2	8	85.2	370.04	4.8	4.84
11	6	85.2	670.04	5	5.15
5	4	585.2	70.04	5.4	5.23
6	8	585.2	70.04	5.5	5.31
13	6	585.2	370.04	5.7	5.57
14	6	585.2	370.04	5.8	5.57
15	6	585.2	370.04	5.6	5.57
7	4	585.2	670.04	6.2	6.18
8	8	585.2	670.04	6.3	6.23
10	6	1085.2	70.04	8.6	7.99
3	4	1085.2	370.04	9.1	8.55
4	8	1085.2	370.04	9.2	8.7
12	6	1085.2	670.04	9.4	8.91

The input corrosion current density was then analyzed by RSM using Minitab 19 to obtain a mathematical model of the relationship of the research variables. Based on estimates of the parameters and experimental results, RSM revealed an empirical relationship between the response and the input variables, as expressed by the fitted regression model below. The regression second-order polynomial equation for corrosion current density developed using BBD is shown in the equation below in terms of uncoded factors,

$$i_{corr} (\mu A/cm^2) = 5.840 - 0.542 \cdot pH - 0.001217 \cdot [Cl^-] + 0.001539 \cdot [SO_4^{2-}] + 0.0438 \cdot [pH]^2 \\ + 0.000004 \cdot [Cl^-]^2 - 0.000000 \cdot [SO_4^{2-}]^2 + 0.000050 pH \cdot [Cl^-] - 0.00000\, pH \cdot [SO_4^{2-}] + \\ 0.00000\, [Cl^-] \cdot [SO_4^{2-}], \quad (3)$$

The parameters $[SO_4^{2-}]^2$, $pH \cdot [SO_4^{2-}]$, and $[Cl^-] \cdot [SO_4^{2-}]$ in Equation (3) are independent of corrosion current density because their coefficients are zero. For the convenience of observation, insignificant parameters can be omitted and Equation (3) can be rewritten with uncoded factors, as follows,

$$i_{corr} (\mu A/cm^2) = 5.840 - 0.542 \cdot pH - 0.001217 \cdot [Cl^-] + 0.001539 \cdot [SO_4^{2-}] + 0.0438 \cdot [pH]^2 \\ + 0.000004 \cdot [Cl^-]^2 + 0.000050\, pH \cdot [Cl^-], \quad (4)$$

From Equation (4), the predicted results shown in Table 4 were calculated. In addition to the linear effect of the parameter for the corrosion current density of carbon steel in soil, Equation (4) also provided an understanding of the quadratic and interaction effects of the parameters. Verification of the equation's reliability will be clarified in the following sections.

3.2. Validity Evaluation of the Fitted Model

Analysis of variance (ANOVA) was conducted to investigate the relationship between a response variable and one or more predictor variables [26].

This study used standard $F_{critical} = F (9, 5, 0.05) = 4.7725$, where 9 is the degree of freedom of regression, 5 is the degree of freedom of residual error, and 0.05 is the level of significance. Table 5 shows the model had a high calculated F-value of 812.85. The calculated F-value is much higher than $F_{critical}$, indicating that it can reject the null hypothesis that all the coefficients are zero, and this model is statistically significant [27]. The F statistic must be used in combination with the *p*-value when deciding if the overall results are significant. The *p*-value is determined by the F statistic and is the probability that the results could have happened by chance. If the *p*-value is less than the alpha level, then the results are not significant and the null hypothesis cannot be rejected. In the sciences, the alpha level is usually set to be less than 0.05, which indicates model validation. Therefore, when the *p*-value output from the ANOVA of this study was 0.000, there was a significant difference between variables.

Table 5. Analysis of variance (ANOVA) for the fitted equation.

Source	Degree of Freedom	Adj. Sum of Square	Adj. Mean Square	F-Value	$F_{critical}$	*p*-Value	Remarks
Model	9	43.8940	4.8771	812.85	4.7725	0.000	Significant
Error	5	0.0300	0.0060	-	-	-	-
Null hypothesis: All the coefficients are zero $\beta_1 = \beta_2 = \beta_3 = \beta_{11} = \beta_{22} = \beta_{33} = \beta_{12} = \beta_{13} = \beta_{23} = 0$							
Lack-of-Fit	3	0.0100	0.0033	0.33	19.1643	0.808	Reasonable
Pure Error	2	0.0200	0.0100	-	-	-	-
Total	14	43.5200	-	-	-	-	-
Null hypothesis: Model is an appropriate fit for the data→No lack of fit							
R^2: 99.93%		R^2 (adj.): 99.81%			R^2 (pred.): 99.53%		

Another important value in ANOVA is lack of fit; the F-value for lack of fit with the null hypothesis that the model is an appropriate fit for the data. Therefore, in contrast to the F-value of the model, the null hypothesis should be "can reject", F value of lack of fit with the null hypothesis should be "cannot be rejected". In this study, the calculated F-value of lack of fit is 0.33 and the *p*-value is 0.808, with standard $F_{critical} = F(3, 2, 0.05) = 19.1643$, where 3 is the degree of freedom of lack of fit, 2 is the degree of freedom of pure error, 0.05 is the level of significance. In Table 5, F-value of lack of fit is much smaller than $F_{critical}$; the *p*-value is higher than 0.05. Thus, the null hypothesis cannot be rejected, the relationship assumed in the model is reasonable, and there is no lack of fit in this model.

Moreover, the value of R^2 is 99.93%, which indicates a good relationship between the experimental and predicted values of the response, with only 0.07% of the total variation not explained by the empirical model. The adjusted value (R^2 adj) is 99.81%, which suggests that the total variation of 0.19% for corrosion current density can be attributed to the independent parameters. The closeness between R^2 and adjusted R^2 implies a high significance for the model. The analysis and observations revealed a strong correlation between the experiments results and the values predicted by the statistical model, demonstrating the model success.

3.3. Preliminary Study of the Effects of pH, Chloride, and Sulfate Concentrations on the Corrosion Current Density in a Soil Environment

An F-test can show if a group of variables is jointly significant and a T-test can indicate if a variable is statistically significant. To precisely determine the effect of each factor on the corrosion current density (i.e., which influence is reliable and which is not), the Pareto chart in Figure 4 shows the standardized effects of pH, chloride, and sulfate concentrations on the corrosion current density evaluation. The T-value, which is used to measure how large the coefficient is in relationship to its standard error, is equal to the coded coefficient divided by its standard error. The Pareto chart was used to determine the magnitude and importance of the effects. Each bar represents the T-value for a type of factor; the height of the bar represents any important factors. Therefore, the biggest influence on the corrosion current density of carbon steel is chloride concentration. On the Pareto chart, bars that cross the reference line are statistically significant. In Figure 4, the bars that represent factors B, BB, C, and AA cross the reference line at 2.57. These factors are statistically significant at the 0.05 level in terms of the current model. Because the Pareto chart displays the absolute value of the effect, it determines large effects; however, it cannot determine effects that increase or decrease the response. A normal probability plot of the standardized effect is used to examine the magnitude and direction of the effects on one plot. In this case, all parameters have a significant influence on the corrosion current density when their *p*-value is less than 0.05. If the *p*-value is greater than 0.10, the model terms are insignificant. In this case, the *p*-value of pH is 0.403, $[SO_4^{2-}] \cdot [SO_4^{2-}]$ is 0.562, pH·$[Cl^-]$ is 0.253, pH·$[SO_4^{2-}]$ is 1.000 and $[Cl^-] \cdot [SO_4^{2-}]$ is 1.000, which are inefficient model terms as shown in Table 6. However, the *p*-value of the model shown in Table 5 is 0.000; therefore, removing them is not necessary. The Pareto chart and Table 6 are in agreement with another study regarding separate factors. The correlation between pH and corrosion rates is very low in this model, and it was inconclusive regarding the effective correlation between pH and corrosion current density. There are other studies with different results on this issue [28–30]. Of the three factors investigated in this study, it can be concluded that the sulfate concentration effect is minor while the chloride concentration has a major influence on the corrosion rate of carbon steel in a soil environment.

3.4. Representation of Model: Response Surface Plotting and Contour Plot of Corrosion Current Density with Each Factor of Carbon Steel in the Soil Environment

Using Statistica software, two-dimensional (2D) surface contour plots and three-dimensional (3D) plots were generated to estimate the value of variation response to better understand the effect of each factor on the corrosion current density. Surface plots of the response function were useful for understanding both the individual and combined effects of pH and chloride. This study examined a combination of the effects of three different factors, but the 3D plots could only represent two factors and the response corrosion current density of carbon steel. Therefore, one factor must be given a fixed value (i.e., a hold value) to represent the other two factors.

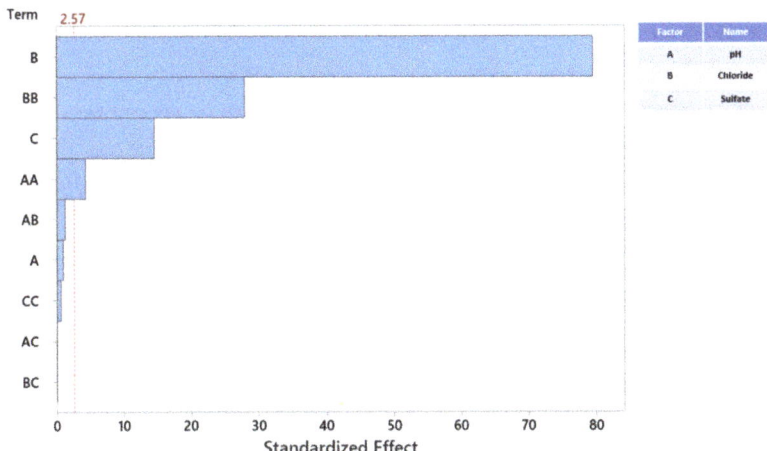

Figure 4. Pareto chart for the standardized effects of pH, chloride concentration, and sulfate concentration on the corrosion current density of carbon steel in a soil environment.

Table 6. Estimated coefficients with coded units for corrosion current density (i_{corr}).

Term	Coefficient	Standard Error Coefficient	T for H_0 [a] Coefficient = 0	p-Value
Constant	5.7000	0.0447	127.46	0.000
pH	0.0250	0.0274	0.91	0.403
$[Cl^-]$	2.1750	0.0274	79.42	0.000
$[SO_4^{2-}]$	0.4000	0.0274	14.61	0.000
pH · pH	0.1750	0.0403	4.34	0.007
$[Cl^-]·[Cl^-]$	1.1250	0.0403	27.91	0.000
$[SO_4^{2-}]·[SO_4^{2-}]$	−0.0250	0.0403	−0.62	0.562
pH ·$[Cl^-]$	0.0500	0.0387	1.29	0.253
pH ·$[SO_4^{2-}]$	0.0000	0.0387	0.00	1.000
$[Cl^-]·[SO_4^{2-}]$	0.0000	0.0387	0.00	1.000

3.5. Interactive Effect of pH and Chloride Concentration (ppm)

The obtained results were represented using a 3D representation of the response surface plot and 2D contour plots. Figure 5 presents the 3D plots demonstrating the effects of pH and chloride on corrosion current density under predefined conditions. In Figure 5, with a hold value of 370.04 ppm for sulfate, the effect of chloride increased by a considerable amount for a corrosion current density from 4 µA/cm² to 9.2 µA/cm². Meanwhile, the corrosion current density seemed to have very small variations in value (almost unchanging) as the pH value increased, which suggests that the corrosion current density is pH-independent.

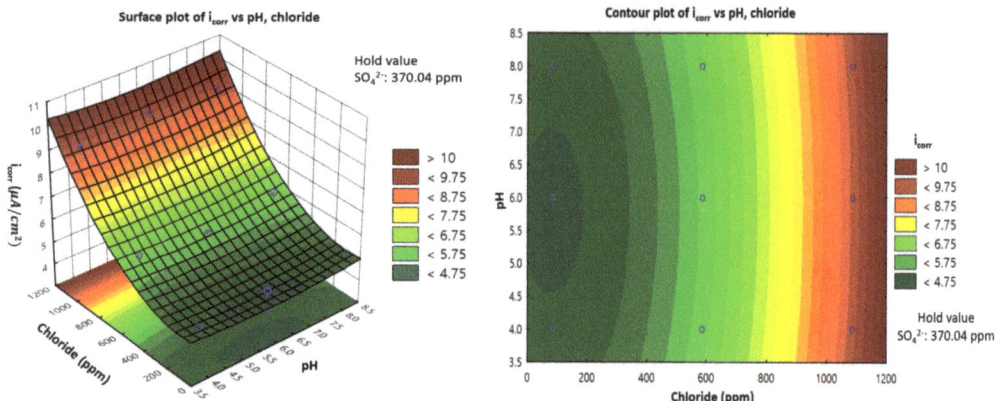

Figure 5. Response surface and contour plots showing the effect of pH and chloride concentration on the corrosion current density (i_{corr}) of carbon steel in a soil environment.

3.6. Interactive Effect of pH and Sulfate Concentration (ppm)

Figure 6 shows the response surface and contour plots for the effect of pH and sulfate on the corrosion current density. With a hold value of 585.02 ppm for chloride, the corrosion current density increased gradually when sulfate concentration increased, from 4 to 6.2 µA/cm². However, with an increase in pH, there was a slight variation in corrosion current density, which was more significant than that in Figure 5. However, the pH-related variation was still considered to be negligible, and the pH seemed to be independent of the corrosion current density.

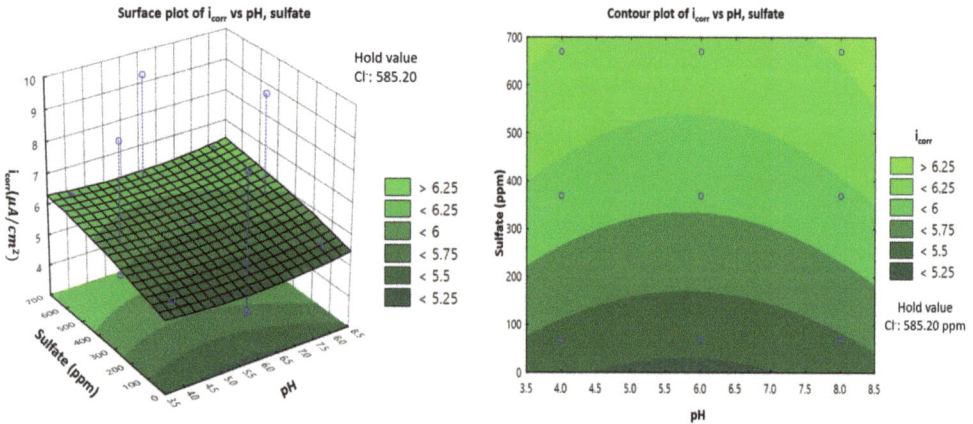

Figure 6. Response surface and contour plots showing the effect of pH and sulfate concentration on the corrosion current density (i_{corr}) of carbon steel in a soil environment.

3.7. Interactive Effect of Chloride and Sulfate Concentrations (ppm)

Figure 7 shows the effect of chloride and sulfate concentrations on the corrosion current density. The surface 2D contour plots and 3D response demonstrate the promotion efficiency. At a hold value of 6 for pH, an increased chloride or sulfate concentration was associated with an increased corrosion current density of carbon steel. From the minimum to maximum values for sulfate concentration, the corrosion current density increased from

4.2 to 5.5 µA/cm². Meanwhile, from the minimum value of chloride concentration to the maximum value of sulfate concentration, the corrosion current density increased from 4.2 to more than 9 µA/cm², which indicates quite a large influence. Several other studies have investigated this behavior and agree with these results [31–33].

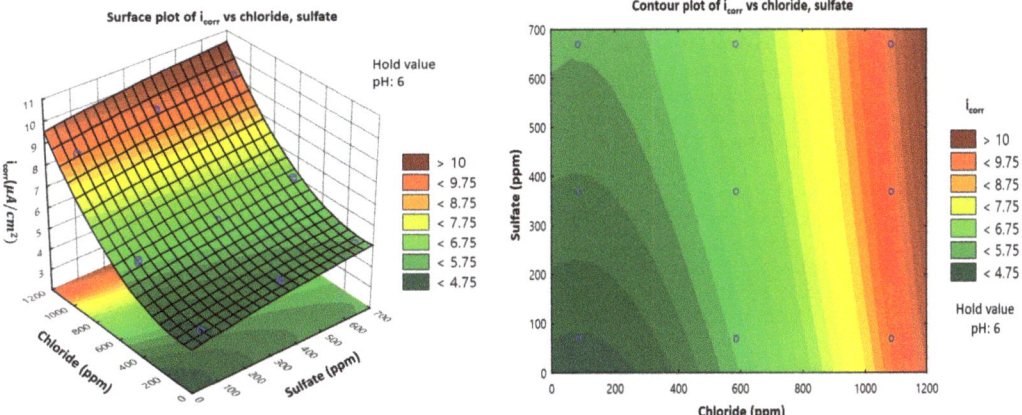

Figure 7. Response surface and contour plots showing the effect of chloride and sulfate concentrations on the corrosion current density (i_{corr}) of carbon steel in a soil environment.

4. Conclusions

Based on the results, the following conclusions could be drawn:

- The effects of pH, chloride concentration and sulfate concentration on the corrosion behavior of a carbon steel pipeline in a soil environment were investigated by statistical method RSM. Research results could be concluded that chloride and sulfate concentrations are a negative influence, pH seemed to be independent of the corrosion current density. A useful mathematical model was suggested for use in exploring methods to protect the buried pipeline.
- The effect level of independent variables on the corrosion rate was found to follow an increasing sequence of pH < sulfate concentration < chloride concentration.

The results show that the model was successful; however, it has a limitation because it can only be used within the experimental ranges. As chloride was the important factor influencing corrosion, further research with expanded chloride concentration ranges is necessary.

Author Contributions: Conceptualization, Y.-S.S.; methodology, N.T.C.; validation, N.T.C. and J.-G.K.; formal analysis, N.T.C.; investigation, N.T.C.; resources, N.T.C. and W.-C.K.; data curation, N.T.C.; writing—original draft preparation, N.T.C.; writing—review and editing, Y.-S.S. and J.-G.K.; supervision, J.-G.K.; project administration, W.-C.K., Y.-S.S. and J.-G.K. All authors have read and agreed to the published version of the manuscript.

Funding: This research was supported by a program for fostering next-generation researchers in engineering from the National Research Foundation of Korea (NRF) funded by the Ministry of Science and ICT (2017H1D8A2031628). This research was supported by the Korea District Heating Corporation (No.0000000014524).

Institutional Review Board Statement: Not applicable.

Informed Consent Statement: Not applicable.

Data Availability Statement: Data sharing is not applicable to this article.

Conflicts of Interest: The authors declare no conflict of interest. The funders had no role in the design of the study; in the collection, analyses, or interpretation of data; in the writing of the manuscript, or in the decision to publish the results.

References

1. Honegger, D.; Wijewickreme, D. Seismic risk assessment for oil and gas pipelines. In *Handbook of Seismic Risk Analysis and Management of Civil Infrastructure Systems*; Woodhead Publishing: Sawston, UK, 2013; pp. 682–715.
2. Chen, X.; Zhao, Y. Research on corrosion protection of buried steel pipeline. *Engineering* **2017**, *9*, 504. [CrossRef]
3. King, F.; Ahonen, L.; Taxén, C.; Vuorinen, U.; Werme, L. *Copper Corrosion under Expected Conditions in a Deep Geologic Repository*; Technical Report; Swedish Nuclear Fuel and Waste Management Co.: Stockholm, Sweden, 2001.
4. Szala, M.; Łukasik, D. Pitting corrosion of the resistance welding joints of stainless steel ventilation grille operated in swimming pool environment. *Int. J. Corros.* **2018**, *2018*, 9408670. [CrossRef]
5. Stefanoni, M.; Angst, U.M.; Elsener, B. Kinetics of electrochemical dissolution of metals in porous media. *Nat. Mater.* **2019**, *18*, 942–947. [CrossRef] [PubMed]
6. Chung, N.-T.; Hong, M.-S.; Kim, J.-G. Optimizing the Required Cathodic Protection Current for Pre-Buried Pipelines Using Electrochemical Acceleration Methods. *Materials* **2021**, *14*, 579. [CrossRef]
7. Saupi, S.; Sulaiman, M.; Masri, M. Effects of soil properties to corrosion of underground pipelines: A review. *J. Trop. Resour. Sustain. Sci.* **2015**, *3*, 14–18. [CrossRef]
8. He, B.; Han, P.; Lu, C.; Bai, X. Effect of soil particle size on the corrosion behavior of natural gas pipeline. *Eng. Fail. Anal.* **2015**, *58*, 19–30. [CrossRef]
9. Hou, Y.; Lei, D.; Li, S.; Yang, W.; Li, C.-Q. Experimental investigation on corrosion effect on mechanical properties of buried metal pipes. *Int. J. Corros.* **2016**, *2016*, 5808372. [CrossRef]
10. Song, Y.; Jiang, G.; Chen, Y.; Zhao, P.; Tian, Y. Effects of chloride ions on corrosion of ductile iron and carbon steel in soil environments. *Sci. Rep.* **2017**, *7*, 6865 . [CrossRef]
11. Wu, Y.H.; Liu, T.M.; Luo, S.X.; Sun, C. Corrosion characteristics of Q235 steel in simulated Yingtan soil solutions. *Mater. Werkst.* **2010**, *41*, 142–146. [CrossRef]
12. Barbalat, M.; Lanarde, L.; Caron, D.; Meyer, M.; Vittonato, J.; Castillon, F.; Fontaine, S.; Refait, P. Electrochemical study of the corrosion rate of carbon steel in soil: Evolution with time and determination of residual corrosion rates under cathodic protection. *Corros. Sci.* **2012**, *55*, 246–253. [CrossRef]
13. Talib, N.S.R.; Halmi, M.I.E.; Gani, S.S.A.; Zaidan, U.H.; Shukor, M.Y.A. Artificial neural networks (ANNs) and response surface methodology (RSM) approach for modelling the optimization of chromium (VI) reduction by newly isolated Acinetobacter radioresistens strain NS-MIE from agricultural soil. *BioMed Res. Int.* **2019**, *2019*, 5785387.
14. Aydin, M.; Uslu, S.; Çelik, M.B. Performance and emission prediction of a compression ignition engine fueled with biodiesel-diesel blends: A combined application of ANN and RSM based optimization. *Fuel* **2020**, *269*, 117472. [CrossRef]
15. Khatti, T.; Naderi-Manesh, H.; Kalantar, S.M. Application of ANN and RSM techniques for modeling electrospinning process of polycaprolactone. *Neural. Comput. Appl.* **2019**, *31*, 239–248. [CrossRef]
16. Gunst, R.F.; Myers, R.H.; Montgomery, D.C. *Response Surface Methodology: Process and Product Optimization Using Designed Experiments*; John Wiley & Sons: Hoboken, NJ, USA, 2016.
17. Sharma, Y.; Srivastava, V.; Singh, V.; Kaul, S.; Weng, C.-H. Nano-adsorbents for the removal of metallic pollutants from water and wastewater. *Environ. Technol.* **2009**, *30*, 583–609. [CrossRef]
18. Rajkumar, K.; Muthukumar, M. Response surface optimization of electro-oxidation process for the treatment of CI Reactive Yellow 186 dye: Reaction pathways. *Appl. Water Sci.* **2017**, *7*, 637–652. [CrossRef]
19. Goh, K.-H.; Lim, T.-T.; Chui, P.-C. Evaluation of the effect of dosage, pH and contact time on high-dose phosphate inhibition for copper corrosion control using response surface methodology (RSM). *Corros. Sci.* **2008**, *50*, 918–927. [CrossRef]
20. Box, G.E.; Draper, N.R. *Empirical Model-Building and Response Surfaces*; John Wiley & Sons: Hoboken, NJ, USA, 1987.
21. Montgomery, D.P.; Plate, C.A.; Jones, M.; Jones, J.; Rios, R.; Lambert, D.K.; Schumtz, N.; Wiedmeier, S.E.; Burnett, J.; Ail, S.; et al. Using umbilical cord tissue to detect fetal exposure to illicit drugs: A multicentered study in Utah and New Jersey. *Am. J. Perinatol.* **2008**, *28*, 750–753. [CrossRef]
22. Breig, S.J.M.; Luti, K.J.K. Response Surface Methodology: A Review on Its Applications and Challenges in Microbial Cultures. *Mater. Today Proc.* **2021**, *42*, 2277–2284. [CrossRef]
23. Kumari, M.; Gupta, S.K. Response surface methodological (RSM) approach for optimizing the removal of trihalomethanes (THMs) and its precursor's by surfactant modified magnetic nanoadsorbents (sMNP)—An endeavor to diminish probable cancer risk. *Sci. Rep.* **2019**, *9*, 18339.
24. Kim, J.-G.; Kim, Y.-W. Cathodic protection criteria of thermally insulated pipeline buried in soil. *Corros. Sci.* **2001**, *43*, 2011–2021. [CrossRef]
25. So, Y.-S.; Hong, M.-S.; Lim, J.-M.; Kim, W.-C.; Kim, J.-G. Calibrating the impressed anodic current density for accelerated galvanostatic testing to simulate the long-term corrosion behavior of buried pipeline. *Materials* **2021**, *14*, 2100. [CrossRef]
26. Ahmadi, M.; Rahmani, K.; Rahmani, A.; Rahmani, H. Removal of benzotriazole by Photo-Fenton like process using nano zero-valent iron: Response surface methodology with a Box-Behnken design. *Pol. J. Chem. Technol.* **2017**, *19*, 104–112. [CrossRef]

27. Di Leo, G.; Sardanelli, F. Statistical significance: *p* value, 0.05 threshold, and applications to radiomics—Reasons for a conservative approach. *Eur. Radiol. Exp.* **2020**, *4*, 1–8. [CrossRef]
28. Tang, D.-Z.; Du, Y.-X.; Lu, M.-X.; Liang, Y.; Jiang, Z.-T.; Dong, L. Effect of pH value on corrosion of carbon steel under an applied alternating current. *Mater. Corros.* **2015**, *66*, 1467–1479. [CrossRef]
29. Prawoto, Y.; Ibrahim, K.; Wan Nik, W.B. Effect of pH and chloride concentration on the corrosion of duplex stainless steel. *Arab. J. Sci. Eng.* **2009**, *34*, 115.
30. Ismail, M.; Noor, N.M.; Yahaya, N.; Abdullah, A.; Rasol, R.M.; Rashid, A.S.A. Effect of pH and temperature on corrosion of steel subject to sulphate-reducing bacteria. *J. Environ. Sci. Technol.* **2014**, *7*, 209–217. [CrossRef]
31. Wasim, M.; Shoaib, S. Influence of chemical properties of soil on the corrosion morphology of carbon steel pipes. In *Metals in Soil-Contamination and Remediation*; IntechOpen: London, UK, 2019.
32. Al-Sodani, K.A.A.; Maslehuddin, M.; Al-Amoudi, O.S.B.; Saleh, T.A.; Shameem, M. Efficiency of generic and proprietary inhibitors in mitigating corrosion of carbon steel in chloride-Sulfate Environments. *Sci. Rep.* **2018**, *8*, 11443, [CrossRef]
33. Arzola, S.; Palomar-Pardavé, M.; Genesca, J. Effect of resistivity on the corrosion mechanism of mild steel in sodium sulfate solutions. *J. Appl. Electrochem.* **2003**, *33*, 1233–1237. [CrossRef]

Article

The Influence of the Type of Electrolyte in the Modifying Solution on the Protective Properties of Vinyltrimethoysilane/Ethanol-Based Coatings Formed on Stainless Steel X20Cr13

Aleksandra Kucharczyk [1], Lidia Adamczyk [1,*] and Krzysztof Miecznikowski [2]

[1] Department of Materials Engineering, Faculty of Production Engineering and Materials Technology, Czestochowa University of Technology, Aleja Armii Krajowej 19, 42-200 Czestochowa, Poland; aleksandra.kucharczyk@pcz.pl

[2] Department of Inorganic and Analytical Chemistry, Faculty of Chemistry, University of Warsaw, Ludwika Pasteura 1, 02-093 Warsaw, Poland; kmiecz@chem.uw.pl

* Correspondence: lidia.adamczyk@pcz.pl

Citation: Kucharczyk, A.; Adamczyk, L.; Miecznikowski, K. The Influence of the Type of Electrolyte in the Modifying Solution on the Protective Properties of Vinyltrimethoysilane/Ethanol-Based Coatings Formed on Stainless Steel X20Cr13. *Materials* **2021**, *14*, 6209. https://doi.org/10.3390/ma14206209

Academic Editor: Vít Křivý

Received: 20 September 2021
Accepted: 15 October 2021
Published: 19 October 2021

Publisher's Note: MDPI stays neutral with regard to jurisdictional claims in published maps and institutional affiliations.

Copyright: © 2021 by the authors. Licensee MDPI, Basel, Switzerland. This article is an open access article distributed under the terms and conditions of the Creative Commons Attribution (CC BY) license (https://creativecommons.org/licenses/by/4.0/).

Abstract: The paper reports the results of the examination of the protective properties of silane coatings based on vinyltrimethoxysilane (VTMS) and ethanol (EtOH), doped with the following electrolytes: acetic acid (AcOH), lithium perchlorate LiClO$_4$, sulphuric acid (VI) H$_2$SO$_4$ and ammonia NH$_3$. The coatings were deposited on stainless steel X20Cr13 by the sol–gel dip-coating method. The obtained VTMS/EtOH/Electrolyte coatings were characterized in terms of corrosion resistance, surface morphology and adhesion to the steel substrate. Corrosion tests were conducted in sulphate media acidified up to pH = 2 with and without chloride ions Cl$^-$, respectively. The effectiveness of corrosion protection was determined using potentiometric curves. It has been demonstrated that the coatings under study slow down the processes of corrosion of the steel substrate, thus effectively protecting it against corrosion.

Keywords: vinyltrimethoxysilane; silane; corrosion; sol–gel

1. Introduction

The majority of metals in contact with atmospheric air (electrochemical corrosion) form a protective layer—an oxide (passive) film of their surface. The phenomenon of passivation provides a basis for the natural corrosion resistance of some metals and construction alloys, such as aluminium or stainless steels. However, it is not fully sufficient in more aggressive media, for example, in locations where the metal is exposed to the action of chloride, bromide, or fluorine ions. The effects of corrosion processes are usually associated with additional, often considerable costs; therefore, various methods for protection against corrosion are used [1,2]. As mentioned above, the process of electrochemical corrosion proceeds in the environment of aggressive solutions of electrolytes. Therefore, this phenomenon affects the correct functioning of different types of current sources and materials (such as concrete, steel, etc.), which many people tend to forget.

Stainless steels, or iron alloys containing 12–18% chromium, have for many years enjoyed great popularity because of their corrosion resistance, availability and price. They have unique characteristics, thanks to which they find increasingly wide application in today's technology. Their corrosion resistance, mechanical or engineering properties enable them to be used in particularly demanding working environments [3]. The martensitic stainless steel X20Cr13 used in the present study, intended for toughening, exhibits high mechanical properties, such as strength, ductility and machinability, while retaining sufficient corrosion resistance. X20Cr13 does not show resistance to chlorine, salts and intercrystalline corrosion. Stainless steel is suitable for operating in the environments of water vapour, low concentrated inorganic acids, solvents and pure water [4]. The corrosion resistance of stainless steels can be enhanced by introducing to them alloy additions, such as nickel or molybdenum, but also by forming protective coats on their surface [5,6].

The modification of metal surface belongs to corrosion prevention methods. Apart from classical metallic, inorganic, or paint coatings, also coatings of organosilicon compounds (siloxane compounds) obtained from modifying solution, attract a lot of attention. These are coatings composed of siloxane bonds formed as a result of the reactions of hydrolysis and condensation. One of the simpler and the most common methods of depositing silane-based coatings is the sol–gel method [7,8]. Silane coatings are used especially for protecting poorly passivating metals, such as Al, Fe, Zn, Mg, Ti and stainless steel [3,9].

Silanes are compounds characterized by low toxicity and, most importantly, offer protection against corrosion and good adhesion to a substrate (e.g., steel). During the deposition of silane-based protective coatings, strong covalent bonds form [5,6]. It has also been shown that organosilanes provide effective corrosion protection of materials, such as aluminium and steel [10,11]. Silanes are used in corrosion protection chiefly as interlayers as they provide the adhesion of coatings to metals [12–16].

The protective properties of silane coatings (structure, stability in time, tightness, corrosion resistance) are related to the parameters of the silane solution (silane type, composition, concentration, pH), as well as with the method of application and the process of drying of the deposited coating at a specific temperature. In many publications [17–20] the authors, prior to depositing a sol–gel coating, used pre-treatment of the steel surface with acids with the aim of enhancing the adhesion and anticorrosive properties. Coatings based on organosilicon compounds are deposited on the surface of substrate elements, usually from modifying solutions being sol–gel solutions [21–32]. In the majority of cases, acetic acid and ammonia were included in the composition of modifying solutions [26–32].

The precursor of the reaction of synthesis in the sol–gel method are various alcoholates of metals, salts or nitrates. After immersing the metal in a diluted silane solution, particles are adsorbed on the metal surface through hydrogen bonds. The key reactions are hydrolysis and condensation, whereby a compact protective coating forms at the silane/metal interface. The hydrolysis and condensation (polycondensation) reactions occur simultaneously within the whole volume of the solution. The properties of the end product and the rate of the process are strongly influenced by, e.g., the R≡H_2O: silane mole ratio, medium pH, solvent type, the nature and concentration of catalysts, and temperature; for example, individual stages of the sol–gel process run faster when an appropriate (acidic or basic) catalyst is used [33–35].

The present study is devoted to the structural examination and corrosion testing of coatings composed of vinyltrimethoxysilane (VTMS), ethanol and electrolytes, deposited on stainless steel X20Cr13. Within the study, the effect of the addition of an acidic and a basic electrolyte on the structural properties and corrosion protection of the investigated stainless steel was examined. The aim of the investigation was to obtain vinyltrimethoxysilane-based coatings by the dip-coating method of the best possible physical, chemical and anticorrosive protection, which could be used for the corrosion protection of applied metals and their alloys (Fe, Al, Zn, Cu and Cu).

Over a dozen or so years, many papers on the protection of metal surface with silanes have been published; it should be emphasized, however, that those publications did not address the effect of modifiers, i.e., electrolytes of varying pH values, on the process of protecting metals covered with silane coatings against corrosion.

2. Materials and Methods

Analytically pure reagents and deionized water were used in experiments. Sol–gel solutions were prepared by mixing vinyltrimethoxysilane (VTMS) of the molecular formula CH_2=CHSi(OC_2H_5)$_3$ (supplied by Sigma Aldrich), anhydrous ethyl alcohol EtOH (supplied by Sigma Aldrich) and electrolyte (acidic and basic, respectively). The volumetric VTMS:EtOH:Electrolyte ratio of the obtained coating was 4.84:2.16:3.0.

The following electrolytes were selected for testing:

- acetic acid (AcOH) by Chempur (the method of depositing the coating based on VTMS and acetic acid was developed in previous studies [36]);
- lithium perchlorate (LiClO$_4$) by Fluka Chemika, as a strongly oxidizing anion of inhibiting properties [37–42];
- sulphuric acid (VI) (H$_2$SO$_4$) by Chempur; an acid medium, with a passive film forming on the metal surface, containing sparingly soluble thermodynamically stable oxidation products [43–46];
- ammonia (NH$_3$)- P.P.H. by Polskie Odczynniki Chemiczne; a basic medium, for controlling the hydrolysis and condensation rates [47–52].

2.1. The Influence of the Reaction Environment on the Sol–Gel Process

Four (4) electrolytes (acetic acid CH$_3$COOH (AcOH), lithium perchlorate LiClO$_4$, sulphuric acid (VI) H$_2$SO$_4$ and ammonia NH$_3$) were chosen for the sol–gel process on account of the hydrolysis reaction: acidic and basic hydrolysis.

Table 1 shows the effect of the medium (electrolyte) on individual stages of the sol–gel process. Table 2 gives examples of substances that speed up this process.

Table 1. The influence of the reaction environment on the speed of the sol–gel process.

Reaction Environment	Hydrolysis	Condensation	Gelation
alkaline	speed drop	high speed	fast
inert	the slowest	faster	easier at pH $\geq$ 7
acidic	high speed	0–2 pH—high speed 2–3 pH—the slowest 3–5 pH—speed drop	slow

Table 2. Substances accelerating the sol–gel process.

Stages of the Sol–Gel Process	Hydrolysis	Condensation
Catalysts	Aqueous solution HCl, HNO$_3$, HF, icy CH$_3$COOH	NH$_3$, NaOH

The sol–gel process can be run by two methods. The first method is one-stage basic or acidic catalysis. In the case of basic catalysis, hydrolysis proceeds with the participation of the hydroxyl anion (OH$^-$). This ion reacts directly with the silicon atom, leading to the formation of silanol and the RO$^-$ group. It enables semi-transparent gels of high porosity to be obtained. In the case of acidic catalysis, on the other hand, the hydrolysis process is initiated by acid. Protons H$^+$ react with the oxygen atoms bonded with the silicon atom in the –OR or OH group. This causes the electron cloud to shift towards the oxygen atom in the Si-O bond. As a consequence, this results in an increase in positive charge on the silicon atom. The water molecule combines with the silicon atom, which is followed by disintegration and the formation of silanols and alcohol. In acidic catalysis, we obtain transparent gels of low porosity [50].

For technological reasons, the sol–gel process is most favourably carried out by using the second of the above-mentioned methods, i.e., two-stage acidic-basic catalysis. This shortens the time necessary for obtaining the gel. The method comprises the first stage—hydrolysis at pH < 7, the second stage—raising the medium reaction up to a pH of approximately 7, whereby we slow down the hydrolysis process and accelerate the gelation process [51].

2.2. Preparation of Test Material

Stainless steel X20Cr13 of the following composition (in wt%): C-0.17; Cr-12.6; Si-0.34; Ni-0.25; Mn-0.30; V-0.04; P-0.024; and S < 0.005 was used for testing. Test samples were in the shape of 5 mm-diameter cylinders. Their walls were isolated in polymethyl methacrylate frames using epoxy resin. The geometric working surface area of the samples was 0.196 cm^2. Prior to experiments, the samples were each time mechanically polished on abrasive papers with a decreasing grit size up to grade 2000, and then rinsed with distilled water and ethyl alcohol. Before applying a coating, each sample was flushed with acetone to degrease its surface. On the prepared electrodes, coatings were deposited by the dip-coating method following the procedure developed in the paper [36]. The composition and the stirring parameters for the test coatings are shown in Table 3.

Table 3. Composition of coatings and mixing parameters.

Number	C$_m$ VTMS [mol dm^{-3}]	EtOH [mL]	H$_2$O [mL]	Electrolyte	Electrolyte Concentration [mol dm^{-3}]	Mixing Time [h]	Immersion Time of the Steel Sample [min]	Rotations/min
1	3.16	2.16	2.9833	CH$_3$COOH 0.0167 mL	0.003	48	20	400–1000
2	3.16	2.16	2.8936	LiClO$_4$ 0.1064 g	0.1	48	20	400–1000
3	3.16	2.16	2.5	2 mol dm^{-3} H$_2$SO$_4$ 0.5 mL	0.1	24	20	400–1000
4	3.16	2.16		1–2 drops of 25% ammonia were added to 100 mL of water, then the pH was measured with a litmus paper	pH = 8–9	24	20	400–1000

The anticorrosive properties of the obtained coatings were assessed in two corrosion media: 0.5 mol dm^{-3} Na$_2$SO$_4$ (pH = 2) and 0.5 mol dm^{-3} Na$_2$SO$_4$ + 0.5 mol dm^{-3} NaCl (pH = 2) using a potentiodynamic technique with the use of a scanning potential from—0.8 V to 1.6 V at a polarization rate of 10 mVs^{-1}. The values of all potentials were measured and expressed relative to the saturated calomel electrode.

The composition and surface appearance of the coatings deposited on the investigated steel were assessed using a JEOL JSM-6610 LV scanning electron microscope with an EDS-type X-ray microanalyzer (JEOL, Tokyo, Japan). Microstructural examination was carried out with a KEYENCE VHX 7000 digital microscope and an Olympus GX41 optical microscope (Keyence, Mechelen, Belgium). The surface roughness of the coatings was measured using an Hommel Tester T1000 profilometer (JENOPTIK Industrial Metrology, Jena, Germany). The microscopic surface maps were made using an AFM (Atomic Force Microscope) NanoScope V MultiMode 8 (BRUKER, Bremen, Germany). The characteristics of the coatings were determined using Attenuated Total Reflection Fourier Transform Infrared Spectroscopy (ATR- FTIR) Bruker Optics-Vertex 70 V (BRUKER, Bremen, Germany). The coatings thicknesses were measured using a DT-20 AN 120 157 meter (ANTICORR, Gdańsk, Poland). Electrochemical measurements were taken in the classical three-electrode system using a CHI 706 measuring station (CH Instruments, Austin, TX, USA). The working electrodes were steel X20Cr13 coated and uncoated as well as glassy carbon, an auxiliary electrode (platinum), and a reference electrode (the saturated calomel electrode, SCE).

The adhesion of the test coatings was assessed by a qualitative test using Scotch^{TM} Tape (Scotch^{TM} Brand, St. Paul, MI, USA).

2.3. The Preparation of Samples for Structural Examination (Sample Cross-Section)

Samples for structural examination (sample cross-section) were prepared in four steps: application of a coating on steel, cutting the sample through in the plane perpendicular to the surface, then the samples were embedded in epoxy resin with graphite. So embedded specimens were subjected to grounding and polishing to obtain a mirror surface.

2.4. Corrosion Resistance Test in a Potassium Hexacyanoferrate (III) Solution (Ferroxyl Test)

For performing a quick test to indicate the barrier nature of the proposed protective coatings, it was applied the well-known and widely used ferroxyl test (ferroxyl indicator) in a modified form. For this purpose, the tested plate was immersed in a solution containing potassium ferricyanide. In the presence of iron ions (formed in the process of corrosion in an acid medium), insoluble blue iron (II) hexacyanoferrate (III) (Prussian Blue) appears, indicating that the material dissolving in anodic locations. A 2-mmol dm^{-3} potassium hexacyanoferrate (III) K$_3$[Fe(CN)$_6$] solution was utilized for testing. An X20Cr13 steel sample uncovered and covered with a VTMS/EtOH/AcOH coating in a VTMS concentration of 3.16 mol dm^{-3} were immersed in potassium hexacyanoferrate (III) solution. Afterward, electrochemical measurements were taken by the cyclic voltammetry in the potential range from −0.6 V to 1.2 V vs. SCE [49,50].

3. Results

3.1. Microstructural Observations and Chemical Analysis

The topography of obtained coatings was assessed using a light microscope. Figures 1 and 2 show the morphology of the investigated coatings deposited on the X20Cr13 stainless steel surface. All of the four coating types uniformly cover the entire surface of the electrodes without any free structural spaces. A structure of the metallic substrate with polishing traces is visible in the photographs (Figure 1). As shown in Figure 1, the coatings morphology is compact, smooth, shiny and transparent over the entire sample surface.

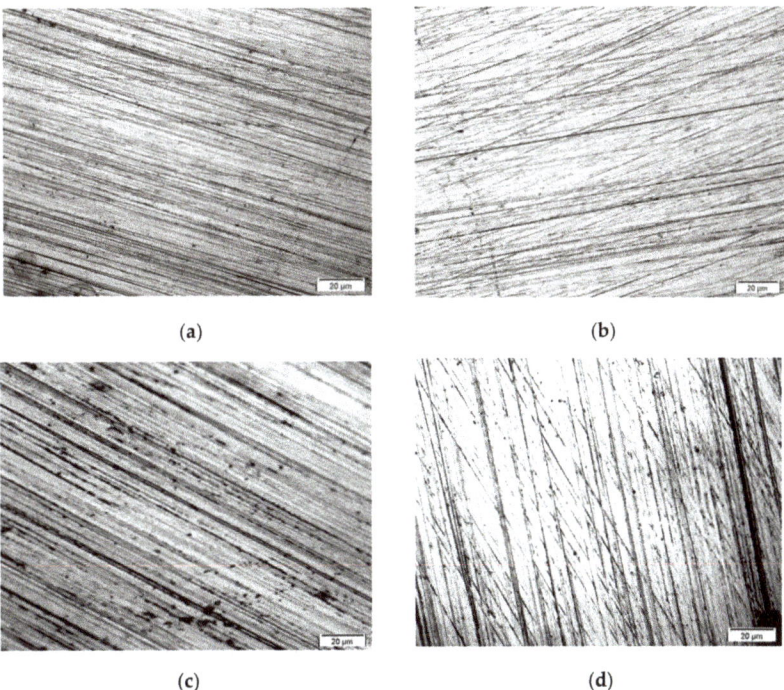

Figure 1. Coatings topography: VTMS/EtOH/AcOH (**a**), VTMS/EtOH/LiClO$_4$ (**b**), VTMS/EtOH/H$_2$SO$_4$ (**c**), VTMS/EtOH/NH$_3$ (**d**), (×200).

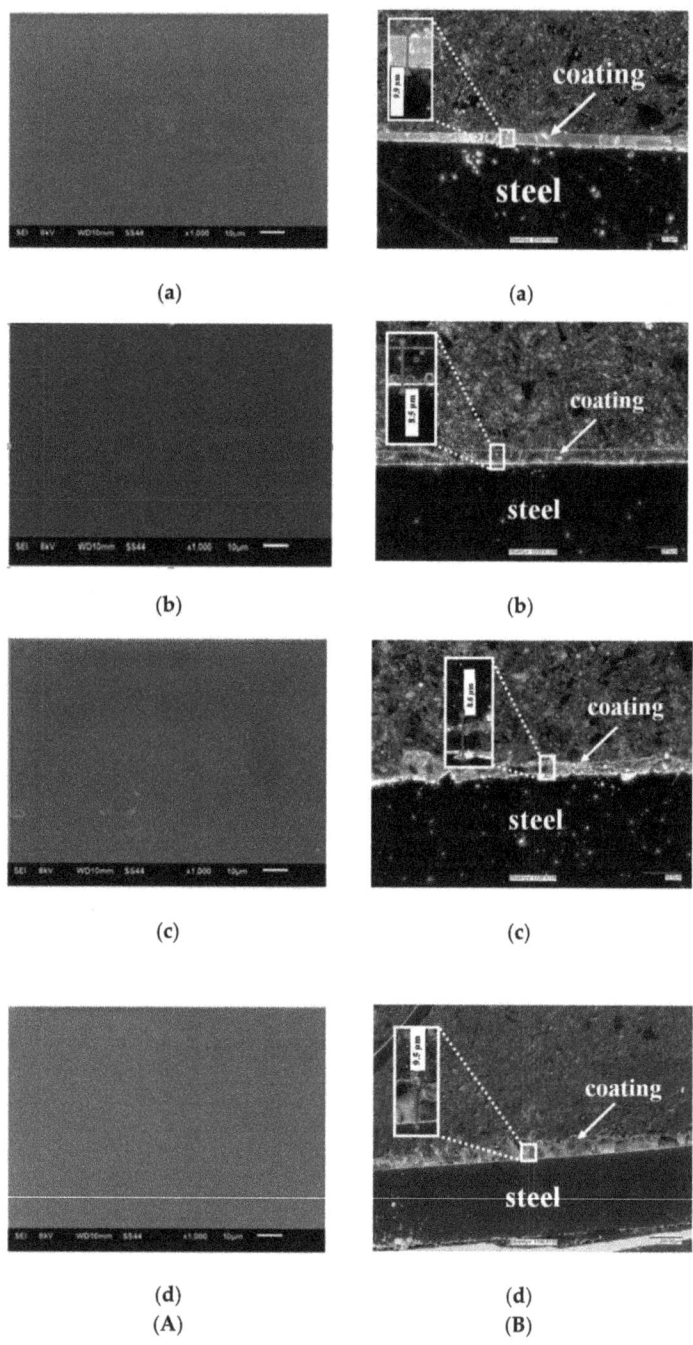

Figure 2. (**A**) Topography of VTMS coatings deposited on steel X20Cr13 (SEM Jeol JSM-6610 LV), (**B**) Cross-section of coatings deposited on steel X20Cr13 (digital microscope KEYANCE VHX 7000, ×200). Coatings: VTMS/EtOH/AcOH (**a**), VTMS/EtOH/LiClO$_4$ (**b**), VTMS/EtOH/H$_2$SO$_4$ (**c**), VTMS/EtOH/NH$_3$ (**d**).

Figure 2 represents the topography of VTMS/EtOH/AcOH, VTMS/EtOH/LiClO$_4$, VTMS/EtOH/H$_2$SO$_4$, VTMS/EtOH/NH$_3$ coatings, revealed using a scanning electron microscope. The obtained results confirm the previous observations: regardless of the added electrolyte, the morphology of VTMS coatings covers uniformly the sample surface, forming a compact, tight and homogeneous structure.

Figure 2B shows the cross-section of the obtained VTMS/EtOH/Electrolyte coatings deposited on the X20Cr13 stainless steel. The recorded profile indicates that the coatings are uniformly deposited on the steel surface. The structural observations confirm that the coatings are free from any cracking and their surface roughness is negligible, which is a huge asset from the point of view of the corrosion resistance of steel. The average coatings thickness (as measured in four locations on the sample) were the following for respective coatings: VTMS/EtOH/AcOH 11.4 µm (a); VTMS/EtOH/LiClO$_4$ 8.05 µm (b); VTMS/EtOH/H$_2$SO$_4$ 8.65 µm (c); VTMS/EtOH/NH$_3$ 12.8 µm (d). Based on the cross-section, it was demonstrated that the modification of the silane solution with various electrolytes had a significant effect on the coating thickness.

3.1.1. Chemical Composition of the VTMS/EtOH/Electrolyte Coatings

The chemical composition of the produced coatings was determined using an electron scanning microscope equipped with an EDS X-ray chemical analyzer. Based on the performed chemical analysis, the contents of silicon for respective coatings are as follows: VTMS/EtOH/AcOH 32.47%; VTMS/EtOH/LiClO$_4$ 29.05%; VTMS/EtOH/H$_2$SO$_4$ 30.26%; VTMS/EtOH/NH$_3$ 25.87%. The rest was made up of the elements C and O.

3.1.2. Testing for Adhesion to the Substrate

Immediately after depositing VTMS/EtOH/Electrolyte coatings, their adhesion to the X20Cr13 stainless steel substrate was tested using Scotch^{TM} Tape. The produced coatings are characterized by good adhesion to the steel substrate.

3.1.3. Surface Roughness of Obtained Coatings

The measurement of the surface roughness of the VTMS/EtOH/Electrolyte coatings deposited on the X20Cr13 stainless steel taken with a profilometer is represented in Figure 3. Both the topography and profile of the coatings confirm that the coatings are free from cracking, and their surface roughness is little (low Ra values). The testing results differ, depending on the electrolyte used; the closest Ra values can be observed for VTMS/EtOH/AcOH and VTMS/EtOH/NH$_3$ coatings (Ra = 0.40–0.43 µm). Table 4 shows the results of the measurement of parameter Ra.

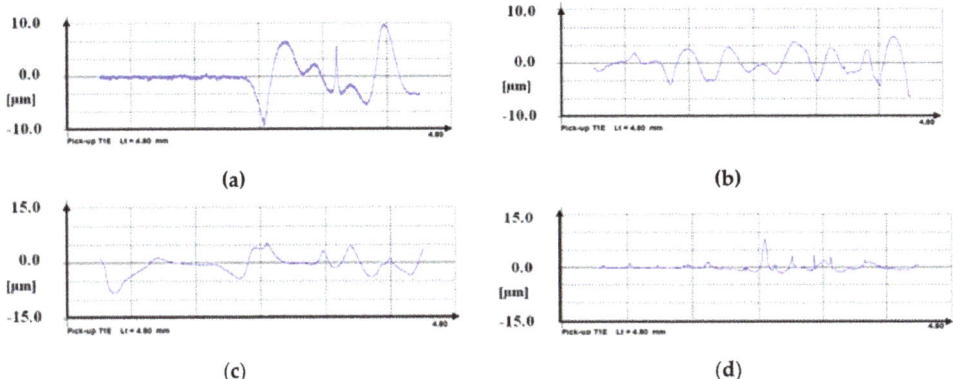

Figure 3. Roughness measurement Ra for coatings deposited on X20Cr13 steel: VTMS/EtOH/AcOH (a), VTMS/EtOH/LiClO$_4$ (b), VTMS/EtOH/H$_2$SO$_4$ (c), VTMS/EtOH/NH$_3$ (d). Profilometer Hommel Tester T1000.

Table 4. Roughness parameter Ra for individual coatings deposited on steel X20Cr13.

Coating	Ra [μm]
VTMS/EtOH/AcOH	0.40
VTMS/EtOH/LiClO$_4$	0.87
VTMS/EtOH/H$_2$SO$_4$	1.32
VTMS/EtOH/NH$_3$	0.43

The examination made using an AFM microscope confirms the previous findings that the addition of electrolyte has an effect on the coating surface roughness. The surface morphologies of VTMS/EtOH/Electrolyte coatings produced on the metal surface, with a varying electrolyte addition, are illustrated in Figure 4. The recorded values of parameter Ra for respective coatings are as follows: VTMS/EtOH/AcOH 0.381 μm; VTMS/EtOH/LiClO$_4$ 0.908 μm; VTMS/EtOH/H$_2$SO$_4$ 1.45 μm; VTMS/EtOH/NH$_3$ 0.389 μm.

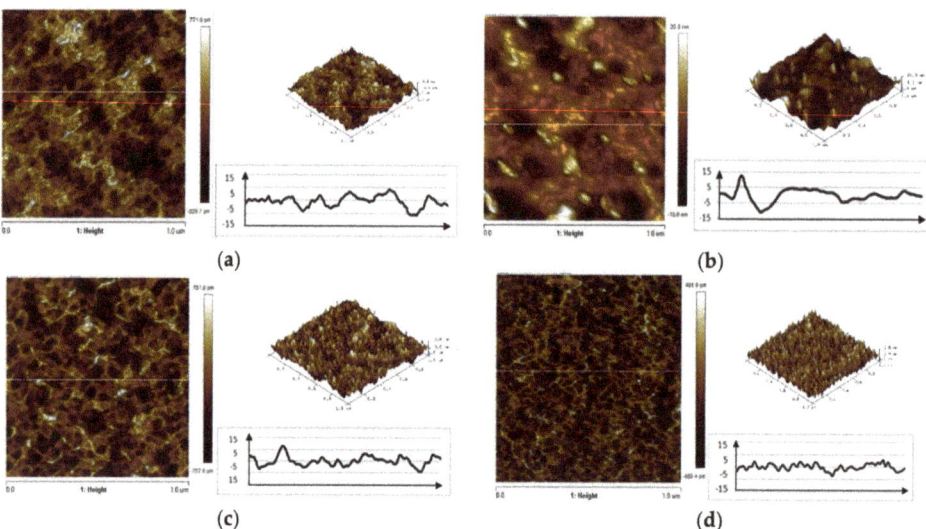

Figure 4. AFM images of the surface of coatings deposited od steel X20Cr13: VTMS/EtOH/AcOH (**a**), VTMS/EtOH/LiClO$_4$ (**b**), VTMS/EtOH/H$_2$SO$_4$ (**c**), VTMS/EtOH/NH$_3$ (**d**). Pictures were taken using an AFM NanoScope V MultiMode 8 Bruker.

Protective coatings are generally porous layers; after some time, the surface of steel or metal will come into contact with an aggressive electrolyte solution, water, or oxygen molecules. A discontinuity in the coating may initiate pitting or crevice corrosion. Obtained protective coatings, as compared to the protected elements, are usually extremely thin.

3.1.4. Thickness of Obtained Coatings

One of the key parameters influencing the corrosion resistance of elements is the thickness of their protective coatings. In the present study, this parameter has been analyzed using three examination methods. Based on profile examination (Figure 2B), the thickness of obtained coatings was analyzed. The thickness of each coating is the average of 4 measurements: VTMS/EtOH/AcOH 11.4 μm (a); VTMS/EtOH/LiClO$_4$ 8.05 μm (b); VTMS/EtOH/H$_2$SO$_4$ 8.65 μm (c); VTMS/EtOH/NH$_3$ 12.8 μm (d).

The recorded thicknesses measured with a profilometer are given in Table 5.

Table 5. Thickness measurement results for individual coatings on steel X20Cr13.

Coating	Coating Thickness [µm]
VTMS/EtOH/AcOH	10.3
VTMS/EtOH/LiClO$_4$	7.9
VTMS/EtOH/H$_2$SO$_4$	8.8
VTMS/EtOH/NH$_3$	11.4

To compare the thicknesses of the coatings, in addition to the methods described above, thickness measurements were taken using a DT-20 Testan meter with an integrated probe designed for measuring on ferro- and non-ferromagnetic substrates. A series of 10 measurements (at different locations on the sample) was done; Table 6 provides recorded thickness values for VTMS/EtOH/Electrolyte coatings. The obtained coatings thickness values are consistent with those produced with a digital microscope and a profilometer.

Table 6. Coating thickness measurement results using a gauge Testan DT-20.

	Coatings Thickness [µm]			
	VTMS/EtOH/ AcOH	VTMS/EtOH/ LiClO$_4$	VTMS/EtOH/ H$_2$SO$_4$	VTMS/EtOH/ NH$_3$
Average	9.6	8.5	8.4	11.5

Based on the performed measurements using three instruments (a digital microscope, profilometer, and a thickness meter), the mean coatings thickness was determined (Table 7).

Table 7. The average thickness of the coatings calculated by measurements from three instruments (a digital microscope, profilometer, and a thickness meter).

Coating	Digital Microscope Average [µm]	Profilometer Average [µm]	Thickness Meter Average [µm]	Thickness of Coatings (Averegae of the Three Devices) [µm]
VTMS/EtOH/AcOH	11.4	10.3	9.6	10.4
VTMS/EtOH/LiClO$_4$	8.05	7.9	8.5	8.2
VTMS/EtOH/H$_2$SO$_4$	8.65	8.8	8.4	8.6
VTMS/EtOH/NH$_3$	12.8	11.4	11.5	11.9

The differences in coatings thickness between individual methods were negligible, which confirms the usefulness of the applied methods for coatings thickness measurement.

3.2. Analysis of Coatings Composition

The characteristics of vinyltrimethoxysilane- based coatings deposited on the X20Cr13 steel substrate were determined using Attenuated Total Reflection Fourier Transform Infrared Spectroscopy (ATR-FTIR). The ATR-FTIR spectra of VTMS/EtOH/Electrolyte coatings are shown respectively in Figure 5. Characteristic absorption peaks were observed for VTMS in the range of 4000–400 cm^{-1}. The absorption bands observed at the values of 2953 cm^{-1}, 2857 cm^{-1} and 1290 cm^{-1} correspond to the asymmetric tensile and bending vibrations of the C-H bond belonging to the –Si-(OCH$_3$) group. Subsequent peaks were noted at the values of 1602 cm^{-1} and 1410 cm^{-1}, which correspond to the tensile vibrations of the C=C bond of the CH$_2$=CH- group. The bands of values of 1000 cm^{-1}, 883 cm^{-1} and 750 cm^{-1} correspond to the vibrations of Si-O-C. The wide band occurring at about 1100 cm^{-1} corresponds to the asymmetric tensile vibrations of the Si-O-Si bond. The peak at the value of 704 cm^{-1} corresponds to the Si-C bond. The wide absorption band at about 3400 cm^{-1} was caused by the –OH groups. The peak at 964 cm^{-1} was ascribed to the asymmetric bending vibrations of the Si-OH bond, whereas the band observed at the wavelength of 3053 cm^{-1} matches the vibrations in the alkene group.

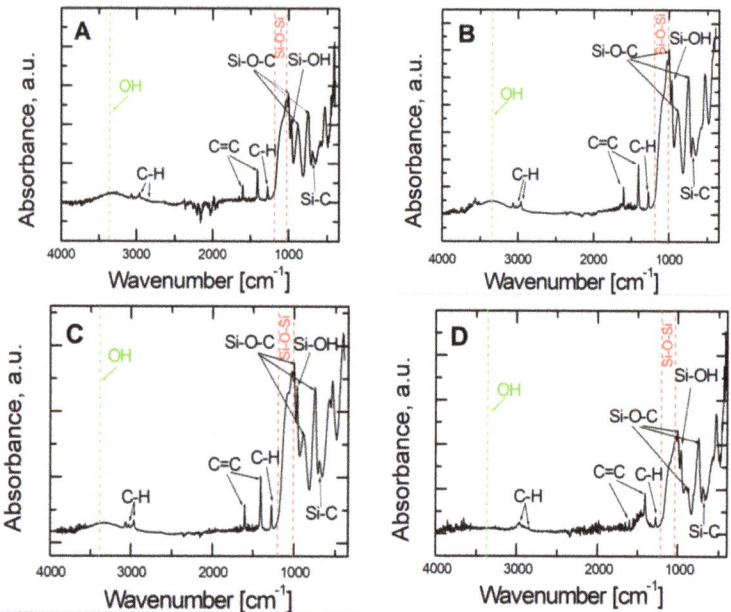

Figure 5. ATR- FTIR spectra obtained for VTMS coatings with a concentration of 3.16 mol dm^{-3} deposited on steel X20Cr13: VTMS/EtOH/AcOH (**A**), VTMS/EtOH/LiClO$_4$ (**B**), VTMS/EtOH/H$_2$SO$_4$ (**C**), VTMS/EtOH/NH$_3$ (**D**).

3.3. Electrochemical Analysis

To make the assessment of the kinetic tendency to either general or pitting corrosion, measurements of open circuit potential (OCP) were taken for steel uncovered and covered with a VTMS coating, respectively, with the addition of electrolyte in the form of either: acetic acid (b), lithium perchlorate (c) sulphuric acid (VI) (d), or ammonia (e).

As shown by the open circuit potential measurements illustrated in Figure 6A,B, the steel not covered with a coating directly after being immersed in the test solutions exhibits a potential of approximately −0.4 V, which decreases for longer exposure times and takes on the value of corrosion potential for steel (−0.5 V). For steel X20Cr13 covered with the coatings: VTMS/EtOH/LiClO$_4$, VTMS/EtOH/H$_2$SO$_4$ and VTMS/EtOH/NH$_3$, Figure 6A, the OCP potential increases for the initial 24 h until reaching a value of 0.4 V. As can be seen from Figure 6A (line b), the steel covered with a VTMS/EtOH/AcOH coating directly after being immersed in the corrosion solution shows a potential of 0.2 V, which increases up to 0.4 V after 24 h exposure. The values of potential for all steels covered with coatings after prolonged immersion in the corrosion solution show potential from the passive range, so more positive than E$_{kor}$ (0.5 V).

The dependence of the open circuit potential of uncoated and coated steel on the time of holding in the chloride ion-containing corrosion solution is represented in Figure 6B. The uncoated X20Cr13 steel undergoes active dissolution after approximately 50 h of immersion in the corrosion solution. By contrast, the steel covered with VTMS-based coatings, upon immersion in the corrosion solution, exhibits a potential from the passive range. The potential of the steel covered with VTMS/EtOH/AcOH coatings increases, for the initial 24 h, up to a value of approximately 0.45 V and stays on this level for another 13.5 days; for VTMS/EtOH/H$_2$SO$_4$, the potential is −0.25 V and remains for 350 h;

for VTMS/EtOH/NH₃, after 150 h, it amounts to −0.35 V and holds on this level for subsequent 200 h; and for VTMS/EtOH/LiClO₄, the potential stays at the level of 0.35 V for 240 h and then dramatically decreases to a value of 0.0 V.

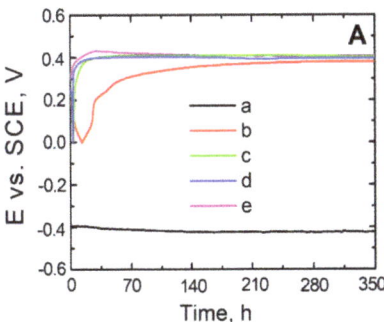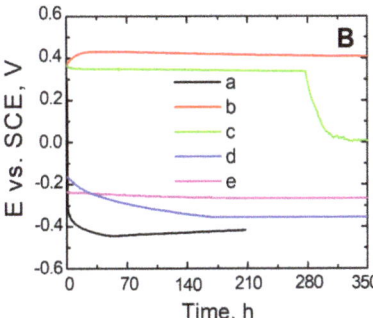

Figure 6. Potential measurement in open circuit potential OCP from exposure time in solution: 0.5 mol dm⁻³ Na₂SO₄ mol dm⁻³ pH = 2 (**A**) and 0.5 mol dm⁻³ Na₂SO₄ + 0.5 mol dm⁻³ NaCl pH = 2 (**B**) for steel X20Cr13 uncovered (a) and covered with coatings VTMS/EtOH: CH₃COOH (b), LiClO₄ (c), H₂SO₄ (d), NH₃ (e).

It is worth noting that the stationary potential value of the coated steel, despite the log time of exposure in the chloride ion-containing corrosion solution, is more positive than the stationary potential value of steel. Microscopic observations after the measurement did not reveal any local corrosion effects under the VTMS/EtOH/AcOH coating, which indicates significant substrate protection.

To establish the most effective influence of electrolytes on the anticorrosion properties of the produced VTMS silane coatings deposited on the X20Cr13 steel, the assessment of their capacity for inhibiting general and pitting corrosion was made using potentiodynamic curves. The experiment was conducted in two solutions:

- for general corrosion: 0.5 mol dm⁻³ Na₂SO₄ pH = 2 (Figure 7A),
- for pitting corrosion: 0.5 mol dm⁻³ Na₂SO₄ + 0.5 mol dm⁻³ NaCl pH = 2 (Figure 7B).

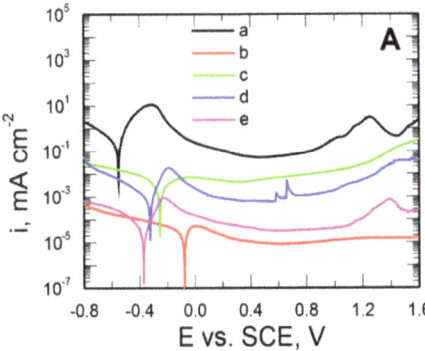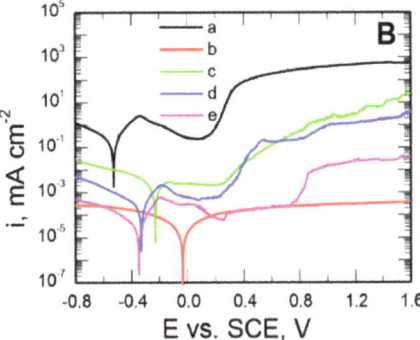

Figure 7. Potentiodynamic polarization curves recorded in the solution: 0.5 mol dm⁻³ Na₂SO₄ pH = 2 (**A**) and 0.5 mol dm⁻³ Na₂SO₄ + 0.5 mol dm⁻³ NaCl pH = 2 (**B**) for uncoated steel X20Cr13 (a) and covered with coatings VTMS concentrations in a 3.16 mol dm⁻³ solution and the addition of an electrolyte: CH₃COOH (b), LiClO₄ (c), H₂SO₄ (d), NH₃ (e). Polarization rate 10 mVs⁻¹, solutions in contact with air.

The potential range of −0.8–1.6 V for the X20Cr13 steel uncoated and coated, respectively.

As follows from Figure 7A, the produced VTMS/EtOH/Electrolyte coatings inhibit the cathodic and anodic processes and shift the corrosion potential of the steel by approximately 0.5 V (the VTMS/EtOH/AcOH coating). The anodic current densities for the steel covered with VTMS/EtOH/Electrolyte coatings in the passive range are smaller by 1–4 times than those for the uncoated steel.

To assess the capacity of the produced coatings to inhibit pitting corrosion, similar potentiodynamic curves were plotted for a sulphate solution acidified to pH = 2, containing an addition of 0.5 mol dm^{-3} of chloride ions (Figure 7B).

The corrosion potential of the X20Cr13 steel for all coatings is shifted by approximately 0.1–0.5 V towards positive values relative to the corrosion potential values recorded for the uncoated steel (E_{kor} = -0.527 V). Lower values of cathodic and anodic current densities were also observed for the steel covered with these coatings, compared to the uncoated steel.

The shape of the polarization curves shows that the pitting nucleation potential (E_{pit}) amounts to, respectively: for the uncoated steel 0.12 V; for the steel covered with the coatings: VTMS/EtOH/$LiClO_4$ 0.18 V; VTMS/EtOH/H_2SO_4 0.19 V; VTMS/EtOH/NH_3 0.64 V. The thermodynamic susceptibility to pitting is similar for the coatings VTMS/EtOH/$LiClO_4$ and VTMS/EtOH/H_2SO_4. As shown by Figure 7B (line b), in the case of employing the VTMS/EtOH/AcOH coating for steel protection, no puncture potential of the passive film (pitting nucleation potential) was observed. The silane coating modified with acetic acid effectively hinders the access of aggressive anions to the steel substrate, thus protecting the substrate against pitting corrosion. Microscopic observations after the measurement did not reveal any local corrosion effects under the VTMS/EtOH/AcOH coating. Figure 7B implies that the application of coatings on steel protects the substrate against local corrosion.

To verify the resistance of coatings deposited on steel to pitting corrosion, the chronoamperometric method was employed. In this method, variations in current density are recorded as a function of time after applying a constant potential to the working electrode. From chronoamperometric curves, one can infer the nucleation of pits.

To determine the stability of the applied coats, the time of holding the test samples in the corrosion solution containing chloride ions and the value of current density were compared at a preset potential. Chronoamperometric curves were recorded in a 0.5 mol dm^{-3} solution of Na_2SO_4 + 0.5 mol dm^{-3} NaCl with pH = 2 at a potential of 0.1 V for uncoated and coated steel, respectively.

Figure 8 shows the chronoamperometric curves for steel plotted at a potential of 0.1 V red out from the polarization curves, Figure 7B. As can be observed, the initiation of pit formation on the steel occurs within several seconds, after which the value of current density dramatically increases. In the case of applying the following coating types, VTMS/EtOH/AcOH, VTMS/EtOH/$LiClO_4$, and VTMS/EtOH/H_2SO_4, the highest corrosion resistance was achieved. The increase in current density for the above-mentioned coatings occurred within a time span ranging from 250 to 312 h. The best capacity to block the transport of chloride ions responsible for pitting corrosion is shown by the VTMS/EtOH/AcOH coating (312 h).

3.4. Corrosion Resistance Test in a Potassium Hexacyanoferrate (III) Solution (Ferroxyl Test)

To demonstrate the corrosion resistance of coatings deposited on the X20Cr13 steel, electrochemical tests were carried out in a 2 mmol dm^{-3} solution of $K_3[Fe(CN)_6]$.

Figure 9 shows a typical voltammetric response of the glassy carbon electrode (A) and the VTMS/EtOH/AcOH–coated X20Cr13 steel electrode (B) in the presence of $Fe(CN)_6^{3-}$ sampler ions. In the case of the pure glassy carbon electrode (Figure 9A), we observe a well-developed and quasi-reversible pair of ferrocyanide ions. By contrast, Figure 9B illustrates the voltammetric response of the VTMS/EtOH/AcOH coating on the X20Cr13 steel substrate, on which it does not observe any electrochemical response in the investigated potential range. This is associated primarily with the fact that ferrocyanide ions do not cross through the produced VTMS/EtOH/AcOH layer (pores in the layer are smaller than the size of the ferrocyanide ion). Additionally, no formation of the blue colouring (Prussian Blue formation) was observed on the steel surface, confirming definitely that the obtained coating provides a compact and tight protective barrier. Moreover, the VTMS/EtOH/AcOH layer formed on the X20Cr13 steel blocked the transport of electrons to ferrocyanide ions, has manifested itself by the attenuation of the redox currents (Figure 9B) [49,50].

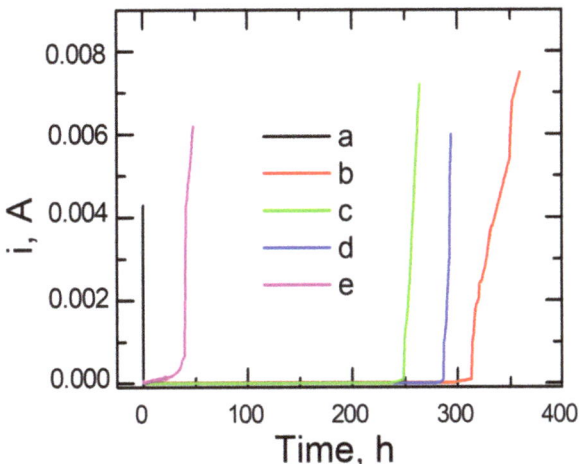

Figure 8. Chronoamperometric curves recorded in a chloride solution (0.5 mol dm^{-3} Na$_2$SO$_4$ + 0.5 mol dm^{-3} NaCl pH = 2) for X20Cr13 steel not covered with the coating (a) and coated with VTMS in 3.16 mol dm^{-3} solution and addition of electrolyte: CH$_3$COOH (b), LiClO$_4$ (c), H$_2$SO$_4$ (d), NH$_3$ (e).

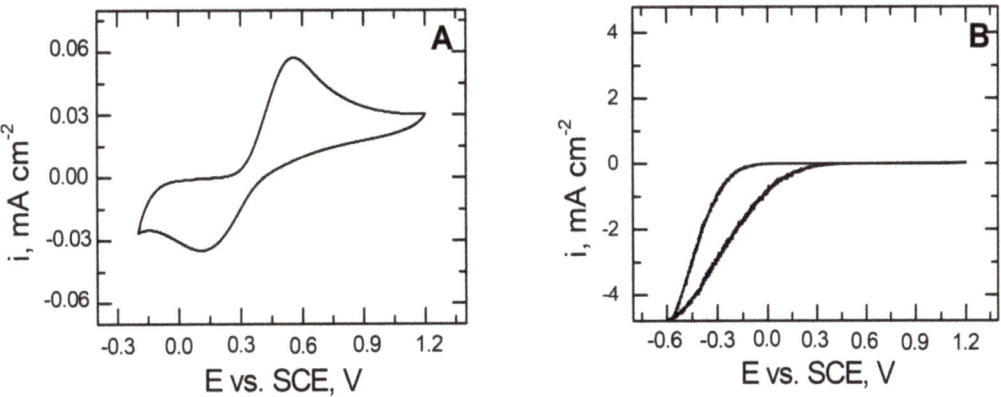

Figure 9. Voltammetric response for: glassy carbon (**A**) and coated X20Cr13 steel with VTMS/EtOH/AcOH (**B**). Electrolyte: 2 mmol dm^{-3} K$_3$[Fe(CN)$_6$]. Polarization rate 10 mVs^{-1}.

4. Conclusions

The investigation of VTMS/EtOH/Electrolyte coatings has shown that the sol–gel process can be used for producing protective layers on stainless steel X20Cr13.

The selection of the appropriate electrolyte has a significant impact on the corrosion and structural properties of VTMS coatings (a uniform surface with no visible defects in the structure). The produced coatings exhibit good adhesion to the substrate and, in addition, extend the duration of steel resistance to the action of chloride and sulphate ions in an acid medium. The best ability to block the transport of chloride ions responsible for the pitting corrosion of steel is shown by the VTMS/EtOH/AcOH coating. The surface roughness and thickness of the coating may be influenced by the size of the doped electrolyte ion. Acetic acid-doped silane coatings deposited on the X20Cr13 steel, with low surface roughness and a small thickness of the coating, exhibit the anticorrosion properties.

Data obtained from potentiodynamic measurements show that the produced VTMS/EtOH/Electrolyte coatings provide stainless steel's anodic and barrier protection. An experiment using a potassium hexacyanoferrate (III) solution has confirmed that the VTMS/EtOH/AcOH coating forms a uniform, tight structure and blocks the transfer of electrons to ferrocyanide ions.

Author Contributions: Conceptualization and idea of this study, A.K. and L.A.; Methodology, L.A. and K.M.; Formal analysis, L.A. and K.M.; Writing—original draft preparation, A.K., L.A. and K.M.; Writing—review and editing, L.A. and A.K.; Visualization, L.A. and K.M.; Supervision, L.A.; Project administration, L.A.; Funding acquisition, L.A. All authors have read and agreed to the published version of the manuscript.

Funding: This research received no external funding.

Institutional Review Board Statement: Not applicable.

Informed Consent Statement: Not applicable.

Data Availability Statement: Not applicable.

Conflicts of Interest: The authors declare no conflict of interest.

References

1. Deflorian, F.; Rossi, S.; Fedrizzi, L. Silane pre-treatments on copper and aluminium. *Electrochem. Acta* **2006**, *51*, 6097–6103. [CrossRef]
2. Bajat, J.B.; Milošev, I.; Jovanovic, Z.; Jancic-Heinemann, R.M.; Dimitrijevic, M.; Miškovic´-Stankovic, V.B. Corrosion protection of aluminium pretreated by vinyltriethoxysilane in sodium chloride solution. *Corros. Sci.* **2010**, *52*, 1060–1069. [CrossRef]
3. Adamczyk, L.; Giza, K.; Dudek, A. Electrochemical preparation of composite coatings of 3,4-etylenodioxythiophene (EDOT) and 4-(pyrrole-1-yl) benzoic acid (PyBA) with heteropolyanions. *Mater. Chem. Phys.* **2014**, *144*, 418–424. [CrossRef]
4. Ryan, M.; Williams, D.; Chater, R.; Hutton, B.M.; McPhail, D.S. Why stainless steel corrodes. *Nature* **2002**, *415*, 770–774. [CrossRef]
5. Asri, R.I.M.; Harun, W.S.W.; Samykano, M.; Lah, N.A.C.; Ghani, S.A.C.; Tarlochan, F.; Raza, M.R. Corrosion and surface modification on biocompatible metals: A review. *Mater. Sci. Eng.* **2017**, *77*, 1261–1274. [CrossRef] [PubMed]
6. Baddoo, N.R. Stainless steel in construction: A review of research, applications, challenges and opportunities. *J. Constr. Steel Res.* **2008**, *64*, 1199–1206. [CrossRef]
7. Brinker, C.J.; Scherer, G.W. *Sol-Gel Science: The Physics and Chemistry of Sol-Gel Processing*; Academic Press: San Diego, CA, USA, 1993.
8. Pierre, A.C. *Introduction to Sol-Gel Processing*; Kluver Academic Publishers: Boston, MA, USA, 1998.
9. Riskin, J.; Khentov, A. *Electrocorrosion and Protection of Metals*, 2nd ed.; Elsevier: Amsterdam, The Netherlands, 2019.
10. Zhu, D.; van Ooij, W.J. Corrosion protection of AA 2024-T3 by bis-[3-(triethoxysilyl)propyl]tetrasulfide in sodium chloride solution.: Part 2: Mechanism for corrosion protection. *Corros. Sci.* **2003**, *45*, 2177–2197. [CrossRef]
11. Plueddemann, E.P. *Silane Coupling Agents*, 2nd ed.; Plenum Press: New York, NY, USA, 1991.
12. Sathyanarayana, M.N.; Yaseen, M. Role of promoters in improving adhesion of organic coatings to a substrate. *Prog. Org. Coat.* **1995**, *26*, 275–313. [CrossRef]
13. Palanivel, V.; Zhu, D.; van Ooij, W.J. Nanoparticle-filled silane films as chromate replacements for aluminum alloys. *Prog. Org. Coat.* **2003**, *47*, 384–392. [CrossRef]
14. Tianlan, P.; Ruilin, M. Rare earth and silane as chromate replacers for corrosion protection on galvanized steel. *J. Rare Earth* **2009**, *27*, 159–163.
15. Balgude, B.; Sabnis, A. Sol–gel derived hybrid coatings as an environment friendly surface treatment for corrosion protection of metals and their alloys. *J. Sol-Gel Sci. Technol.* **2012**, *64*, 124–134. [CrossRef]
16. Liu, B.; Fang, Z.; Wang, H.; Wang, T. Effect of cross linking degree and adhesion force on the anti-corrosion performance of epoxy coatings under simulated deep sea environment. *Prog. Org. Coat.* **2013**, *12*, 1814–1818. [CrossRef]
17. Tavangar, R.; Naderi, R.; Mahdavian, M. Acidic surface treatment of mild steel with enhanced corrosion protection for silane coatings application: The effect of zinc cations. *Prog. Org. Coat.* **2021**, *158*, 106384. [CrossRef]
18. Rouzmeh, S.S.; Naderi, R.; Mahdavian, M. A sulphuric acid surface treatment of mild steel for enhancing the protective properties of an organosilane coating. *Prog. Org. Coat.* **2017**, *103*, 156–164. [CrossRef]

19. Gopi, D.; Saranswathy, R.; Kavitha, L.; Kim, D.K. Electrochemical synthesis of poly(indole-co-thiophene) on low-nickel stainless steel and its anticorrosive performance in 0.5 mol L^{-1} H$_2$SO$_4$. *Polym. Int.* **2014**, *63*, 280–289. [CrossRef]
20. Mekhalif, Z.; Cossement, D.; Hevesi, L.; Delhalle, J. Electropolymerization of pyrrole on silanized polycrystalline titanium substrates. *Appl. Surf. Sci.* **2008**, *254*, 4056–4062. [CrossRef]
21. He, J.; Zhou, L.; Soucek, M.D.; Wollyung, K.M.; Wesdemiotis, C. UV- Curable Hybrid Coatings Based on Vinylfunctionalized Siloxane Oligomer and Acrylated Polyester. *J. Appl. Polym. Sci.* **2007**, *105*, 2376–2386. [CrossRef]
22. Zhang, D.; Williams, B.L.; Shrestha, S.B.; Nasir, Z.; Becher, E.M.; Lofink, B.J.; Santos, V.H.; Patel, H.; Peng, X.; Sun, L. Flame retardant and hydrophobic coatings on cotton fabrics via sol-gel and self-assembly techniques. *J. Colloid Interface Sci.* **2017**, *505*, 892–899. [CrossRef]
23. Latella, B.A.; Ignat, M.; Barbe, C.h.; Cassidy, D.J.; Bartlett, J.R. Adhesion behaviour of organically-modified silicate coatings on stainless steel. *J. Sol-Sel Sci. Technol.* **2003**, *26*, 765–770. [CrossRef]
24. Zucchi, F.; Grassi, V.; Frignani, A.; Trabanelli, G. Inhibition of copper corrosion by silane coatings. *Corros. Sci.* **2004**, *46*, 2853–2865. [CrossRef]
25. Subramanian, V.; van OoIj, W.J. Silane based metal pretreatments as alternatives to chromating. *Surf. Eng.* **1999**, *15*, 2. [CrossRef]
26. Rouzmeh, S.S.; Naderi, R.; Mahdavian, M. Steel surface treatment with free different acid solutions and its effect on the protective properties of the subsequent silane coating. *Prog. Org. Coat.* **2017**, *112*, 133–140. [CrossRef]
27. Wojciechowski, J.; Szubert, K.; Peipmann, R.; Fritz, M.; Schmidt, U.; Bund, A.; Lota, G. Anti-corrosive properties of silane coatings deposited on anodised aluminium. *Electrochim. Acta* **2016**, *220*, 1–10. [CrossRef]
28. Pantoja, M.; Diaz-Benito, B.; Velasco, F.; Abenojar, J.; del Real, J.C. Analysis of hydrolysis of γ-methacryloxypropyltrimethoxysilane and its influence on the formation of silane coatings on 6063 aluminium alloy. *Appl. Surf. Sci.* **2000**, *255*, 6386–6390. [CrossRef]
29. Jeyaram, R.; Elango, A.; Siva, T.; Ayeshamariam, A.; Kaviyarasu, K. Corrosion protection of silane based coatings on mild steel I an aggressive chloride ion environment. *Surf. Interfaces* **2020**, *18*, 100423. [CrossRef]
30. Park, S.K.; Kim, K.D.; Kim, H.T. Preparation of silica nanoparticles: Determination of the optimal synthesis conditions for small and uniform particles. *Colloids Surf. A Physicochem. Eng. Asp.* **2002**, *197*, 7–17. [CrossRef]
31. Kim, K.D.; Kim, H.T. Formation of silica nanoparticles by hydrolysis of TEOS using a mixed semi-batch/batch method. *J. Sol-Gel Sci. Technol.* **2002**, *25*, 183–189. [CrossRef]
32. Bogush, G.H.; Zukoski, C.F. Preparation of monodisperse silica particles: Control of size and mass fraction. *J. Non Cryst. Solids* **1998**, *104*, 95–106. [CrossRef]
33. Meixner, D.L.; Dyer, P.N. Influence of sol-gel synthesis parameters on the microstructure of particulate silica xerogels. *J. Sol-Gel Sci. Technol.* **1999**, *14*, 223–232. [CrossRef]
34. Ran, C.; Lu, W.; Song, G.; Ran, C.; Zhao, S. Study on prolonging the working time of silane solution during the silylation process on carbon steel. *Anti Corros. Methods Mater.* **2011**, *58*, 328. [CrossRef]
35. Metroke, T.; Wang, Y.M.; van Ooij, W.J.; Schaefer, D.W. Chemistry of mixtures of bis-[trimethoxysilylpropyl]amine and vinyltriacetoxysilane: An NMR analysis. *J. Sol-Gel Sci. Technol.* **2009**, *51*, 23–31. [CrossRef]
36. Kucharczyk, A.; Adamczyk, L. The influence of the concentration of ingredients in the immersion deposition process on the protective properties of silan coatings made on stainless steel. *Ochr. Przed Korozją* **2020**, *63*, 290–294.
37. Innocenzi, P.; Brusation, G.; Guglielmi, R.; Bertani, R. New synthetic route to (3-glycidoxypropyl)trimethoxysilane-based hybryd organic-inorganic materials. *Chem. Mater.* **1999**, *11*, 1672–1679. [CrossRef]
38. Milosev, I.; Kapum, B.; Rodic, P.; Iskra, J. Hybrid sol-gel coating agents based on zirconium(IV) propoxide and epoxysilane. *J. Sol-Gel Sci. Technol.* **2015**, *74*, 447–459. [CrossRef]
39. Schmidt, H.; Seiferling, B. Chemistry and applications of inorganic-organic polymers (organically modified silicates). *Mater. Res. Soc. Proc.* **1986**, *73*, 739–750. [CrossRef]
40. Fedorchuk, A.; Walcarius, A.; Laskowska, M.; Vila, N.; Kowalczyk, P.; Cpałka, K.; Laskowski, Ł. Synthesis of vertically aligned Poros silica thin films functionalized by silver ions. *Int. J. Mol. Sci.* **2021**, *22*, 7505. [CrossRef] [PubMed]
41. Adamczyk, L.; Pietrusiak, A.; Bala, H. Corrosion resistance of stainless steel covered by 4-aminobenzoic acid films. *Cent. Eur. J. Chem.* **2012**, *10*, 1657–1668. [CrossRef]
42. Adamczyk, L.; Kulesza, P.J. Fabrication of composite coatings of 4-(pyrrole-1-yl) benzoate-modified poly-3,4-ethylenedioxythiophene with phosphomolybdate and their application in corrosion protection. *Electrochim. Acta* **2011**, *56*, 3649–3655. [CrossRef]
43. Sakakibara, M.; Saito, N.; Nishihara, H.; Aramaki, K. Corrosion of iron in anhydrous methanol. *Corros. Sci.* **1993**, *34*, 391–402. [CrossRef]
44. Kawai, T.; Nishihara, H.; Aramaki, K. Corrosion of iron in electrolytic anhydrous methanol solutions containing ferric chloride. *Corros. Sci.* **1995**, *37*, 823–831. [CrossRef]
45. Shintani, D.; Fukutsuka, T.; Matsuo, Y.; Sugie, Y.; Ishida, T. Passivation films beahavior of stainless steel under high temperature and pressure anhydrous methanol solution containing chlorid. In Proceedings of the CORROSION 2008, New Orleans, LA, USA, 16–20 March 2008.
46. Wang, L.; Gao, H.; Fang, H.; Wang, S.; Sun, J. Effect of methanol on the electrochemical behaviour and surface conductivity of niobium carbide-modified stainless steel for DMFC bipolar plate. *Int. J. Hydrogen Energy* **2016**, *41*, 14864–14871. [CrossRef]
47. Kamitami, K.; Uo, M.; Inoue, H.; Makishima, A. Synthesis and spectroscopy of TPPS-doped silica sol gels by the sol-gel process. *J. Sol-Gel Sci. Technol.* **1993**, *1*, 85–92. [CrossRef]

48. Saka, S. *Handbook of Sol-Gel Science and Technology, Processing, Characterization and Applications*; Springer International Publishing: Midtown Manhattan, NY, USA, 2004.
49. Miecznikowski, K.; Cox, J.A.; Lewera, A.; Kulesza, P.J. Solid state voltammetric characterization of iron hexacyanoferrate encapsulated in silica. *J. Solid State Electrochem.* **2000**, *4*, 199–204. [CrossRef]
50. Miecznikowski, K.; Cox, J.A. Electroanalysis based in stand-alone matrices and electrode-modifying films with silica sol-gel frameworks: A review. *J. Solid State Electrochem.* **2020**, *24*, 2617–2631. [CrossRef]
51. Owczarek, E.; Adamczyk, L. Electrochemical and Anticorrosion Properties of Bilayer Polyrhodanine/Isobutyltriethoxysilane Coatings. *J. Appl. Electrochem.* **2016**, *46*, 635–643. [CrossRef]
52. Furuhashi, T.; Yamada, Y.; Hayashi, M.; Ichihara, S.; Usui, H. Corrosion-resistant nickel thin films by electroless deposition in foam of electrolyte. *MRS Commun.* **2019**, *9*, 352–359. [CrossRef]

Communication

Hydrogen-Induced Cracking Caused by Galvanic Corrosion of Steel Weld in a Sour Environment

Jin Sung Park [1], Jin Woo Lee [2] and Sung Jin Kim [1,*]

[1] Department of Advanced Materials Engineering, Sunchon National University Jungang-ro, Suncheon 540-742, Korea; pjs1352@naver.com
[2] POSCO Technical Research Laboratories, Pohang 790-704, Korea; sjhrte1156@nate.com
* Correspondence: sjkim56@scnu.ac.kr; Tel.: +82-61-750-3557

Abstract: This study examined the hydrogen-induced cracking (HIC) caused by galvanic corrosion of an ASTM A516-65 steel weld in a wet sour environment using a combination of standard immersion corrosion test, electrochemical analyses, and morphological observation of corrosion damage. This study showed that the weld metal has lower open circuit potential, and higher anodic and cathodic reaction rates than the base metal. The preferential dissolution and much higher density of localized corrosion damage were observed in the weld metal of the welded steel. On the other hand, the presence of weldment can make steel more susceptible to HIC, specifically, in areas of the base metal but not in the weld metal or heat affected zone, which is in contrast to typical expectations based on metallurgical knowledge. This can be explained by galvanic corrosion interactions between the weldment and the base metal, acting as a small anode and a large cathode, respectively. This type of galvanic couple can provide large surface areas for infusing cathodically-reduced hydrogen on the base metal in wet sour environments, increasing the susceptibility of welded steel to HIC.

Keywords: ASTM A516-65 steel; weld; hydrogen-induced cracking; galvanic corrosion; sour environment

1. Introduction

Hydrogen degradation of ferrous alloys has attracted considerable attention in the scientific and engineering community for more than 100 years [1–3]. Among the degradation phenomena, hydrogen-induced cracking (HIC) and sulfide stress corrosion cracking (SSCC) are major technical problems that remain to be addressed in the petrochemical industries. Premature failure caused by hydrogen embrittlement (HE) occurs mainly at the welded joints of the steel structures [4–6]. This is because welds are formed under a welding thermal cycle with rapid heating and cooling processes, and they are normally comprised of dendritic and heterogeneous structures with several metallurgical defects [7,8]. Hence, the welded joint is considered the most critical and problematic part of steel structures.

Sour corrosion is the deterioration that occurs on a metal surface in a highly acidic environment containing H_2S. This type of corrosion is expected to occur preferentially at the welds when the welded steel structures are exposed to wet sour environments. The sour corrosion process can be briefly summarized as anodic metal dissolution ($M \rightarrow M^{n+} + e^-$) followed by the infusion of cathodically-reduced hydrogen ($H^+ + e^- \rightarrow H$) in steel [9–11]. The poisoning effect by H_2S facilitates the absorption of atomic hydrogen, and the hydrogen could be trapped at certain metallurgical defects in steel [9,12,13]. According to internal pressure theory [14], which has been widely accepted as a mechanism of HE in steels, the continuous trapping of atomic hydrogens tends to recombine into molecular hydrogen ($H + H \rightarrow H_2$) and leads to significant volume expansion, resulting in the nucleation of cracks.

The heat affected zone (HAZ) with a coarse grain size can be the most inferior part of the welded steel, and numerous papers have reported the (hydrogen-induced) mechanical degradation of the HAZ depending on the welding heat input [15,16]. In the case of weld metal (WM), welding consumables can be one of the factors controlling the resistance to

hydrogen-assisted cracking (HAC) failures. On the other hand the welding consumable is adopted considering mostly the mechanical properties of the base metal (BM). This welding consumable does not guarantee the WM will have high resistance to corrosion or corrosion-induced HAC. In this regard, Pagotto et al. [17] reported a much higher anodic dissolution current at welds relative to the BM of carbon steel in a neutral aqueous environment, employing a scanning vibrating electrode technique (SVET).

In contrast to the common metallurgical aspects described above, the authors found that the HIC ratio of the BM was higher than that of the WM when the welded steel was exposed to a wet sour environment. Moreover, the cracking problem of the BM became worse in the presence of the WM compared to the unwelded steel sample. This is closely associated with the formation of a galvanic couple with the WM and BM in an acidic aqueous environment. In this study, the National Association of Corrosion Engineers (NACE) standard HIC test, diffusible hydrogen measurement, and several electrochemical evaluations were conducted to clarify the mechanistic reason for the more serious damage by HIC in the BM, which can be protected galvanically.

2. Experimental Procedure

The test materials under investigation were equivalent to an ASTM A516-65 grade pressure vessel steel plate with a 15 mm thickness produced by an industrial rolling process. The chemical compositions of the two types of steel samples used in this study, termed Steel A and B, are listed in Table 1. The steels were normalized by heating to 910 °C for 8 min and cooled to room temperature in air.

Table 1. Chemical compositions of the two tested steel samples.

Specimens	C	Mn	Si	S	P
Steel A	0.15–0.18	1.1–1.15	0.3–0.4	<0.003	<0.005
Steel B	0.12–0.15	1.1–1.15	0.3–0.4	<0.003	<0.005

To produce the welded samples for Steel A and B, a double X-groove was produced, and the tandem submerged arc welding (SAW) was performed using two solid wires (OE-SD3, 0.07% C-0.9% Mn-0.3% Si) with diameters of 4 mm each and an OP121TT (0.07% C-1.6% Mn-0.3% Si) flux. The total heat input from the sum of two electrodes was approximately 30 kJ/cm, which was calculated using the welding parameters; these are listed in Table 2.

Table 2. Welding parameters and calculated heat input.

	Electrodes	Welding Current (A)	Arc Voltage (V)	Welding Speed (cm/min)	k-Coefficient	Heat Input (kJ/cm)
Inside	Lead (DC)	650	28	65	1	30.1
	Tail (AC)	480	30			
Outside	Lead (DC)	850	30	85		33.6
	Tail (AC)	650	34			

A brief schematic diagram of the double-pass welded sample is presented in Figure 1a.

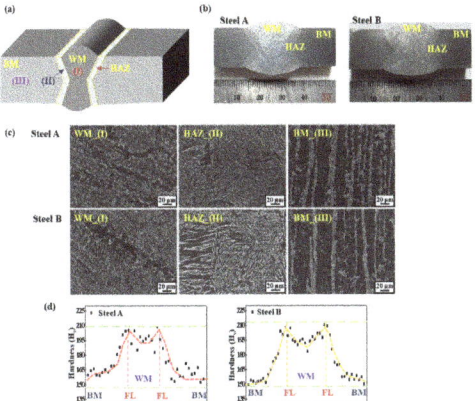

Figure 1. (**a**) Brief schematic diagram of the double-pass welded sample; (**b**) Cross-section view of the two welded samples; (**c**) Microstructures in the WM, HAZ, and BM of the two welded samples; (**d**) Vickers hardness profile of the two welded samples.

Metallographic observations of the WM, HAZ, and BM by optical microscopy (OM) (Zeiss, Jena, Germany) and field-emission scanning electron microscopy (FE-SEM) (Hitachi, Tokyo, Japan) were conducted after the steel samples had been polished up to a 1 μm surface finish and etched in a 3% Nital solution. After the macro- and micrographic observations, a Vickers hardness test of the two welded samples was performed with a constant force of 500 gf for 10 s, and the hardness distributions in different zones of the welded joints were obtained.

The HIC sensitivity was evaluated according to the NACE TM0284 standard method [18], and the sensitivity indices of the crack length ratio (CLR (%)) and crack area ratio (CAR (%)) were determined using an ultrasonic detection method. To ensure reproducibility, three samples obtained from the two tested materials (Steel A and B) were evaluated. After the HIC tests, the diffusible hydrogen contents introduced in the samples were measured using a glycerin volumetric method in reference to JIS Z3113 standard [19]. For this, immediately after the HIC tests, the samples were inserted into the glycerin column that was maintained at 45 °C, using liquid nitrogen as a medium to prevent hydrogen diffusion out of the samples. After three days, the volume of hydrogen collected at the top of the glycerin column was measured.

For the mechanistic study, three types of electrochemical measurements (open circuit potential (OCP), potentiodynamic (PD) polarization, and galvanic current) were conducted in a simulated wet sour environment (5% NaCl + 0.5% CH_3COOH + 0.05 M Na_2S solution). A typical three-electrode cell composed of a steel sample, a Pt grid, and a saturated calomel electrode (SCE), which acted as the working, counter, and reference electrode, respectively, was used for the OCP and PD polarization measurements. Before the tests, the samples were ground to 2400 grit sand-paper and cleaned ultrasonically in ethanol. For the PD polarization, the potential was scanned from −500 mV to 500 mV vs OCP at a scan rate of 0.2 mV/s. A potentiostat (Gamry reference 600, Pennsylvania, America) in zero-resistance ammeter (ZRA) mode was used to measure the variations in the galvanic current flow between the WM and BM with dimensions of 1 cm^2. The distance between the two electrodes was 20 mm. With these electrochemical analyses, the surface and cross-section morphologies of the welded steel sample were observed after immersion in a simulated wet sour solution for seven days.

3. Results and Discussion

Figure 1 shows the macrostructure, microstructure, and hardness profile of the two welded steel samples. The major differences between the two BMs of the two samples lie in the banding index of pearlite and the level of the 2nd phase particles, which was reported previously [20], but they were not the focus of the present study. Under the same welding conditions and consumables, however, there was no significant difference in the macro- and microstructures of weldments of the two samples. The distributions of the hardness profile of the two welded samples showed a similar pattern: the highest and lowest hardness values were measured around the fusion lines and the BM, respectively.

According to common knowledge in welding metallurgy, a HAZ with high hardness and a coarse grain can be considered the most critical and problematic area in a high-strength steel weld [21–23]. For this reason, the SSC of a welded steel sample, which is equivalent to the sample in this investigation, occurred in the HAZ, which has been discussed elsewhere [21,24,25] and is not covered in the present study. The point the authors try to make in this study is to clarify the underlying mechanisms of the changes in the HIC levels with the presence of a narrow weldment in steel samples.

Figure 2 presents the ultrasonically detected HICs of the unwelded and welded steel samples, which had been immersed in a NACE solution saturated with H_2S. The differences in the HIC sensitivity (Table 3) between the two unwelded samples are discussed elsewhere [20]. The focus here was on the changes in the HIC levels and distributions after welding the steel samples.

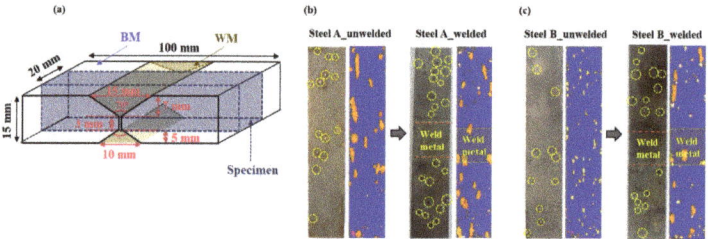

Figure 2. (a) Schematic diagram showing the dimensions of welded sample for HIC test in reference to NACE TM0284 standard method; ultrasonically detected HICs of the unwelded and welded samples after HIC test: (b) Steel A and (c) Steel B.

Table 3. Mean values (μ) and their standard deviations (σ) of ultrasonically detected CLR (%) and CAR (%) values of the two samples after HIC test in reference to NACE TM0284 standard method.

Specimens		CLR (%)		CAR (%)	
		μ	σ	μ	σ
Steel A	unwelded	4.48	1.82	7.88	2.04
	welded	7.46	1.98	7.58	2.35
Steel B	unwelded	1.49	0.88	2.24	0.87
	welded	3.98	1.12	5.81	1.86

In contrast to common expectation, most HICs occurred in the BM in the welded samples, not in the HAZ or WM. Moreover, it is interesting to note that the HIC levels in the BM of both welded steel samples were even higher than those of unwelded samples. Under the same materials in the absence of an externally applied stress, the higher susceptibility to HIC could mainly be due to the higher infusion of hydrogen in the materials [26,27]. This suggests that the presence of a weldment in the tested sample could lead to additional hydrogen uptake in the BM some distance from the weldment. This is supported by the increase in the amount of diffusible hydrogen contents ($[H]_{diff.}$) introduced in the tested samples after welding, as shown in Figure 3. Although the $[H]_{diff.}$ is the sum of the diffusible hydrogen contents obtained from the BM, HAZ, and WM, and each contribution cannot be extracted separately, judging from the HIC occurrence location (Figure 2), most of it may be obtained from the BM.

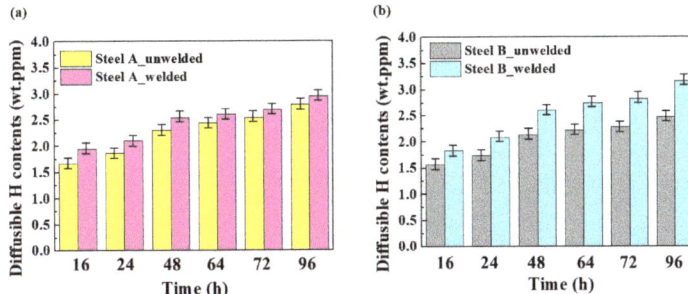

Figure 3. Diffusible hydrogen contents in the unwelded and welded samples: (**a**) Steel A and (**b**) Steel B. (Error bars represent the standard deviations of the mean values).

An electrochemical approach can be adopted to understand the underlying mechanism behind the higher hydrogen infusion and resulting HIC in the BMs of the welded samples. Because the WM and BM are connected electrically, an electrochemical potential difference can be generated through the differences in chemical composition between them, leading to a galvanic current flow from the anode to the cathode. Figure 4a shows that the WM has a slightly lower open circuit potential than the BM, indicating that the WM may be a more active electrode and act as an anode. Although no significant differences in the PD polarization curves (Figure 4b) between the BM and WM of each steel sample were observed, the anodic reaction (Fe $\rightarrow$ Fe^{2+} + 2e$^-$) and cathodic reaction (H$^+$ + e$^-$ $\rightarrow$ H) rates of the WM were slightly higher than those of the BMs. From a practical aspect, the difference in corrosion current density (i_{corr}) between the WM and BM appears to be insignificant. Nevertheless, there could be sufficient driving force for galvanic corrosion between the WM and BM when they are coupled, which can be supported by the galvanic current flow and current level, as shown in Figure 4c. The measured positive current density also indicates that the galvanic current flows from the WM to the BM, and more electrons can be supplied to the BM. In particular, the geometry of the welded steel sample, or even large-sized welded steel structures such as welded pipes, can provide a large cathode (BM) and small anode (WM) ratio. This ratio can be another significant factor expediting galvanic corrosion. From an electrochemical perspective, the larger the cathode area compared to the anode, the greater the galvanic current, which is the more favorable condition for galvanic interactions. Hence, it can be accepted that the anodic steel dissolution and cathodic hydrogen reduction are dominant on the WM and BM, respectively, in the welded sample. The preferential dissolution and much higher density of localized corrosion damage of the WM in the welded sample, shown in Figure 4d, can be the metallographic observation of galvanic corrosion.

The formation of this type of galvanic couple between the WM and BM in an acidic sour solution leads to more hydrogen reduction (H$^+$ + e$^-$ $\rightarrow$ H) in the BM, resulting in the higher infusion of hydrogen in the BM and more vulnerability to HIC. This process is illustrated schematically in Figure 5.

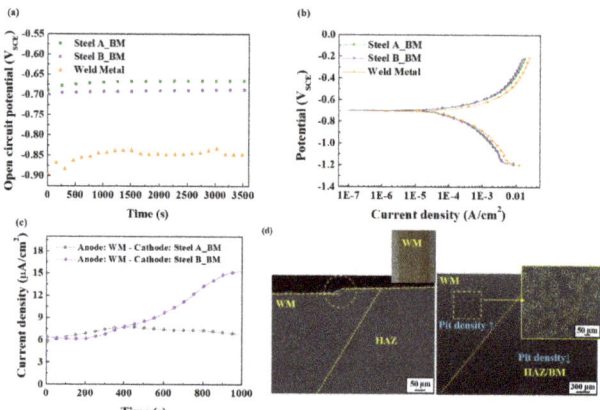

Figure 4. Measurements of (**a**) OCP, (**b**) PD polarization, and (**c**) Galvanic current; (**d**) Metallographic observation of weld metal in Steel A after an immersion test.

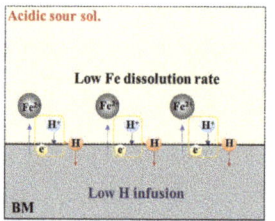
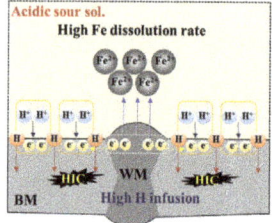

Figure 5. Brief schematic illustration showing the mechanism of galvanic corrosion between the WM and BM, and hydrogen infusion and cracking in the BM.

4. Summary

This work elucidates the preferential occurrence of HIC in the BM in the welded steel under wet sour corrosion with a series of experimental results. The major findings are summarized as follows.

The preferential occurrence of HIC in the BM is caused primarily by the fact that the chemical composition of WM, which was optimized for mechanical properties, is slightly anodic to the BM. Even if the WM is close in chemical composition to the BM, the dendritic and inhomogeneous microstructure of the WM can also produce a potential difference with the BM, leading to galvanic corrosion. This leads to the galvanic current flows from the WM to the BM, suggesting that more electrons can be supplied to the BM. Hence, the cathodic hydrogen reduction is more dominant on the BM resulting in the higher infusion of hydrogen in the BM (i.e., welded samples had 5.73% (Steel A) and 28.45% (Steel B) higher $[H]_{diff.}$ than unwelded samples) and more susceptibility to HIC (i.e., welded samples had 66.51% (Steel A) and 167.11% (Steel B) higher CLR (%) than unwelded samples). Therefore, optimizing the WM so that its chemical composition is very close to that of the BM or slightly nobler than that of the BM is an effective strategy to suppress preferential anodic dissolution and the formation of HIC in the WM and BM, respectively, in the welded steels under an acidic sour environment.

Author Contributions: Conceptualization, J.W.L. and S.J.K.; methodology, J.S.P., J.W.L., and S.J.K.; investigation, J.S.P. and S.J.K.; data curation, J.S.P. and S.J.K.; writing-original draft preparation, J.S.P. and S.J.K.; review and editing, S.J.K. All authors have read and agreed to the published version of the manuscript.

Funding: This research was supported in part by the National Research Foundation of Korea (NRF) grant funded by the Korea government (MSIT), No. 2019R1C1C1005007. In addition, this work was partly funded and conducted under the Competency Development Program for Industry Specialists of the Korean Ministry of Trade, Industry and Energy (MOTIE), operated by the Korea Institute for Advancement of Technology (KIAT), No. P0002019, HRD Program for High Value-Added Metallic Materials Expert.

Data Availability Statement: The data that support the plots and other findings of the current study are available from the corresponding author on reasonable request.

Conflicts of Interest: The authors declare no conflict of interest.

References

1. Johnson, W.H. On some remarkable changes produced in iron and steel by the action of hydrogen and acids. *Proc. R. Soc.* **1875**, *23*, 168–175. [CrossRef]
2. Beachem, C.D. New model for hydrogen-assisted cracking (hydrogen embrittlement). *Hydrog. Adv. Mater.* **2010**, *22*, 1128–1135. [CrossRef]
3. Lynch, S. Discussion of some recent literature on hydrogen-embrittlement mechanisms: Addressing common misunderstandings. *Corros. Rev.* **2019**, *37*, 377–395. [CrossRef]
4. Hardie, D.; Charles, E.A.; Lopez, A.H. Hydrogen embrittlement of high strength pipeline steel. *Corros. Sci.* **2006**, *48*, 4378–4385. [CrossRef]
5. Lindley, C.; Rudd, W.J. Influence of the level of cathodic protection on the corrosion fatigue properties of high strength welded joint. *Mar. Struct.* **2001**, *14*, 397–416. [CrossRef]
6. Świerczyńska, A.; Fydrych, D.; Landowski, M.; Rogalski, G.; Łabanowski, J. Hydrogen embrittlement of X2CrNiMoCuN25-6-3 super duplex stainless steel welded joint under cathodic protection. *Constr. Build. Mater.* **2020**, *238*, 117697. [CrossRef]
7. Banerjee, K.; Chatterjee, U.K. Effect of microstructure on hydrogen embrittlement of weld-simulated HSLA-80 and HSLA-100 steels. *Mater. Trans. A* **2003**, *34*, 1297–1309. [CrossRef]
8. Shiraiwa, T.; Kawate, M.; Briffod, F.; Kasuya, T.; Enoki, M. Evaluation of hydrogen-induced cracking in high-strength steel welded joints by acoustic emission technique. *Mater. Des.* **2020**, *190*, 108573. [CrossRef]
9. Kim, S.J. Effect of the elastic tensile load on the electrochemical corrosion behavior and diffusible hydrogen content of ferritic steel in acidic environment. *Int. J. Hydrog. Energy* **2017**, *42*, 19367–19375. [CrossRef]
10. Kim, S.J.; Kim, K.Y. A review of corrosion and hydrogen diffusion behaviors of high strength pipe steel in sour environment. *J Weld. Join.* **2014**, *32*, 13–20. [CrossRef]
11. Li, X.; Ma, X.; Zhang, J.; Akiyama, E.; Wang, Y.; Song, X. Review of hydrogen embrittlement in metals: Hydrogen diffusion, hydrogen characterization, hydrogen embrittlement mechanism and prevention. *Acta Metall. Sin-Engl.* **2020**, *33*, 759–773. [CrossRef]
12. Plennevaux, C.; Kittel, J.; Fregonese, M.; Normand, B.; Ropital, F.; Grosjean, F.; Cassagne, T. Contribution of CO_2 on hydrogen evolution and hydrogen permeation in low alloy steels exposed to H_2S environment. *Electrochem. Comm.* **2013**, *26*, 17–20. [CrossRef]
13. Monnot, M.; Nogueira, R.P.; Roche, V.; Berthomé, G.; Chauveau, E.; Estevez, R.; Mantel, M. Sulfide stress corrosion study of super martensitic stainless steel in H_2S sour environment: Metallic sulfides formation and hydrogen embrittlement. *App. Surf. Sci.* **2017**, *394*, 132–141. [CrossRef]
14. Zapffe, C.A.; Sims, C.E. Hydrogen embrittlement, internal stress and defects in steel. *Trans. AIME* **1941**, *145*, 225–261.
15. Gáspár, M. Effect of welding heat input on simulated HAZ areas in S960QL high strength steel. *Metals* **2019**, *9*, 1226. [CrossRef]
16. Shi, M.; Zhang, P.; Wang, C.; Zhu, F. Effect of high heat input on toughness and microstructure of coarse grain heat affected zone in Zr bearing low carbon steel. *ISIJ Int.* **2014**, *54*, 932–937. [CrossRef]
17. Pagotto, J.F.; Montemor, M.F.; Recio, F.J.; Motheo, A.J.; Simões, A.M.; Herrasti, P. Visualisation of the galvanic effects at welds on carbon steel. *J. Braz. Chem. Soc.* **2015**, *26*, 667–675. [CrossRef]
18. NACE Standard TM0284. *Standard Test Method for Evaluation of Pipeline and Pressure Vessel Steel for Resistance to Hydrogen Induced Cracking*; NACE International: Houston, TX, USA, 2003.
19. JIS Z3113. *Method for Measurement of Hydrogen Evolved from Deposited Metal*; Japan Standards Association: Tokyo, Japan, 1983.
20. Park, J.S.; Lee, J.W.; Hwang, J.K.; Kim, S.J. Effects of alloying elements (C,Mo) on hydrogen assisted cracking behaviors of A516-65 steels in sour environments. *Materials* **2020**, *13*, 4188. [CrossRef]
21. Cho, D.M.; Park, J.S.; Lee, J.W.; Kim, S.J. Study on hydrogen diffusion and sulfide stress cracking behaviors of simulated heat-affected zone of A516-65 grade pressure vessel carbon steel. *Korean J. Met. Mater.* **2020**, *58*, 599–609. [CrossRef]
22. Kang, Y.J.; Kim, M.J.; Kim, G.D.; Kim, N.K.; Song, S.W. Characteristics of susceptible microstructure for hydrogen-induced cracking in the coarse-grained heat-affected zone of carbon steel. *Metall. Mater. Trans. A* **2020**, *51*, 2143–2153. [CrossRef]
23. Lee, J.A.; Lee, D.H.; Seok, M.Y.; Baek, U.B.; Lee, Y.H.; Nahm, S.H.; Jang, J.I. Hydroge-induced toughness drop in weld coarse-grained heat-affected zones of linepipe steel. *Mater. Charact.* **2013**, *82*, 17–22. [CrossRef]

24. Zafra, A.; Belzunce, J.; Rodríguez, C.; Pariente, I.F. Hydrogen embrittlement of the coarse grain heat affected zone of a quenched and tempered 42CrMo4 steel. *Int. J. Hydrog. Energy* **2020**, *45*, 16890–16908. [CrossRef]
25. Albiter, A. Sulfide stress cracking assessment of carbon steel welding with high content of H_2S and CO_2 at high temperature: A case study. *Engineering* **2020**, *12*, 863–885. [CrossRef]
26. Laureys, A.; Eeckhout, E.V.E.; Petrov, R.; Verbeken, K. Effect of deformation and charging conditions on crack and blister formation during electrochemical hydrogen charging. *Acta Mater.* **2017**, *127*, 192–202. [CrossRef]
27. Trautmann, A.; Mori, G.; Oberndorfer, M.; Bauer, S.; Holzer, C.; Dittmann, C. Hydrogen uptake and embrittlement of carbon steel in various environments. *Materials* **2020**, *13*, 3604. [CrossRef]

Article

Calibrating the Impressed Anodic Current Density for Accelerated Galvanostatic Testing to Simulate the Long-Term Corrosion Behavior of Buried Pipeline

Yoon-Sik So [1], Min-Sung Hong [1], Jeong-Min Lim [1], Woo-Cheol Kim [2] and Jung-Gu Kim [1],*

[1] School of Advanced Materials Science and Engineering, Sungkyunkwan University, Suwon 16419, Korea; soy4718@skku.edu (Y.-S.S.); smith803@skku.edu (M.-S.H.); alsdl0311@skku.edu (J.-M.L.)
[2] Technical Efficiency Research Team, Korea District Heating Corporation, 92 Gigok-ro, Yongin 06340, Korea; kwc7777@kdhc.co.kr
* Correspondence: kimjg@skku.edu

Abstract: Various studies have been conducted to better understand the long-term corrosion mechanism for steels in a soil environment. Here, electrochemical acceleration methods present the most efficient way to simulate long-term corrosion. Among the various methods, galvanostatic testing allows for accelerating the surface corrosion reactions through controlling the impressed anodic current density. However, a large deviation from the equilibrium state can induce different corrosion mechanisms to those in actual service. Therefore, applying a suitable anodic current density is important for shortening the test times and maintaining the stable dissolution of steel. In this paper, to calibrate the anodic current density, galvanostatic tests were performed at four different levels of anodic current density and time to accelerate a one-year corrosion reaction of pipeline steel. To validate the appropriate anodic current density, analysis of the potential vs. time curves, thermodynamic analysis, and analysis of the specimen's cross-sections and products were conducted using a validation algorithm. The results indicated that 0.96 mA/cm^2 was the optimal impressed anodic current density in terms of a suitable polarized potential, uniform corrosion, and a valid corrosion product among the evaluated conditions.

Keywords: galvanostatic test method; underground infrastructure; long-term corrosion; carbon steel

1. Introduction

With the recent development of general industry, the demand for various types of pipeline has increased. Numerous infrastructures have been built in downtown underground areas. Most of these underground infrastructures are aimed at achieving long-term use, since they are generally difficult to maintain or replace. Therefore, it is essential to verify the long-term corrosion behavior of buried metallic structures.

In fact, despite the various developments, underground structural failures continue to occur [1–3]. Here, the corrosion of metallic infrastructures in underground soil is a major issue that presents numerous safety and economic concerns [4–8]. In short, pipelines can be damaged by corrosion, which can lead to the failure of the structures within a soil environment. However, it is difficult to detect the failure of a large underground system, which means understanding the long-term corrosion behavior is crucial to mitigating unpredictable failures. As such, various studies have been conducted on the long-term corrosion mechanism for metals in a soil environment. The majority of these studies involved the use of immersion tests to analyze the corrosive characteristics of the metals [9–12]. However, obtaining the results of this type of test requires a long period of time (at least several months). It is also difficult to maintain the same environmental conditions during the entire test period, making it difficult to yield reproducible results. Since an immersion test is not suitable for evaluating the long-term corrosion properties of materials, an appropriate acceleration test must be considered. Here, electrochemical acceleration methods

present the most effective approach for simulating long-term corrosion. Among the various methods, galvanostatic tests allow for accelerating the surface chemical reactions through controlling the impressed anodic current density. It is thus a suitable method for long-term corrosion studies.

The impressed anodic current density and the time can be calculated using Faraday's law [13]. Applying an appropriate anodic current density is the key factor here, since the expected corrosion reaction cannot be achieved if the impressed density is too high [14]. Meanwhile, accelerated corrosion testing methods must allow for shortening the test time and inducing the same mechanism of degradation as that in actual service. However, there exists no international standard for these tests. With this in mind, this study was aimed at providing an academic standard for anodic current density that can be applied to accelerate and simulate the long-term corrosion of metals in a soil environment. Thus, a potentiodynamic (PD) polarization test was conducted to determine the corrosion current density for carbon steel in synthetic groundwater. To determine the most appropriate anodic current density, galvanostatic (GS) tests were performed at four different anodic current densities using an acceleration period that represented the one-year corrosion reaction of a specific pipeline. Each anodic current density value was determined according to the corrosion current density, which was obtained from the PD polarization measurements and through the use of Faraday's law. Meanwhile, an optical microscope (OM) was used to observe the surface and cross-section morphologies of the specimens, while X-ray diffraction (XRD) analysis was used to confirm the species of each oxide following the galvanostatic experiments.

2. Materials and Methods

2.1. Specimen and Solution

The specimen was cut into cuboid-shaped pieces with dimensions of $10 \times 10 \times 5$ mm^3. The chemical composition of the tested pipeline steel is presented in Table 1 (ASTM A 139), while the composition of the synthetic soil solution is presented in Table 2. The results were obtained from three soil environment sites close to an operating pipeline.

Table 1. Composition of the tested steel specimen.

C	Mn	P	S	Fe
0.25	1.00	0.04	0.04	Balance

Table 2. Composition of the tested solution.

pH	Temperature	CaCl$_2$ (ppm)	MgSO$_4$·7H$_2$O (ppm)	NaHCO$_3$ (ppm)	H$_2$SO$_4$ (ppm)	HNO$_3$ (ppm)
6.4	60 °C	133.2	59	208	85	22.2

2.2. Electrochemical Analysis to Optimize the Impressed Anodic Current Density for GS Testing

To evaluate the corrosion resistance of pipeline steel, a PD polarization test was conducted using a multi-potentiostat/galvanostat instrument (VMP-2, Bio-Logic Science Instruments, Seyssinet-Pariset, France). Meanwhile, a three-electrode cell was constructed using pipeline steel as the working electrode (WE), two pure graphite rods as the counter electrode (CE, Qrins, Seoul, Korea), and a saturated calomel electrode as the reference electrode (RE, Qrins). Prior to conducting the electrochemical tests, the specimens were abraded with 600-grit silicon carbide paper. These prepared steel surfaces were then covered with silicone rubber, leaving an area of 0.25 cm^2 unmasked before they were exposed to a synthetic soil solution at 60 °C under aerated conditions, and then rinsed with ethanol and finally dried using nitrogen gas. Prior to all the electrochemical tests, the specimens were immersed in a test solution for 3 h to attain a stable surface. The PD polarization measurements were performed at a potential sweep of 0.166 mV/s from an initial potential of −250 mV vs. an open circuit potential (OCP) up to a final potential of 0 V$_{SCE}$. A GS test was then performed to accelerate corrosion reaction of steel after 3 h of OCP measurements at four different anodic current densities. Each of the impressed anodic current density values was determined according to the corrosion current density, which was obtained from the PD polarization measurements.

2.3. Surface Analysis

An OM (SZ61TRC, Olympus Korea Co., Seocho-gu, Seoul, Korea) was used to observe the surface morphology and the cross-section of each specimen, while each oxide product was analyzed via XRD (Rigaku Ultima III X-ray diffractometer, Tokyo, Japan) analysis with Cu Kα_1 radiation (λ = 1.54056 Å) over a 2θ range of 20°–70°, using a step-size of 0.017° and a step-time of 1 s, to confirm the species of each oxide following the galvanostatic experiments.

3. Results and Discussion

3.1. Corrosion Behavior of Pipeline Steel in a Synthetic Soil Solution

The PD polarization curve related to a synthetic soil solution at a pH of 6.4 and a temperature of 60 °C is shown in Figure 1, while the PD results are summarized in Table 3.

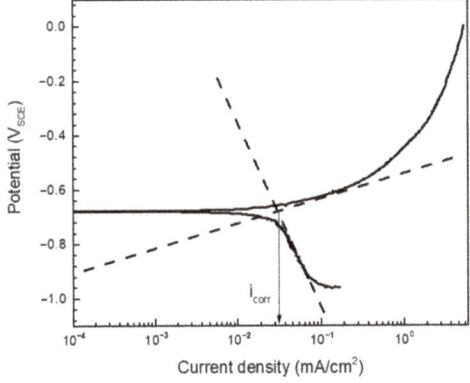

Figure 1. The PD polarization curve of steel in a synthetic soil solution at pH 6.4 and a temperature of 60 °C.

Table 3. Results of the PD (potentiodynamic) polarization measurements.

E$_{corr}$ (mV$_{SCE}$)	i$_{corr}$ (μA/cm^2)	β$_a$ (mV/decade)	β$_c$ (mV/decade)
−678.03	27.96	113.6	579.6

To calculate the corrosion current density, the Tafel extrapolation method (as described in the equation below) was applied. Here, Equation (1) describes the linear relationship between the over-potential and the log scale current density [14,15]:

$$\eta = a \pm \beta_{a,c} \log |i| \tag{1}$$

where a = −β$_a$log(i$_0$) or β$_c$log(i$_0$), β$_a$ ≅ (RT/(1 − α)nF) is the Tafel slope of the anodic polarization curve, β$_c$ ≅ (RT/αnF) is the Tafel slope of the cathodic polarization curve, i$_0$ is the exchange current density, α is the charge transfer coefficient, n is the charge number, R is the gas constant (8.314 J/[mol·K]), and T is the absolute temperature [K]. From Equation (1), a linear relationship was derived, and the corrosion current density was measured as approximately 27.96 μA/cm^2 according to the Tafel extrapolation in the PD polarization curve (Figure 1). No passivation behavior of the anodic polarization curve was observed, which means that in a synthetic soil solution, pipeline steel will be homogeneously corroded.

3.2. Corrosion Acceleration Using the GS Method

The mass loss of pipeline steel can be calculated for each current and experiment time by applying Faraday's law. The mass loss by PD test for the one-year corrosion of pipeline steel is given in the following Equation (2):

$$m = ita/nF = (27.96 \text{ μA/cm}^2 \times 1 \text{ year} \times 55.84)/(2 \times 96500 \text{ C}) = 0.255 \text{ g/cm}^2 \quad (2)$$

where m is mass loss, i is corrosion current density, F is Faraday's constant (96,500 C/ equivalent), n is the number of equivalents exchanged, a is the atomic weight, and t is time [16]. To accelerate the one year of corrosion, the impressed anodic current density and test time were set accordingly. The impressed anodic current densities were selected to investigate a wide range as possible, starting from the maximum output (24 mA, 3435.29 times faster than corrosion rate) of the potentiostat instrument (VMP-2) before being reduced to 0.024 mA (3.43 times faster than corrosion rate) in 1/10 stages. Meanwhile, the exposure times were also determined to maintain the same theoretical metal loss for each test. With the reduction in exposure time, the acceleration coefficient, which is the ratio between impressed anodic current density and corrosion current density, increased sharply. All of these variables are detailed in Table 4.

Table 4. Experimental conditions and calculated variables.

Impressed Anodic Current Density (mA/cm^2)	Exposure Time (h)	Mass Loss by PD Test (g/cm^2)	Anodic Dissolution Rate (mm/y)	Acceleration Coefficient
0.096	2551.35		1.14	3.43
0.96	255.14	0.255	11.44	34.34
9.6	25.51		114.39	343.35
96	2.55		1143.90	3435.29

The logarithmic shape of the curves in Figure 2 follows Equation (3), known as the Sand equation [17]. This equation defines the quantitative relationship between the impressed anodic current density and time:

$$(i\tau^{1/2})/(C_0^*) = (nFD_0^{1/2}\pi^{1/2})/2 \quad (3)$$

where i is the impressed current density, τ is the transition time, C_0^* is the bulk concentration of the reactant, n is the coefficient number, and D_0 is the diffusion coefficient of the reactant. When the impressed current can no longer be supported by the intended metal dissolution reaction, the potential changes to an alternative electron-transfer reaction [17]. However, in most of the accelerated GS corrosion tests, the reactants were the metal itself. Therefore, the probability that the concentration of the reactants reaches zero is extremely low. In Figure 2, since the transition point could not be observed, i.e., the potential vs. time curve exhibited a logarithmic curve shape, the GS tests were appropriately performed under all conditions. Nevertheless, there were differences in the stability of the curve shape for each condition. As shown in Figure 2a, the wavering potential curve was recorded only at the slowest reaction rate of 0.096 mA/cm^2. When the reaction rate is slower, oxides will have more opportunity to adsorb onto the electrode surface. Hence, it can be expected that there will be an obstruction of the reaction area comprising the laminated oxide layer. Consequently, in this experiment, the 0.096 mA/cm^2 condition appeared to be invalid for accelerating a homogeneous corrosion using the GS method.

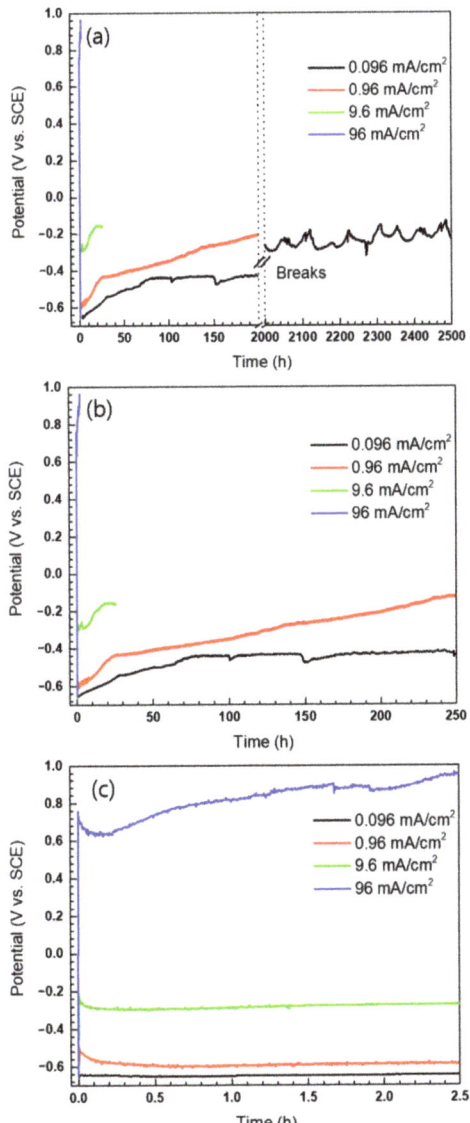

Figure 2. Potential (V_{SCE}) vs. time (h) curves during the GS test at 60 °C. The shape of the graph follows the Sand equation. (**a**) Entire curves with breaks; (**b**) graph from 0 to 250 h; (**c**) graph from 0 to 2.5 h.

Meanwhile, the measured potential of each condition was different depending on the impressed anodic current density. As the impressed anodic current density increased, the WE potential increased to more positive values and was generated at a higher current during a relatively short period of time, with the initial and final values of the measured potentials shown in Figure 3a. Here, it was clear that the measured WE potential depended on the impressed anodic current density. Therefore, thermodynamic analysis was then conducted to verify the electrochemically accelerated reaction. The reactions and the Nernst equations that primarily occur in the Fe-H_2O system at 60 °C are described in terms of

Equations (4)–(6) [13], while these are also presented in terms of a pH-potential diagram in Figure 3b.

$$Fe^{2+} + e^- = Fe, \varepsilon_{[Fe2+/Fe]} = -0.199 + 0.033 \log[Fe^{2+}], [V_{SCE}] \quad (4)$$

$$Fe(OH)_3 + 3H^+ + e^- = Fe^{2+} + 3H_2O, pH = 6.65 - 0.5 \log[Fe^{2+}] \quad (5)$$

$$Fe(OH)_2 + 2H^+ = Fe^{2+} + 2H_2O, \varepsilon_{[Fe2+/Fe(OH)3)]} = 1.298 - 0.177\, pH - 0.066 \log[Fe^{2+}], [V_{SCE}] \quad (6)$$

In Figure 3b, the metal ions appeared to be stable in the range of –0.397 to 0.561 V_{SCE} when the soluble ion activity was 10^{-6} at pH = 6.4 and 60 °C, as shown in the Pourbaix diagram. Meanwhile, as Figure 3a shows, the 0.096 and 0.96 mA/cm² current densities had initial and final potentials of −0.651 and −0.244 V_{SCE}, −0.597 and −0.206 V_{SCE}, respectively, while the 9.6 mA/cm² current density had an initial potential of −0.295 and a final potential of −0.153 V_{SCE}. These three conditions indicated the region where the metal ions were both stable and sensitive to corrosion (see Figure 3b). As such, the corrosion of carbon steel will be accelerated accordingly. However, the 96 mA/cm² current density had extreme potentials, with an initial potential of 0.648 V_{SCE} and a final potential of 0.960 V_{SCE}. In this potential range, a metal-dissolution reaction is no longer stable, meaning the thermodynamic stability of the pipeline steel will change from iron to iron oxide/hydroxide ions, based on the Pourbaix diagram [18]. In addition, an oxygen-evolution reaction may occur close to 1 V_{SCE}. As such, the 96 mA/cm² condition, as a current value for accelerating the corrosion process, cannot be reasonably accepted.

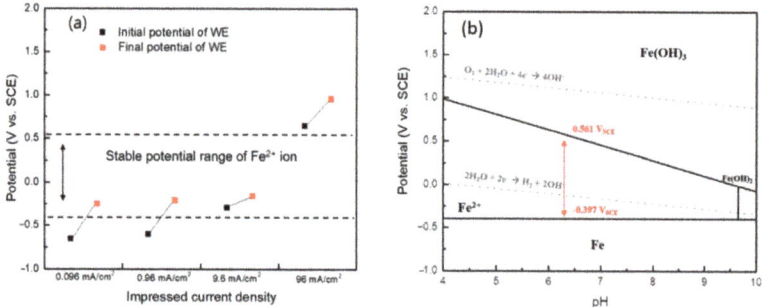

Figure 3. (a) Initial and final potentials during the GS test, and (b) Pourbaix diagram of an Fe-H₂O system (activity of Fe^{2+}: 10^{-6}, 60 °C).

Meanwhile, Table 5 shows the parameters following the accelerated tests for corrosion over a one-year period. During the GS test, the electrode was exposed to a water-based solution, with the primary reaction being the production of electrons via metal dissolution. However, a self-corrosion reaction also occurred [19], which consumed the electrons generated by the dissolution of iron from the WE surface [20]. Due to this self-corrosion reaction, the generated electrons were not completely transported to the CE; rather, they reacted with the water and oxygen at the WE's surface. As a result, the mass loss in the GS acceleration test became larger than the theoretical mass loss calculated using Faraday's law. As such, in electrochemically accelerated tests, the self-corrosion ratio in relation to the corrosion rate could be around 20–30%, as shown in Table 5. Nevertheless, the electrochemical acceleration test had the advantage of a significantly reduced testing time. For example, while with the 0.96 mA/cm² condition, a 28.9% self-corrosion ratio was indicated, the testing time was generally reduced from one year to around 10 days. Thus, the electrochemical acceleration method still has merits despite the self-corrosion aspect.

Table 5. Measured parameters of the one-year corrosion-accelerated specimen as a function of impressed anodic current density.

Impressed Anodic Current Density (mA/cm^2)	Mass Loss by GS Test (g/cm^2)	Corrosion Rate [1] (mm/y)	Self-Corrosion Rate [2] (mm/y)	Self-Corrosion Ratio [3] (%)
0.096	0.364	1.63	0.49	30.06
0.96	0.360	16.09	4.65	28.90
9.6	0.336	150.23	35.85	23.86
96	0.332	1485.05	341.15	22.97

[1] Corrosion rate: calculated values from mass loss by GS test. [2] Self-corrosion rate: difference between corrosion rate by PD test and GS test. [3] Self-corrosion ratio: (self-corrosion rate/corrosion rate) × 100.

3.3. Analysis of Surfaces

Cross-sectional images of the tested specimens obtained using the OM are shown in Figure 4. Most of the images clearly indicated a uniform corrosion. However, at the lowest anodic current density of 0.096 mA/cm^2, different behavior, which indicated localized corrosion, was observed. As shown in Figure 2, an unstable potential was recorded at the anodic current density of 0.096 mA/cm^2. As noted above, when the reaction rate is slower, oxides will have more opportunity to adsorb onto the electrode's surface. Hence, most of the iron oxides produced via the self-corrosion reaction covered the reaction area at the anodic current density of 0.096 mA/cm^2. Consequently, an accelerated corrosion reaction occurred at the localized site of the narrow reaction area, due to the layer of iron oxide formed on the surface.

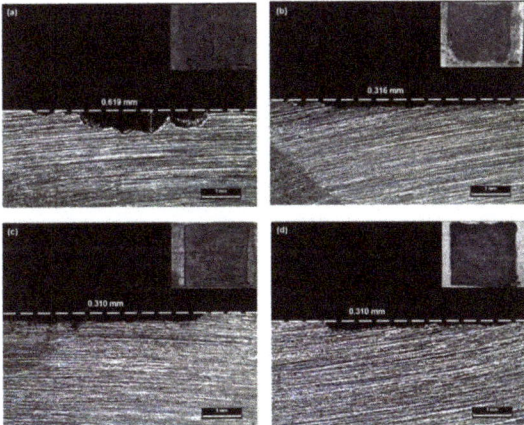

Figure 4. Cross-sectional images following the galvanostatic test at different applied current densities: (**a**) 0.096 mA/cm^2; (**b**) 0.96 mA/cm^2; (**c**) 9.6 mA/cm^2; (**d**) 96 mA/cm^2.

Meanwhile, phase analysis of the oxides under each anodic current density condition was conducted using the XRD instrument, with the major phase of the oxides shown in Figure 5. Here, the corrosion product of the lower anodic current density group was mainly comprised of goethite. However, the higher anodic current density condition (96 mA/cm^2) showed different XRD peaks, which represented ferrihydrite and magnetite. According to the existing literature, the main iron oxide phase in general soil environments is goethite [10,20–24]. Furthermore, the XRD results at the 0.096 and 0.96 mA/cm^2 conditions were similar to those obtained for actual environment iron oxides in previous studies. Therefore, these conditions accurately reflected the long-term corrosion behavior of buried pipeline.

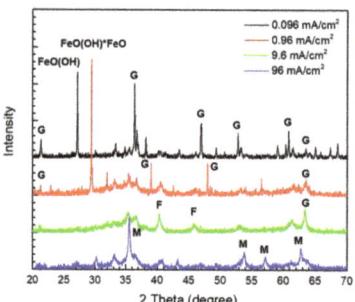

Figure 5. The XRD results following the galvanostatic tests based on current density: G: goethite (α-FeO[OH]); F: ferrihydrite (Fe$_2$O$_3$·0.5H$_2$O); M: magnetite (Fe$_3$O$_4$).

3.4. Validation of the Galvanostatically Accelerated Testing

To establish an appropriate impressed anodic current density for the accelerated test, a validation process was conducted for all tests, based on the process shown in Figure 6. This process was initially implemented to assess the inflection points in the potential vs. time curves observation, and thus to confirm the effect of the environment during the test. In the galvanostatic test, all the curves exhibited a stable logarithmic shape, indicating that the pipeline steel reacted well. Meanwhile, the validation process involved comparing the measured potentials with the Pourbaix diagram. Here, the majority of the conditions are within a thermodynamically stable Fe ion range. However, the 96 mA/cm^2 condition departed from this range and was thus excluded from the valid current density range. Next, the validation process was continued in terms of the cross-sectional images of the steel pipeline specimen to confirm the presence of uniform corrosion, which was indeed clearly confirmed by the majority of the images. However, the lowest anodic current density (0.096 mA/cm^2) indicated different corrosion behavior (localized corrosion) and was thus also excluded from our determination of a valid anodic current density. Finally, the validation process involved analyzing the corrosion product using XRD analysis. Here, 0.96 mA/cm^2 was found to be the optimal anodic current density for reproducing long-term corrosion behavior.

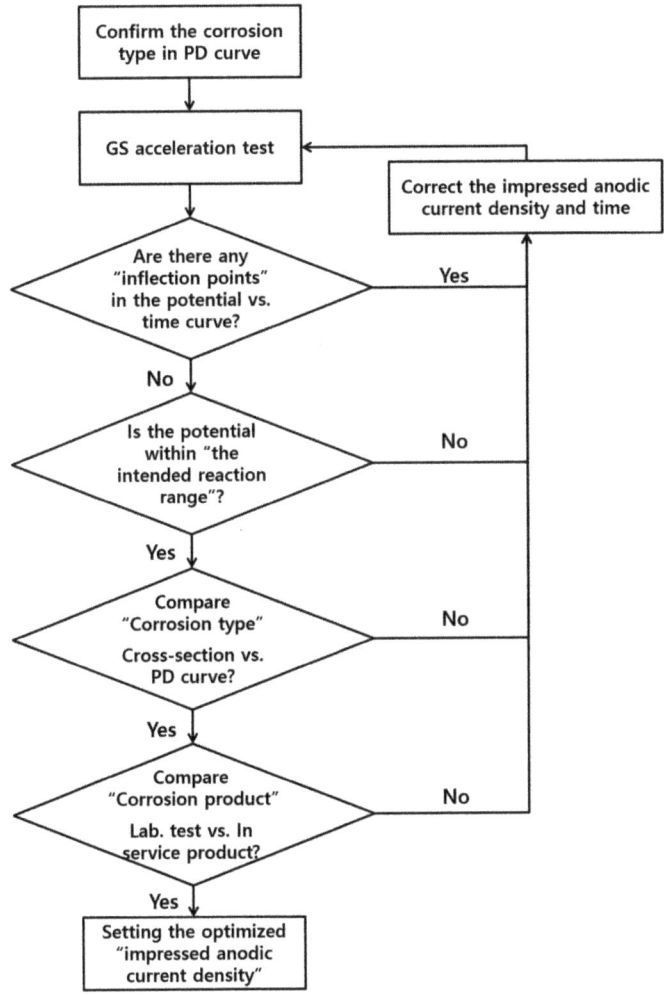

Figure 6. Flowchart showing the validation process for the GS accelerated testing.

4. Conclusions

In this study, the impressed anodic current density for a GS test aimed at determining long-term corrosion behavior was evaluated using an electrochemical test, OM observation, and XRD analyses. To verify the accelerated test, analysis of the potential vs. time curves, thermodynamic analysis, and analysis of the cross-sections and products of the specimen were performed. During the GS test, based on the laminated oxide layer, the most unstable potential form was recorded at the slowest reaction rate (0.096 mA/cm^2). Meanwhile, the highest anodic current density (96 mA/cm^2) demonstrated extreme potentials that were out of the Fe ion's stable range. The majority of the OM images clearly indicated uniform corrosion. However, the slowest condition at the anodic current density of 0.096 mA/cm^2 indicated localized corrosion. The XRD peaks at the 0.096 and 0.96 mA/cm^2 conditions corresponded to a corroded buried pipeline product (goethite), while the 9.6 and 96 mA/cm^2 conditions indicated the presence of different oxides. In conclusion, the anodic current density of 0.96 mA/cm^2 was found to be the most suitable for conducting the GS acceleration testing of carbon steel in a soil environment. Based on this finding, an appropriate validation process was established for an accelerated corrosion test aimed at predicting long-term corrosion lifetimes. Furthermore, it is expected to help determine the reliable impressed anodic current density by applying the validation process to design accelerating metal corrosion.

Author Contributions: Conceptualization, Y.-S.S. and W.-C.K.; methodology, Y.-S.S. and J.-M.L.; validation, M.-S.H.; investigation, J.-M.L. and W.-C.K.; writing—original draft preparation, Y.-S.S.; writing—review and editing, M.-S.H. and J.-G.K.; supervision, J.-G.K. All authors have read and agreed to the published version of the manuscript.

Funding: This work was supported by the program for fostering next-generation researchers in engineering of National Research Foundation of Korea (NRF), funded by the Ministry of Science and ICT (2017H1D8A2031628). This work also was supported by an NRF grant funded by the Korean Government (NRF-2020-Research Staff Program) (NRF-2020R1I1A1A01074866).

Institutional Review Board Statement: Not applicable.

Informed Consent Statement: Not applicable.

Data Availability Statement: Data is contained within the article material.

Acknowledgments: This research was supported by the Korea District Heating Corporation (No. 0000000014524).

Conflicts of Interest: The authors declare no conflict of interest.

References

1. Kim, S.-H.; So, Y.-S.; Kim, J.-G. Fracture behavior of locally corroded steel pipeline in district heating system using the combination of electrochemistry and fracture mechanics. *Met. Mater. Int.* **2019**, *1–8*, 1671. [CrossRef]
2. Kim, Y.-S.; Kim, J.-G. Investigation of Weld Corrosion Effects on the Stress Behavior of a Welded Joint Pipe Using Numerical Simulations. *Met. Mater. Int.* **2019**, *25*, 918–929. [CrossRef]
3. Ismail, A.; El-Shamy, A.J.A.C.S. Engineering behaviour of soil materials on the corrosion of mild steel. *Appl. Clay Sci.* **2009**, *42*, 356–362. [CrossRef]
4. Alamilla, J.; Espinosa-Medina, M.; Sosa, E. Modelling steel corrosion damage in soil environment. *Corros. Sci.* **2009**, *51*, 2628–2638. [CrossRef]
5. Alkhateeb, E.; Ali, R.; Popovska-Leipertz, N.; Virtanen, S. Long-term corrosion study of low carbon steel coated with titanium boronitride in simulated soil solution. *Electrochim. Acta* **2012**, *76*, 312–319. [CrossRef]
6. Feng, Q.; Yan, B.; Chen, P.; Shirazi, S.A. Failure analysis and simulation model of pinhole corrosion of the refined oil pipeline. *Eng. Fail. Anal.* **2019**, *106*, 104177. [CrossRef]
7. Melchers, R.E.; Petersen, R.B.; Wells, T.J. Empirical models for long-term localised corrosion of cast iron pipes buried in soils. *Corros. Eng. Sci. Techn.* **2019**, *54*, 678–687. [CrossRef]
8. Cole, I.S.; Marney, D.J.C.S. The science of pipe corrosion: A review of the literature on the corrosion of ferrous metals in soils. *Corros. Sci.* **2012**, *56*, 5–16. [CrossRef]
9. Neff, D.; Dillmann, P.; Bellot-Gurlet, L.; Beranger, G. Corrosion of iron archaeological artefacts in soil: Characterisation of the corrosion system. *Corros. Sci.* **2005**, *47*, 515–535. [CrossRef]
10. Song, Y.; Jiang, G.; Chen, Y.; Zhao, P.; Tian, Y.J.S.R. Effects of chloride ions on corrosion of ductile iron and carbon steel in soil environments. *Sci. Rep.* **2017**, *7*, 6865. [CrossRef] [PubMed]
11. Tan, M.Y.; Varela, F.; Huo, Y.; Wang, K.; Ubhayaratne, I. An Overview of Recent Progresses in Monitoring and Understanding Localized Corrosion on Buried Steel Pipelines. In Proceedings of the NACE International Corrosion Conference Proceedings, Houston, TX, USA, 6–9 May 2020; pp. 1–12.
12. Hong, M.S.; Kim, S.H.; Im, S.Y.; Kim, J.G. Effect of ascorbic acid on the pitting resistance of 316L stainless steel in synthetic tap water. *Met. Mater. Int.* **2016**, *22*, 621–629. [CrossRef]
13. Jones, D.A. *Principles and Prevention of Corrosion*, 2nd ed.; Prentice-Hall: Upper Saddle River, NJ, USA, 1992; pp. 76–92.
14. Bockris, J.O.M. *Modern Electrochemistry 2B: Electrodics in Chemistry, Engineering, Biology and Environmental Science*, 2nd ed.; Plenum Publishers: New York, NY, USA, 1998; pp. 1646–1652.
15. Byeon, S.I. *The Fundamentals of Corrosion of Metals and Their Application into Practice*; Gyomoon: Seoul, Korea, 2006; pp. 331–341.
16. Muralidharan, S.; Kim, D.-K.; Ha, T.-H.; Bae, J.-H.; Ha, Y.-C.; Lee, H.-G.; Scantlebury, J.J.D. Influence of alternating, direct and superimposed alternating and direct current on the corrosion of mild steel in marine environments. *Desalination* **2007**, *216*, 103–115. [CrossRef]
17. Bott, A.W. Controlled current techniques. *Curr. Sep.* **2000**, *18*, 125.

18. Pourbaix, M.J.N. *Atlas of Electrochemical Equilibria in Aqueous Solution*; National Association of Corrosion Engineeers: Houston, TX, USA, 1974; pp. 307–321.
19. Berntsen, T.; Laethaisong, N.; Seiersten, M.; Hemmingsen, T. Uncovering carbide on carbon steels by use of anodic galvanostatic polarization and its effect on CO_2 corrosion. *Corrosion* **2016**, *72*, 534–546.
20. Stefanoni, M.; Angst, U.M.; Elsener, B. Kinetics of electrochemical dissolution of metals in porous media. *Nat. Mater.* **2019**, *18*, 942–947. [CrossRef] [PubMed]
21. Leban, M.B.; Kosec, T. Characterization of corrosion products formed on mild steel in deoxygenated water by Raman spectroscopy and energy dispersive X-ray spectrometry. *Eng. Fail. Anal.* **2017**, *79*, 940–950. [CrossRef]
22. Wei, B.; Qin, Q.; Bai, Y.; Yu, C.; Xu, J.; Sun, C.; Ke, W. Short-period corrosion of X80 pipeline steel induced by AC current in acidic red soil. *Eng. Fail. Anal.* **2019**, *105*, 156–175. [CrossRef]
23. You, J.S.N.; Zhang, Y. Corrosion behavior of low-carbon steel reinforcement in alkali-activated slag-steel slag and Portland cement-based mortars under simulated marine environment. *Corros. Sci.* **2020**, *175*, 108874. [CrossRef]
24. Yan, M.; Sun, C.; Xu, J.; Ke, W. Anoxic corrosion behavior of pipeline steel in acidic soils. *Ind. Eng. Chem. Res.* **2014**, *53*, 17615. [CrossRef]

Article
Structural Integrity of Steel Pipeline with Clusters of Corrosion Defects

Maciej Witek

Gas Engineering Group, Warsaw University of Technology, 20 Nowowiejska St., 00-653 Warsaw, Poland; maciej.witek@pw.edu.pl

Abstract: The main goal of this paper is to evaluate the burst pressure and structural integrity of a steel pipeline based on in-line inspection results, in respect to the grouping criteria of closely spaced volumetric surface features. In the study, special attention is paid to evaluation of data provided from the diagnostics using an axial excitation magnetic flux leakage technology in respect to multiple defects grouping. Standardized clustering rules were applied to the corrosion pits taken from an in-line inspection of the gas transmission pipeline. Basic rules of interaction of pipe wall metal losses are expressed in terms of longitudinal and circumferential spacing of the features in the colony. The effect of interactions of the detected anomalies on the tube residual strength evaluated according to the Det Norske Veritas Recommended practice was investigated in the current study. In the presented case, groups of closely-spaced defects behaved similarly as individual flaws with regard to their influence on burst pressure and pipeline failure probability.

Keywords: steel pipeline; in-line inspection results; interacting corrosion defects; structural integrity

1. Introduction

Degradation of underground steel structures during their service lives leads to occurrence of volumetric surface defects and reduction of the tube wall thickness, as it is shown in Figure 1. The steel pipelines are buried almost on whole their length, and the properties of the soil are the most important factor of the corrosion; however, there are many other parameters such as an influence of the straight currents. A corrosion rate of high pressure steel pipelines needs to be controlled by their operators during the maintenance. Periodic in-line inspections (ILI), using an axial excitation magnetic flux leakage technology (Figure 2), are usually performed by gas transmission grid operators to detect and size the tube wall metal losses during the certain time intervals. If more than one diagnostic survey is performed at on a steel pipeline, so-called defect matching can be performed in order to evaluate the growth of the corrosion in the specific maintenance conditions [1]. Direct application of theoretical fracture mechanics methods to the assessment of the volumetric features provided from the in-line inspection is not appropriate due to uncertainty of the in-line inspection tool results highlighted by the author in [2]. The steel pipe wall burst was analysed by the author in [3]. However, a lot of studies deal with investigation of strength and structural integrity of steel pipelines with wall metal losses and longitudinally-oriented grooves similar to cracks using different methodologies, for instance, the latest publications applying a finite element method [4] and a linear elastic fracture failure mode [5]. A rupture pressure prediction model for steel tubes affected by the stray current corrosion based on artificial neural network was applied in [6].

Figure 1. A photograph of corrosion colonies on the steel pipe surface. Author's source.

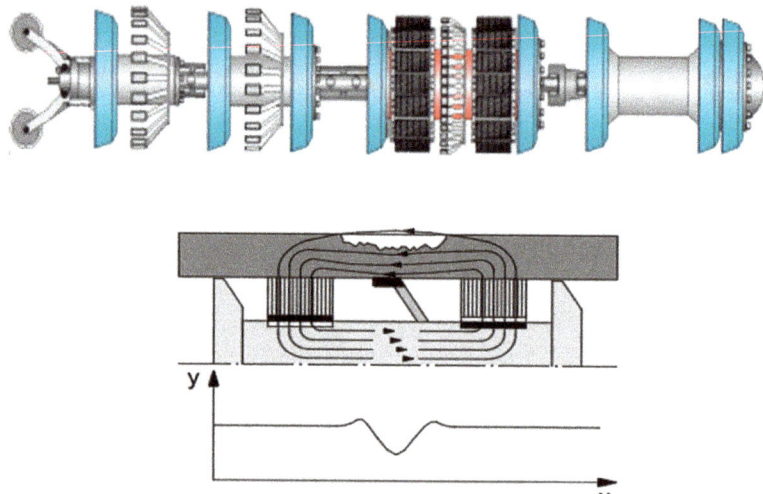

Figure 2. Axial excitation magnetic flux leakage inspection technology: a diagnostics tool (at the top) and a measurement principle (at the bottom). Author's source.

Many research projects, in which the failure behaviour and assessment of cylindrical shells containing adjacent corrosion indications were investigated, have been carried out worldwide over the past 50 years. From those studies the conclusions are as follows. The interaction between each pair of metal losses within a group is governed by several parameters, namely, spacing, tube outside diameter, wall thickness, depth of anomalies, and shapes of defects. Among these factors, the distance between each pair of volumetric flaws is most significant. Rules of interaction of pipe wall metal losses are generally expressed in terms of longitudinal and circumferential spacing of the features in the cluster and are studied in many publications, for example [7]. The authors applied a finite element method to find new interaction rules of corrosion flaws for longitudinal and circumferential aligned metal losses as well as to compare these rules to the available grouping standards. However, the failure pressure of a cluster of closely spaced corrosion pits is generally smaller than the rupture pressures in the case when the defects are considered as isolated. This reduction in the corroded pipe strength occurs due to the possible interaction between the adjacent tube wall metal losses.

The influence of the results of analysis of the anomaly colonies on the rupture of the pipe subjected to the internal pressure was evaluated in the present study. There are two main subjects of interest in the investigation of closely spaced interacting corrosion defects. The first issue is the rules of grouping the metal losses, whereas the other one is prediction of the failure pressure of the pipe with adjacent defects. This research is focused on the assessment of tube wall volumetric surface flaws in respect to the criteria of interaction. In the present paper, DNV-RP-F101:2010 Det Norske Veritas Recommended practice [8] standard is applied to find defects which can be considered as clusters within the population of the external surface metal losses taken from the diagnostics data.

The mechanical reliability of the underground infrastructure within long-term operation, counted in decades, can be analysed as a stochastic process of random degradation of the structure elements. The algorithms calculating the probability of failure can be divided into methods based on functions of random parameters [9] and a theory of random variables which is applied in the current paper similarly as in [10,11]. There are plenty of publications which analyse the time dependent structural integrity relying on the stochastic models and on the results of the in-service diagnostics as well. Many publications, for example, [12], as well as the author's work [13], calculate the failure probability considering defects as single isolated corrosion pits. However, a few publications focus on the interacting corrosion areas and, for this reason, this problem is the subject of the current research. The aim of the present study is to calculate the failure probability of the steel pipeline failure based on the in-line inspection data containing groups of indications. A limited number of publications focus on the multiple colonies of corrosion features and, for this reason, this issue is the subject of the current research. The following words: group, colony, cluster are synonyms and were used exchangingly in the text.

2. Clustering Rules for Tube Wall Metal Losses

There are two main subjects of interest in the investigation of the interacting pipe wall metal losses. The first issue are coincidence rules, whereas the other one is prediction of the failure pressure of a tube with adjacent colonies of defects. Section 2.1 presents the grouping criteria for the pipe wall volumetric anomalies detected by the axial excitation magnetic flux leakage in-line inspection tools. Section 2.2 contains the standardized assessment level 2 methodology for burst pressure calculations of a steel pipe with the interacting metal losses.

2.1. Grouping Criteria for Volumetric Defect Colonies

According to the Benjamin Adilson's classification presented in [14], there are three types of interactions of volumetric features caused by corrosion. Type 1 is a group of metal losses separated circumferentially; however, their individual profiles overlap when projected into the longitudinal plane through the wall thickness. Type 2 is found in colonies in which the flaws are or are not longitudinally aligned and their individual profiles do not overlap when projected onto the longitudinal plane through the wall thickness, i.e., their projected individual profiles are separated by the length of the full pipe wall-thickness. Type 3 is a combination of 2 above mentioned types. The described types are graphically shown in Det Norske Veritas Recommended practice [8].

The rules of interactions establish the limit value of the distance between two individual defects in the colony, beyond which the interaction is negligible. For the purpose of analysis of interactions, the longitudinal spacing s_l and the circumferential distance s_c of each metal loss in the group are usually verified. A majority of the currently available rules of clustering the corrosion flaws adopt expressions containing two following conditions to be met [8]:

(a) longitudinal spacing along the pipe axis is less than $s_{Li} \leq (s_l)_{\lim}$

$$(s_l)_{\lim} = 2.0 \sqrt{Dt} \qquad (1)$$

and
(b) circumferential spacing $s_{ci} \leq (s_c)_{\lim}$

$$(s_c)_{\lim} = \pi \sqrt{Dt} \qquad (2)$$

where: s_{li}—longitudinal spacing of each defect in the colony, [mm]; s_{ci}—circumferential spacing of each anomaly in the group, [mm]; D—tube outside diameter, [mm]; t—pipe wall nominal thickness, [mm].

If the cluster is composed of more than two metal losses, the rules of interaction are applied to all possible pairs of adjacent defects within the group and the above mentioned criteria are verified. In the present study, the evaluation rules applied to the repeated in-line inspection data of gas transmission pipeline are presented.

2.2. Analytical Assessment of Interacting Defects

Det Norske Veritas Recommended practice Corroded pipelines level 2 methodology was developed for the calculation of the burst pressure of thin-walled cylindrical shells with interacting volumetric surface colonies and is applied in the current paper. The combined length of the interacting metal losses is calculated as follows [8]:

$$l_{nm} = l_m + \sum_{i=n}^{i=m-1} (l_i + s_i) \qquad (3)$$

where:
l_i—axial length of each interacting metal loss from n to m, [mm];
s_i—circumferential distance of each interacting metal loss from n to m, [mm].
l_{nm}—combined length of defects in the longitudinal direction [mm].

The effective depth of the combined flaw formed from all n to m of the interacting metal losses is calculated as follows [8]:

$$d_{nm} = \frac{\sum_{i=n}^{i=m} d_i l_i}{l_{nm}} \qquad (4)$$

where:
d_{nm}—effective depth of the metal loss combined from n to m, [mm];
d_i—circumferential depth of each interacting metal loss from n to m, [mm];

The rupture pressure of the steel pipe with the combined corrosion colony formed from all n to m of the interacting metal losses is calculated as follows:

$$P_{burst} = \frac{2tf_u \left(1 - \frac{d_{nm}}{t}\right)}{(D-t)\left(1 - \frac{d_{nm}}{tM_{nm}}\right)} \qquad (5)$$

where:
P_{burst}—burst pressure of the steel pipe with the combined defect formed from all n to m, [MPa];
f_u—ultimate tensile strength of the steel, [MPa];
M_{nm}—Folias factor, [–], corresponding to the combined length of defects in the longitudinal direction is expressed as:

$$M_{nm} = \sqrt{1 + 0.31 \left(\frac{l_{nm}}{\sqrt{Dt}}\right)^2} \qquad (6)$$

3. In-Line Inspection Indications Clustering

In order to illustrate the methodology for the interactions of defects described in Section 2, the clustering criteria were implemented to evaluate the possible material loss colonies along the studied gas transmission grid. The case study considers a 711 mm outer diameter cathodically protected pipeline of the total length of 147.267 km and the tube wall thickness of 10.5 mm. The maximum operating pressure value of all the pipeline sections is $MOP = 5.5$ MPa. The material used is equivalent to L360NE steel grade, according to EN-ISO 3183, with the following average parameters: ultimate tensile strength $\sigma_U = 554.7$ MPa, average yield stress $\sigma_Y = 370$ MPa and average elasticity modulus of steel $E = 202$ GPa. The pipe as well as the mechanical properties of the girth weld were confirmed by the destructive tests conducted on the steel coupons taken from the pipeline after 13 years of operation, directly after the first diagnostic [15]. Two in-line inspections were performed with the use of axial excitation magnetic flux leakage diagnostic tools. A period of time between the first and the second survey was 12 years. In this paper, all analyses are based on the real data for the external metal losses of the tube wall. Relying on the report of the first ILI, the number of 1347 external corrosion pits were found on the outside surface of the wall. Based on the data of the second ILI, the amount of 2838 external surface flaws were found. The number of 72 external closely spaced metal losses taken from the second inspection data were classified as groups for detailed considerations. For the research purpose, three of groups were described below in details and shown in Figures 3–5. The limit of longitudinal spacing along the pipe axis in the studied case needs to be less than $s_L \leq (s_L)_{\lim} = 176.9$ mm and circumferential spacing $s_c \leq (s_c)_{\lim} = 277.8$ mm. Dimensions of the defects from group 1 are summarized in Table 1 and the graphical presentation of features spacing is shown in Figure 3. Description of defects in colonies are as follows: C1D1—cluster 1, defect 1.

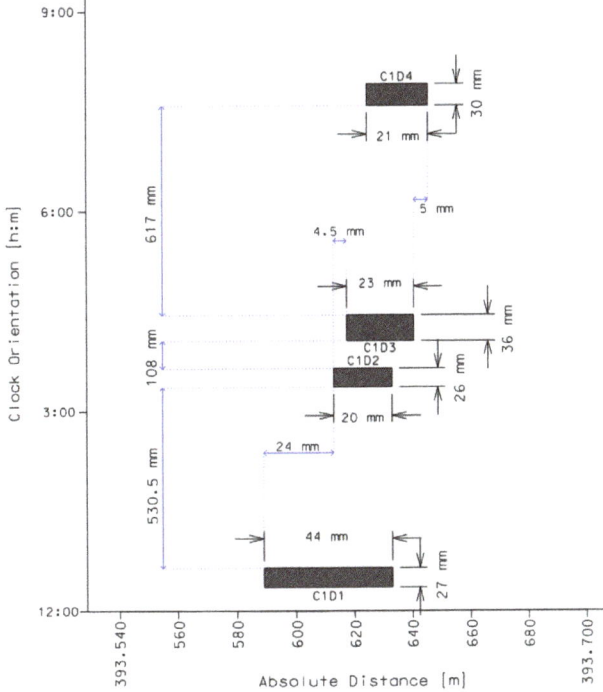

Figure 3. Graphical presentation of volumetric defects in colony 1. Source: Author's analysis.

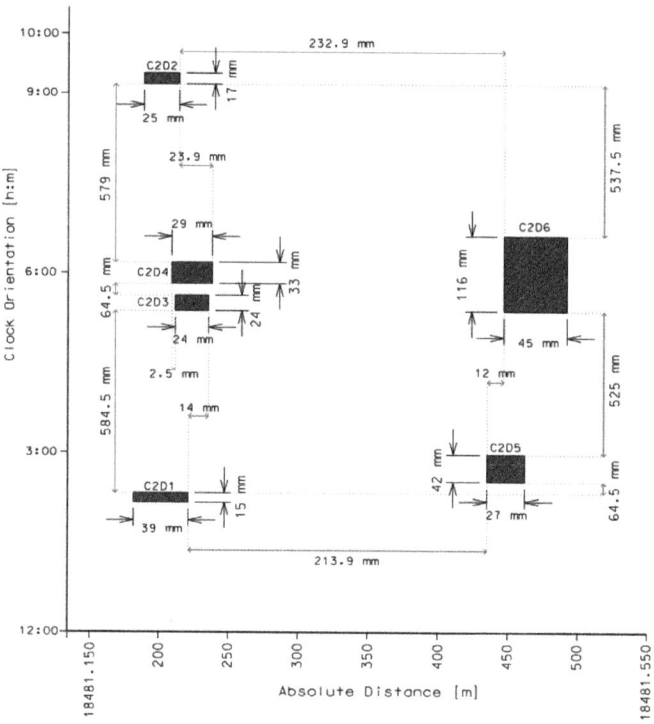

Figure 4. Graphical presentation of metal losses spacing in group 2. Source: Author's analysis.

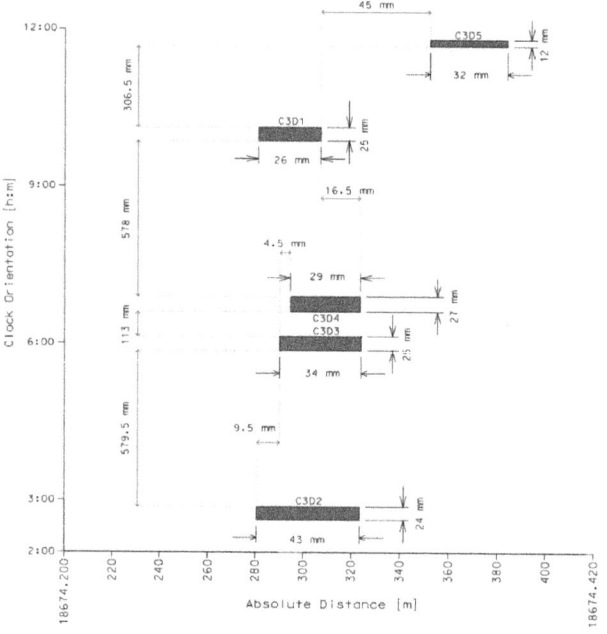

Figure 5. Graphical presentation of defects spacing in group 3. Source: Author's analysis.

Table 1. Anomaly dimensions in cluster 1. Source: Author's analysis.

Defect №	Absolute Distance [m]	Relative Depth (d/t) [%]	Axial Length [mm]	Width [mm]	Clock Orientation [hour:min]
C1D1	393.611	17%	44	27	12:30
C1D2	393.623	22%	20	26	03:30
C1D3	393.629	25%	21	30	07:45
C1D4	393.635	20%	23	36	04:15

Group 1 contains four flaws separated circumferentially; however, their individual profiles overlap when projected onto the longitudinal plane through the wall thickness (interaction type 1), as it is illustrated in Figure 3. Due to the limit of circumferential spacing for each pair of metal losses of the studied pipeline $(s_c)_{lim}$ = 271.5 mm, only two indications, C1D2 and C1D4, can be considered as interacting with the combined length of l_{nm} = 27.5 mm and the maximum depth of 22% of the pipe wall thickness which corresponds to 2.42 mm. The effective depth of the external metal loss combined from C1D2 and C1D4 calculated according to Equation (4) is equal to d_{nm} = 3.43 mm.

Dimensions of the defects from group 2 are summarized in Table 2. Cluster 2 contains six indications divided into two subgroups whose individual profiles overlap in the circumferential as well as in the longitudinal plane through the wall thickness (interaction type 1 and type 2), as it is shown in Figure 4. Due to the limit of circumferential spacing for the studied pipeline $(s_c)_{lim}$ = 271.5 mm and a value of longitudinal spacing along the pipe axis less than $s_L \leq (s_L)_{lim}$ = 172.8 mm of each pair, only two anomalies, C2D3 and C2D4, can be considered as interacting with the combined length of l_{nm} = 29.0 mm and the maximum depth of 31% of wall thickness which corresponds to the value of 3.41 mm. The effective depth of the metal loss combined from C2D3 and C2D4 calculated according to Equation (4) is equal to d_{nm} = 5.10 mm.

Table 2. Metal loss sizing in group 2. Source: Author's analysis.

Defect №	Absolute Distance [m]	Relative Depth (d/t) [%]	Axial Length [mm]	Width [mm]	Clock Orientation [hour:min]
C2D1	18481.201	29%	39	15	02:15
C2D2	18481.201	25%	25	17	09:15
C2D3	18481.222	21%	24	24	05:30
C2D4	18481.222	31%	29	33	06:00
C2D5	18481.447	16%	27	42	02:45
C2D6	18481.468	19%	45	116	06:00

Dimensions of the metal losses from cluster 3 are summarized in Table 3. Colony 3 is the group of five anomalies separated circumferentially; however, four individual profiles overlap when projected onto the longitudinal plane through the wall thickness (interaction type 1), as it is illustrated in Figure 5. Due to the limit of circumferential spacing for the considered pipeline $(s_c)_{lim}$ = 271.5 mm and a value of longitudinal spacing along the pipe axis less than $s_L \leq (s_L)_{lim}$ = 172.8 mm of each pair, only the two flaws C3D3 and C3D4 can be considered as interacting with the combined length of l_{nm} = 33.5 mm and maximum depth of 22% of pipe wall thickness which corresponds to the value of 2.42 mm. The effective depth of the metal loss combined from C3D3 and C3D4 calculated according to Equation (4) is equal to d_{nm} = 4.0 mm.

Table 3. Flaws dimensions in the colony 3. Source: Author's analysis.

Defect №	Absolute Distance [m]	Relative Depth (d/t) [%]	Axial Length [mm]	Width [mm]	Clock Orientation [hour:min]
C3D1	18674.294	17%	26	25	10:00
C3D2	18674.302	22%	43	24	02:45
C3D3	18674.307	19%	34	25	06:00
C3D4	18674.309	22%	29	27	06:45
C3D5	18674.368	27%	32	12	11:45

The results of the plastic burst pressure calculations according to Equations (5) and (6), in the three analysed cases of colonies of the volumetric defects, are summarized in Table 4.

Table 4. The results of the pipe wall plastic collapse calculated according to DNV-RP-F101 standard. Source: Author's analysis.

Colony №	Defect №	Defect Depth [mm]	Axial Length [mm]	Burst Pressure [MPa]	Colony №
C1	C1D2	2.42	20.0	17.875	C1D2
	C1D4	2.20	23.0	17.868	C1D4
C2	C1D2 + C1D4	3.43	27.5	17.796	C1D2 + C1D4
	C2D3	2.31	24.0	17.861	C2D3
C3	C2D4	3.41	29.0	17.784	C2D4
	C2D3+ C2D4	5.10	29.0	17.666	C2D3 + C2D4

From Table 4, it can be concluded that rupture pressure calculated according to Det Norske Veritas Recommended practice for every analysed group of flaws is in the difference range of 2 bar compared to the assessment of the individual features. The results of analytical calculation of burst pressure of the defected pipes, taking into consideration interactions of metal losses obtained from the in-line inspection, show a little influence of closely spaced corrosion flaws on the burst pressure of the studied case.

4. Probability of Pipeline Rupture

In order to estimate a failure probability in a long operation time, the limit state based on a pressure difference between the plastic collapse of the remaining pipe wall thickness and an expected value of the gas working pressure as a random variable were employed.

4.1. Calculation Methodology

The limit state function of the plastic collapse of the *j*-th pipeline section affected by a part-wall reduction caused by corrosion is expressed as:

$$g(\vec{X}) = P_{fj} - OP_{max} \tag{7}$$

where:

$g(\vec{X})$—limit state function of the tube wall plastic collapse;
$\vec{X}$—vector of random variables related to the pipeline segment;
P_{fj}—failure pressure of the j-th pipeline section affected by corrosion, [MPa];
OP_{max}—maximum operating pressure of the segment, [MPa].

The time-dependent theoretical failure pressure for a straight pipe with a part-wall volumetric surface defect is a function of the following variables:

$$P_{fj}(T) = f(t, d, l, D, c_d, \alpha, c_L, f_u, T) \tag{8}$$

where:

T—time period, [year];

c_L—axial corrosion rate for the defect length, [mm/year];
c_d—proportionality coefficient of a power law function for the defect depth, [mm/year$^\alpha$];
α—exponential coefficient of a power law function for the flaw depth;
f_u—ultimate tensile strength of the material used in design. [MPa].

Due to active corrosion of the pipe wall, the pipeline reliability decreases with time of the system operation. Increments of dimensions of metal losses during operation are described for the feature depth in radial direction with the following functions:

$$d(T) = d_{mean}(T_0) + c_d \cdot T^\alpha \tag{9}$$

and for the length of the longitudinally-oriented tube external surface features in the axial direction:

$$L(T) = L(T_0) + c_L \cdot T \tag{10}$$

Failure probability for a pipeline with corrosion grooving with time (T) can be calculated as follows:

$$Pof_j\left(\vec{X}, T\right) = P[g(\vec{X}, T) \leq 0] \tag{11}$$

$$P[g(\vec{X}, T) \leq 0] = \int_{g(\vec{X},T) \leq 0} f(x_i, T) dx_i \tag{12}$$

where:

Pof_j—probability of failure of the j-th pipeline segment with an active corrosion defect, [1/year].

The failure occurs when $g\left(\vec{X}, T\right) \leq 0$. For a specific time period, Monte Carlo numerical simulation is conducted by random generating of numbers for variables P_{fj}, with respect to statistical distribution of the input parameters specific for a segment. For each evaluation of limit state function (7), occurrence of $g\left(\vec{X}, T\right) 0$ is counted.

Probability of rupture of the j-th section of the pipeline, with the assumption of independence of n individual failures Pof_{jt} of tubes connected in series, is calculated as

$$Pof_j\left(\vec{X}, T\right) = 1 - \prod_{t=1}^{n}(1 - Pof_{jt}(\vec{X}, T)) \tag{13}$$

where:

n—number of pipe wall metal losses, [–].

For each external corrosion flaw, the total number of failure events N_f is determined after Monte Carlo samples were generated and rupture probability of a single defect as a function of time can be obtained using the following Equation.

$$Pof_{jt}\left(\vec{X}, T\right) = \frac{N_f}{MC} \tag{14}$$

where:

N_f—total number of failure events when $g\left(\vec{X}, T\right) \leq 0$;

MC—total number of Monte Carlo trials at the specific time step for calculations of the j-th pipeline segment failure probability.

4.2. Input Data for Structural Integrity Evaluation

For the input parameters specified below, the pipe diameter and the wall thickness are modelled as random variables relying on the tube manufacturer's certificates. The random variables listed below have been obtained from the two repeated diagnostics on the same pipeline. For ILI of oil and gas pipelines, it is a common practice to track the same anomaly in different inspections (i.e., so-called defect matching) based on the longitudinal and circumferential positions of the indication reported by ILI tools. Taken into consideration the clustering criteria, only six pairs of flaws from the total amount of 138 fully matched external corrosion pits were classified as interacting. Eventually, a number of 132 external metal losses are selected for structural reliability estimation taken into consideration the above mentioned interacting criteria.

The external pipe surface corrosion coefficients for the cathodically protected underground structures are derived from the literature [16] and from the author's other papers [13]. The reliability estimations of the pipeline were conducted with coefficients of corrosion growth rates according to Equations (9) and (10): $c_d = 0.164$ mm, $\alpha = 0.78$, $c_l = 1.4$ mm/year. The linear length increment of the metal loss of the initial length $L = 174$ mm with a function of service time is presented in Figure 6.

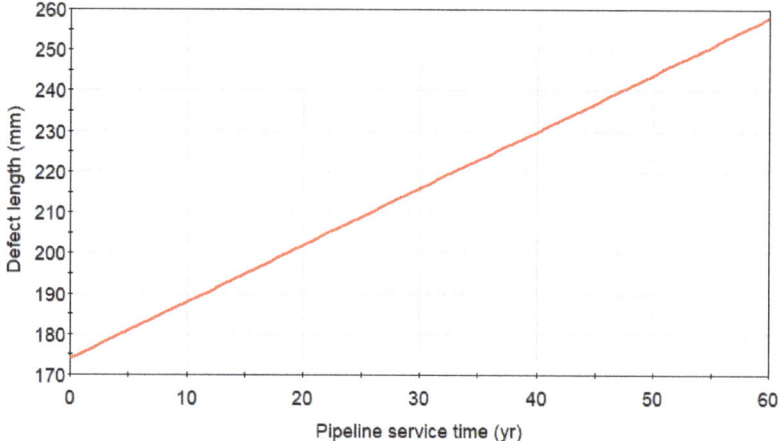

Figure 6. The linear length increment of the detected metal loss of the initial length l = 174 mm and a growth rate of 1.4 mm/year within the predicted pipeline service time. Source: Author's analysis.

The list of statistical distribution of input parameters for a structural integrity evaluation is presented in Table 5. Computations of failure probability were conducted with the use of an academic license of GoldSim software.

Table 5. Statistical distribution of input parameters for reliability evaluation. Source: Author's analysis.

No.	Parameter	Unit	Mean Value	Uncertainty Coefficients	Distribution Type
1.	Yield tensile strength (f_y)	MPa	370.6	StD[f_y] = 12.2 COV[f_y] = 3.3 %	Lognormal
2.	Ultimate tensile strength (f_u)	MPa	554.7	StD[f_u] = 19.4 COV[f_u] = 3.5 %	Lognormal
3.	Pipe wall thickness (t)	mm	11.0	StD[t] = 0.5 COV[t] = 4.5 %	Normal
4.	Tube outside diameter (D)	mm	711.0	StD[D] = 20.3 COV[D] = 2.8 %	Normal

Table 5. Cont.

No.	Parameter	Unit	Mean Value	Uncertainty Coefficients	Distribution Type
5.	Maximum operating pressure (MOP)	MPa	5.5	s = 0.3 COV[MOP] = 5.5 %	Gumbel
6.	Metal loss depth (d)	mm	defect specific	StD[d] = 0.6 COV[d] = 26.6 %	Normal
7.	Flaw length (L)	mm	defect specific	StD[L] = 34.6 COV[L] = 76.9 %	Lognormal
8.	Defect depth growth rate $d(T)$ as a power law function with parameters c_d, α.	mm/year	c_d = 0.164, α = 0.78	Parameters c_d, α deterministic	Parameters cd, α deterministic
9.	Metal loss length growth rate $L(T)$ as a power law function with parameters c_l.	mm/year	c_l = 1.4	Parameter c_l deterministic	Parameter cl deterministic

5. Pipeline Structural Reliability Considering Flaws Grouping

Integrity computations, taking into consideration the plastic collapse of the remaining wall thickness of the tube, were carried out relying on the in-line inspections data for the considered case. Due to a corrosion increment of the steel, the structural reliability decreases with time of the underground pipeline operation. The plot of the burst pressure (green line, right axis) and a probability of rupture logarithmic graph (red line, left axis) for the longest single metal loss of initial length L = 322 mm and maximum depth of 12% of the wall thickness, detected during the second ILI, as a function of service time for the considered case, is presented in Figure 7. The failure probability calculated for service time within 60 years starting from the second in-line inspection, even for non-repaired longest defect, is low and remain lower than a related code-based target value set for a pipeline safety class high as not higher than 10^{-3} per annum [8]. In the later years of maintenance, e.g., when the operation life of the studied pipeline is more than 40 years, a rate of the failure probability increase is strong, which means the rapid aging process of the steel underground structure.

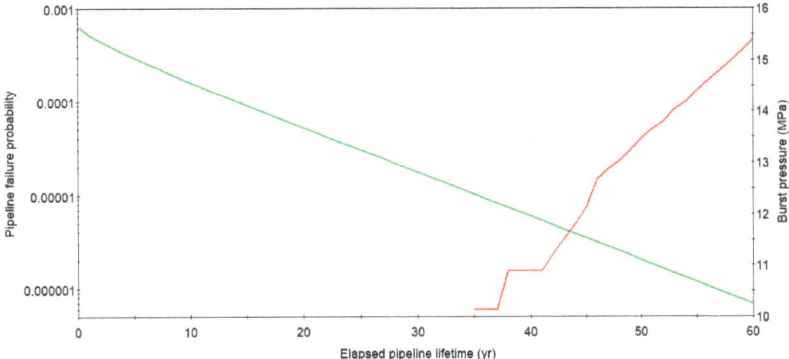

Figure 7. Change of rupture pressure and failure probability with time for X52 DN 700 MOP = 5.5 MPa pipeline with the longest detected defect and the initial depth of 12%. Source: Author's analysis.

The plot of the burst pressure and a probability of rupture logarithmic graph for cluster 1 with the combined length of l_{nm} = 27.5 mm and the maximum depth of 22% of the wall thickness as a function of pipeline service time for the considered case is presented in Figure 8. The failure probability calculated for service time within 60 years starting from the second diagnostics, even for non-repaired defect colony 1, is low and remain lower than a related code-based target value set as not higher than 10^{-3} per annum. In the later years of pipeline service, exceeding 40 years, the failure probability increase is strong, showing the rapid aging effect of the steel buried structure.

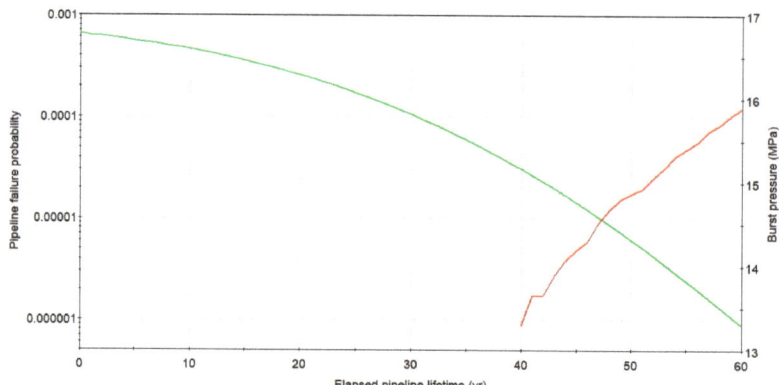

Figure 8. Change of rupture pressure and failure probability with time for X52 DN 700 MOP = 5.5 MPa pipeline with defect group1. Source: Author's analysis.

The graph of the burst pressure and a probability of rupture for flaw group 2 with the combined length of l_{nm} = 29.0 mm and the maximum depth of 3.41 mm of the pipe wall thickness as a function of the operating time for the considered case is presented in Figure 9. The failure probability calculated for pipeline service in time, after 53 maintenance years, remains higher than a related code-based target value set for a safety class high as not exceeding 10^{-3} per annum.

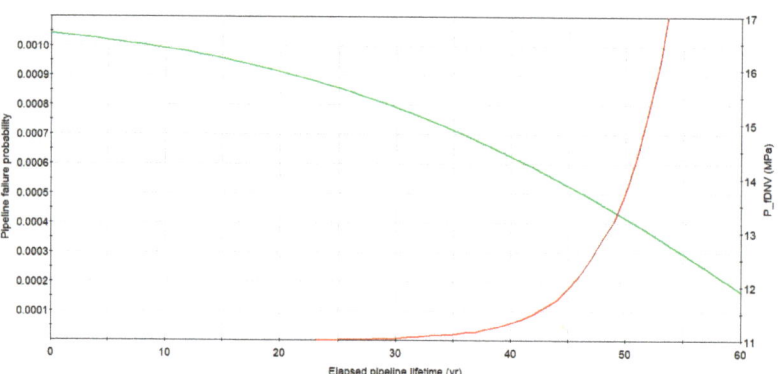

Figure 9. Change of burst pressure and failure probability with time for L360NE DN 700 MOP = 5.5 MPa pipeline with metal loss group 2. Source: Author's analysis.

The plot of the burst pressure and a probability of rupture logarithmic graph for cluster 3 with the total length of l_{nm} = 33.5 mm and the maximum depth of 22% of the wall thickness which corresponds to the value of 2.42 mm is shown in Figure 10. The probability of burst calculated for a pipeline operating period not more than 60 years, is low and remains lower than a related code-based target value set for a safety class high as not exceeding than 10^{-3} per annum. If the operation period of the studied buried pipeline is more than 40 years, a rate of the failure probability increase is strong, which means the rapid aging effect of the steel.

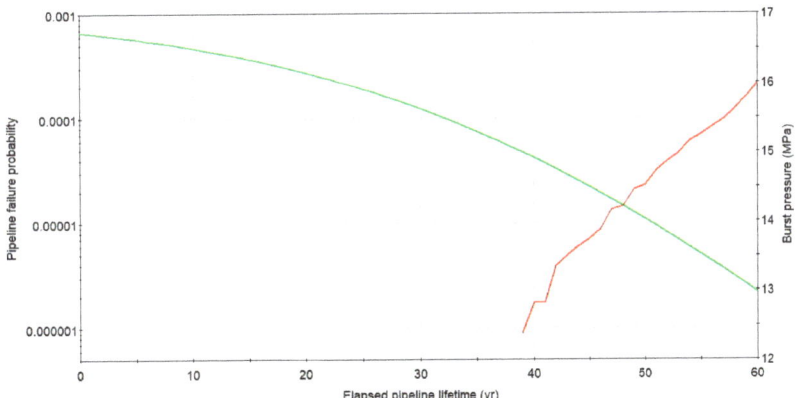

Figure 10. Change of burst pressure and failure probability with time X52 DN 700 MOP = 5.5 MPa pipeline with defect group 3. Source: Author's analysis.

6. Conclusions

The calculation results of the burst pressure of the defected pipes relying on Det Norske Veritas Recommended practice Corroded pipelines, taking into consideration interactions of metal losses obtained from in-line inspections, show a little influence of closely spaced indications on the rupture pressure of the considered steel pipeline.

The probability of pipeline burst in the studied corrosion colonies cases do not differ significantly from the corresponding case when features are assessed as isolated. In the case considered in the present paper, grouping of closely spaced defects is almost consistent compared to the assessment of the individual flaws in respect of the burst pressure calculations and the probability estimations. The burst probability computations of the studied pipeline are independent of the results of corrosion grouping indications due to both small areas and the depth of the defects, which are the most important impact factors. The failure probability calculated for the pipeline service time within 50 years, starting from the second in-line inspection, even for non-repaired corrosion clusters, is low and remains lower than a related code-based target value set for a safety class high as not exceeding 10^{-3} per annum. In the later years of the studied pipeline operation, beyond 40 years, the structural integrity decrease is strong, which means the rapid aging degradation of the steel underground structure.

Funding: The author declare that the research has not received external funding. This paper was co-financed under the research grant of the Warsaw University of Technology supporting the scientific activity in the discipline of Environmental Engineering, Mining and Power Engineering.

Institutional Review Board Statement: Not applicable.

Informed Consent Statement: Not applicable.

Data Availability Statement: The data presented in this study are available on request from the corresponding author.

Conflicts of Interest: The authors declare that there are no conflict of interest that are directly or indirectly related to the research.

References

1. Witek, M.; Batura, A.; Orynyak, I.; Borodii, M. An integrated risk assessment of onshore gas transmission pipelines based on defect population. *Eng. Struct.* **2018**, *173*, 150–165. [CrossRef]
2. Witek, M. Validation of in-line inspection data quality and impact on steel pipeline diagnostic intervals. *J. Nat. Gas Sci. Eng.* **2018**, *56*, 121–133. [CrossRef]
3. Witek, M. Life cycle estimation of high pressure pipeline based on in-line inspection data. *Eng. Fail. Anal.* **2019**, *104*, 247–260. [CrossRef]

4. Kong, F.T.; Wordu, A.H. Burst strength analysis of pressurized steel pipelines with corrosion and gouge defects. *Eng. Fail. Anal.* **2020**, *108*, 104347.
5. Guillal, A.; Seghier, M.E.A.B.; Nourddine, A.; Correia, J.A.; Mustaffa, Z.B.; Trung, N.-T. Probabilistic investigation on the reliability assessment of mid- and high-strength pipelines under corrosion and fracture conditions. *Eng. Fail. Anal.* **2020**, *118*, 104891. [CrossRef]
6. Liu, X.; Xia, M.; Bolati, D.; Liu, J.; Zheng, Q.; Zhang, H. An ANN-based failure pressure prediction method for buried high-strength pipes with stray current corrosion defect. *Energy Sci. Eng.* **2019**, *8*, 248–259. [CrossRef]
7. Li, X.; Bai, Y.; Su, C.; Li, M. Effect of interaction between corrosion defects on failure pressure of thin wall steel pipeline. *Int. J. Press. Vessel. Pip.* **2016**, *138*, 8–18. [CrossRef]
8. DNV-RP-F101. *Det Norske Veritas Recommended Practice Corroded Pipelines*; 2010; Available online: http://www.opimsoft.com/download/reference/rp-f101_2010-10.pdf (accessed on 8 February 2021).
9. Zhang, S.; Zhou, W. System reliability of corroding pipelines considering stochastic process-based models for defect growth and internal pressure. *Int. J. Press. Vessels Pip.* **2013**, *111*, 120–130. [CrossRef]
10. Sahraoui, Y.; Khelif, R.; Chateauneuf, A. Maintenance planning under imperfect inspection of corroded pipelines. *Int. J. Press. Vessel. Pip.* **2013**, *104*, 76–82. [CrossRef]
11. Li, S.-X.; Yu, S.-R.; Zeng, H.-L.; Li, J.-H.; Liang, R. Predicting corrosion remaining life of underground pipelines with a mechanically-based probabilistic model. *J. Pet. Sci. Eng.* **2009**, *65*, 162–166. [CrossRef]
12. Pesinis, K.; Tee, K.F. Bayesian analysis of small probability incidents for corroding energy pipelines. *Eng. Struct.* **2018**, *165*, 264–277. [CrossRef]
13. Witek, M. Gas transmission failure probability estimation and defect repairs activities based on in-line inspection data. *Eng. Fail. Anal.* **2016**, *70*, 255–272. [CrossRef]
14. Adilson, C.; Benjamin, J.; Luiz, F.; Freire, R.; Vieira, D.; Divino, J.S. Interaction of corrosion defects in pipelines e Part. 1: Fundamentals. *Int. J. Press. Vessel. Pip.* **2016**, *144*, 56–62.
15. Szteke, W.; Biłous, W.; Wasiak, J.; Hajewska, E.; Przyborska, M.; Wagner, T. Badania Wycinka Rury ze Stali G-355 z Gazociągu Po 15-Letniej Eksploatacji Część II.: Badania Metodami Niszczącymi. Available online: https://www.researchgate.net/publication/320798876_The_investigation_of_the_gas_pipeline_from_G355_steel_after_15-years_of_exploitation_Part_I_The_investigation_with_nondestructive_methods (accessed on 8 February 2021).
16. Sahraoui, Y.; Chateauneuf, A. The effects of spatial variability of the aggressiveness of soil on system reliability of corroding underground pipelines. *Int. J. Press. Vessel. Pip.* **2016**, *146*, 188–197. [CrossRef]

Article

Effects of Ti and Cu Addition on Inclusion Modification and Corrosion Behavior in Simulated Coarse-Grained Heat-Affected Zone of Low-Alloy Steels

Yuhang Wang, Xian Zhang *, Wenzhui Wei, Xiangliang Wan, Jing Liu and Kaiming Wu *

The State Key Laboratory of Refractories and Metallurgy, Hubei Province Key Laboratory of Systems Science in Metallurgical Process, Collaborative Innovation Center for Advanced Steels, Wuhan University of Science and Technology, Wuhan 430081, China; wangyuhang@wust.edu.cn (Y.W.); weiwenzhui@126.com (W.W.); wanxiangliang@wust.edu.cn (X.W.); liujing2015@wust.edu.cn (J.L.)
* Correspondence: xianzhang@wust.edu.cn (X.Z.); wukaiming@wust.edu.cn (K.W.)

Abstract: In this paper, the effects of Ti and Cu addition on inclusion modification and corrosion behavior in the simulated coarse-grained heat-affected zone (CGHAZ) of low-alloy steels were investigated by using in-situ scanning vibration electrode technique (SVET), scanning electron microscope/energy-dispersive X-ray spectroscopy (SEM/EDS), and electrochemical workstation. The results demonstrated that the complex inclusions formed in Cu-bearing steel were (Ti, Al, Mn)-O_x-MnS, which was similar to that in base steel. Hence, localized corrosion was initiated by the dissolution of MnS. However, the main inclusions in Ti-bearing steels were modified into TiN-Al_2O_3/TiN, and the localized corrosion was initiated by the dissolution of high deformation region at inclusion/matrix interface. With increased interface density of inclusions in steels, the corrosion rate increased in the following order: Base steel ≈ Cu-bearing steel < Ti-bearing steel. Owing to the existence of Cu-enriched rust layer, the Cu-bearing steel shows a similar corrosion resistance with base steel.

Keywords: low-alloy steel; SEM; inclusion; anodic dissolution; pitting corrosion

1. Introduction

Low-alloy steel has been widely used as a construction material in the marine environment, owing to its remarkable mechanical properties and low cost [1–3]. In the harsh marine environment, low-alloy steels are susceptible to localized corrosion due to the existence of aggressive anions, such as Cl^- and SO_4^{2-} [4–6]. Moreover, localized corrosion induced by inclusions is usually in conjunction with a high local corrosion rate, which can result in a structural failure [7].

The weldability of low-alloy steel had a significant impact on its application. In the construction of bridges, ships, and steel structures, high heat input welding was used to improve welding productivity. However, the grain coarsening appeared in the heat-affected zone during the welding process, leading to degraded mechanical properties [8]. Moreover, residual stress generated during the solidification and shrinkage process can facilitate the formation of cold cracking in the heat-affected zone [9,10]. To improve the mechanical property of the coarse-grained heat-affected zone (CGHAZ) induced by high heat input welding, alloying elements, such as Ti and Cu, were added to refine the microstructure during the welding process [11–13]. The addition of Ti generates the dispersive and fine TiN precipitates, which effectively hinders the migration of grain boundaries. Thus, the addition of a small amount of Ti significantly inhibited the prior austenite grain growth [14,15]. It has been reported [16,17] that Cu addition effectively improves the acicular ferrite fraction, which would lead to a superior impact toughness.

However, the addition of alloying elements can significantly affect the corrosion resistance of the steel, owing to its influence on protective rust layer formation, inclusion

number density, size distribution, and chemical composition [18–20]. In Cu and Cr containing steel, an elemental Cu- and Cr-enriched layer would generate on the steel surface, this compact rust layer can significantly inhibit both anodic and cathodic reactions, lowering the corrosion rate of steel [18,19,21]. With the addition of alloying elements, such as S and Al, the number density and average size of MnS, MnO, and Al_2O_3 obviously increased, which would lead to the decrease of corrosion resistance [20,22,23]. On the contrary, the addition of RE elements can improve the corrosion resistance of the steel, owing to the much smaller average size of the modified inclusions than that of normal inclusions [24–27].

In our previous work, the mechanism of pitting initiation and propagation process induced by $(Zr-Ti-Al)-O_x$ inclusions in Zr-Ti deoxidized low-alloy steel were thoroughly investigated [28]. In previous studies, the impact of Ti and Cu addition on microstructure and toughness in simulated CGHAZ of low-alloy steels [15,16] and the localized corrosion behavior induced by inclusions, such as Al_2O_3, MnS, were discussed in detail [29–31]. However, with addition of Ti, Cu, and Mn, the correlation between inclusion modification and corrosion behavior in simulated CGHAZ of low-alloy steel has not yet been established.

In the present work, the effect of Ti and Cu addition on inclusion modification and corrosion behavior in CGHAZ of low-alloy steel was investigated. First, the inclusion number density, size distribution, and chemical composition with the addition of Ti and Cu were characterized by field-emission scanning electron microscopy/energy-dispersive spectrometry (FE-SEM-EDS). In addition, an immersion test coupled with in situ scanning vibrating electrode technique (SVET) and SEM/EDS was used to investigate the pit initiation and the propagation process induced by inclusions. Moreover, potentiodynamic polarization measurement was employed to analyze the corrosion resistance in the CGHAZ of Ti, Cu addition, and base low-alloy steels. Finally, the impact of inclusion number density, average diameter, and chemical composition on corrosion behavior was clarified.

2. Materials and Experimental

2.1. Sample Preparation

Three experimental steels, micro-alloyed with different Cu and Ti content, were prepared in a 10 kg vacuum melt induction furnace (Wuhan University of Science and Technology, Hubei, China), indicated as base (X70 pipeline) steel, Cu-bearing steel, and Ti-bearing steel, respectively. The chemical compositions of the investigated steel are listed in Table 1. The cylindrical ingots with 120 mm in diameter and 100 mm in length were reheated to 1250 ± 20 °C and forged into the cuboid with cross-section of 30 mm × 30 mm. A thermal simulator (Gleeble 3800, DSI, Bolingbrook, IL, USA) was used to simulate the CGHAZ. The samples were machined into cuboid with the dimensions of 11 mm × 11 mm × 100 mm. To simulate the submerged-arc welding at heat input of 100 kJ·cm^{-1}, the thermal cycle simulation was proceeded with a peak temperature of 1350 °C, a heating rate of 300 Ks^{-1}, and a holding time of 3 s. Besides, the cooling time from 800 to 500 °C was 52.8 s [16]. The heat input and the cooling time from 800 to 500 °C satisfy the Equation (1):

$$t = \left(0.67 - 5 \times 10^{-4} T_0\right) \times E \times \left(\frac{1}{500 - T_0} - \frac{1}{800 - T_0}\right) \quad (1)$$

where T_0 is the initial temperature 20 °C. Then, it can be obtained that the cooling times from 800 °C to 500 °C was 52.8 s, which were approximately equivalent to welding at the heat input of 100 kJ·cm^{-1}. The absorbed energy in the simulated CGHAZ of these three types of steels at −20 °C was measured by Charpy V-notch test (Table 2). Samples cut from the simulated CGHAZ with a size of 10 mm × 5 mm × 5 mm were sequentially ground with 2000 grit SiC paper and then polished with diamond paste down to 2.5 μm, ultrasonically degreased in ethanol.

Table 1. Chemical composition of the experimental steels (wt.%). Adapted with permission from ref. [16]. 2021 Taylor & Francis.

Samples	C	Si	Mn	Nb	V	Ti	Cu	Al	Fe
Base (X70) steel	0.055	0.21	1.61	0.038	0.022	0.012	0	0.022	Balance
Ti-bearing steel	0.057	0.23	1.63	0.039	0.021	0.061	0	0.025	Balance
Cu-bearing steel	0.056	0.21	1.60	0.040	0.021	0.010	0.32	0.024	Balance

Table 2. Impact toughness in the simulated CGHAZ of the experimental steels.

Samples	−20 °C Absorbed Energy (J)	
	Mean	Deviation
Base steel	78	±12
Ti-bearing	6	±0.6
Cu-bearing	241	±18

2.2. Surface Characterization

The prior austenite grain diameters of three types of steels were calculated by Nano Measurer, and about nine optical micrographs with 200 magnification were measured for each sample. The number density, size distribution, and chemical composition of inclusions were measured by an SEM (EVO MA10, ZEISS, Oberkochen, Germany). The morphology and elemental distribution of the inclusions were observed by an FE-SEM (Apreo S HiVac, FEI, Hillsboro, OR, USA) equipped with an EDS (AZteclive Ultim Max 100, OXFORD Instruments, Abingdon, UK).

To observe the localized corrosion process induced by inclusions of these steel, immersion tests (10 min) were implemented in 0.5 wt.% NaCl solution. After the immersion test, approximately ten inclusions in these samples were characterized via SEM/EDS to obtain the corrosion morphology and elemental distribution, which can be used to analyze the mechanism of localized corrosion induced by inclusions in CGHAZ.

2.3. In Situ Scanning Vibration Electrode Technique

A SVET system (Versascan, Ametek, Berwyn, PA, USA) was used to conduct the in situ micro-electrochemical measurement, which can produce the distribution and magnitude of local current intensities on the corroding surface of the samples in 0.5% NaCl solution. On the corroding surface of the steel, oxidation and reduction reactions usually occurred in separate regions. Owing to the nature of the difference of each reaction, an electric field was generated in the solution [32]. These extremely small potential variations over the corroding surface could be detected by the scanning vibration probe. For the test, the vibrating electrode was an insulated Pt-Ir probe with a diameter of approximately 5 μm, and its vibration frequency was 80 Hz and a vibration amplitude of 30 μm. According to the noise frequencies, the lock-in amplifier could filter out the electrical noise and then transformed the potential vibration into local current density by Ohm's law: $I = -\Delta E/R$, where $R = d/k$, d represent the amplitude of the microelectrode (30 μm), and k represent the conductivity of the solution (8.42 mS·cm^{-1} for 0.5% NaCl solution) [23].

2.4. Electrochemical Tests

The potentiodynamic polarization tests were conducted in 0.5% NaCl solution at 25 ± 1 °C via an electrochemical workstation (E4, Zahner, Kronach, Germany). During the tests, a three-electrode cell was composed by a counter electrode of platinum plate, a saturated calomel (SCE) reference electrode and the steel sample as the working electrode. The scan range of potentiodynamic polarization tests was from −300 mV$_{vs.OCP}$ to 300 mV$_{vs.OCP}$ and a scan rate of 0.5 mV·s^{-1}. The potentiodynamic polarization tests for each sample were conducted at least three times for reproducibility.

3. Results

3.1. Microstructure Characterization

The micrographs of the simulated CGHAZ of the three types of steels after thermal cycles are shown in Figure 1. In the simulated CGHAZ, the prior austenite grain diameter of base steel, Ti-bearing steel, and Cu-bearing steel was 66, 61, and 60 μm, respectively. The microstructures in the simulated CGHAZ of the three types of steels were composed of bainite, acicular ferrite, and martensite-austenite (M-A) constituents (Figure 1). The fractions and distribution of the M-A constituents and acicular ferrite in these three types of steels were slightly different, which was thoroughly investigated in previous studies [15,16]. The focus of this work is the influence of element addition on the formation of inclusions and corrosion behavior of the steels.

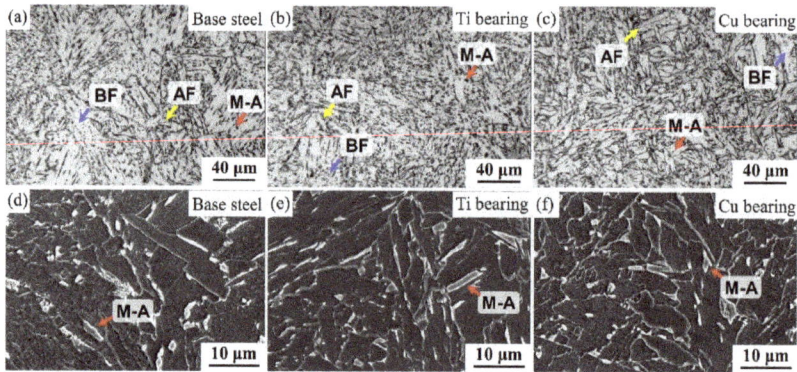

Figure 1. Optical micrographs and SEM micrographs in the simulated CGHAZ of three types of steels. BF, Bainite ferrite; AF, Acicular ferrite; (**a**) Optical micrographs of base steel; (**b**) Optical micrographs of Ti bearing steel; (**c**) Optical micrographs of Cu bearing steel; (**d**) SEM micrographs of base steel; (**e**) SEM micrographs of Ti bearing steel; (**f**) SEM micrographs of Cu bearing steel.

3.2. Characterization of Inclusions

The elemental composition, number density, and average diameter of the complex inclusions in CGHAZ of three types of steels were statistically analyzed via an SEM/EDS, and the result is shown in Table 3. In base steel, the main types of inclusions were MnS, (Ti, Al)-O_x and (Ti, Al, Mn)-O_x-S_y. In Ti-bearing steel, the main types of inclusions were Al_2O_3, TiN and TiN-Al_2O_3. In Cu-bearing steel, the main types of inclusions were MnS, (Ti, Al, Mn)-O_x and (Ti, Al, Mn)-O_x-S_y. Figure 2 shows the morphologies and chemical composition of extracted inclusions with the highest number density in three types of steels. Different from the globular feature of inclusions in base steel and Cu-bearing steel, the shape of inclusions in Ti-bearing steel was obviously irregular and acute.

Table 3. Statistical analysis of inclusions in three types of steels.

Sample	Types	Number Density (mm^{-2})		Average Diameter (μm)	
		Individual	Overall	Individual	Overall
Base steel	MnS	4.16		1.31	
	(Ti, Al)-O$_x$	4.65	18.53	1.92	1.54
	(Ti, Al, Mn)-O$_x$-S$_y$	9.72		1.46	
Ti-bearing	Al$_2$O$_3$	5.76		1.53	
	TiN	90.54	99.77	1.77	1.75
	TiN-Al$_2$O$_3$	3.47		1.55	
Cu-bearing	MnS	8.27		1.39	
	(Al, Ti, Mn)-O$_x$	18.00	49.70	1.86	1.71
	(Ti, Al, Mn)-O$_x$-S$_y$	23.43		1.70	

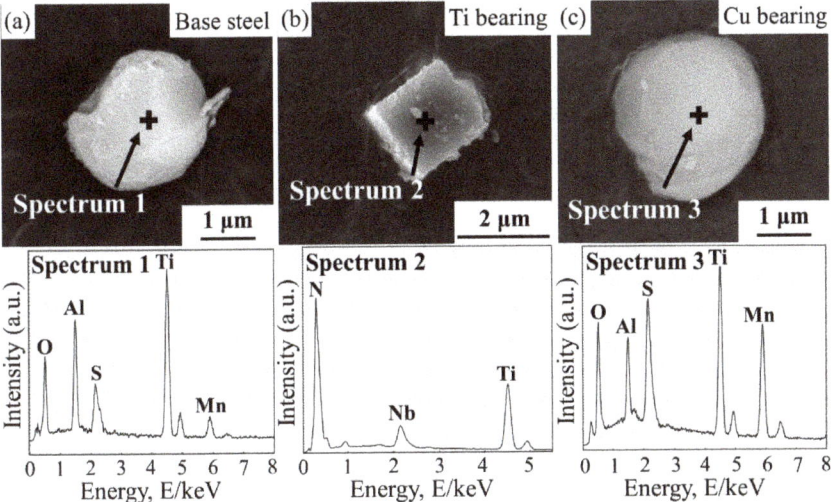

Figure 2. SEM images and corresponding EDS results of extracted inclusions in three types of steels. (**a**) Base steel. (**b**) Ti-bearing steel. (**c**) Cu-bearing steel.

The ability of inclusions to induce localized corrosion was not only limited by the chemical composition but also influenced by the dimension and density of inclusions [25,27,33,34]. Hence, the number density and size distribution of inclusions in three types of low-alloy steels were statistically analyzed via an SEM/EDS (Figure 3a,b). With the increased content of Ti and Cu elements, the density of inclusions obviously increased in two types of steels (Figure 3b). The number density of inclusions in Ti-bearing steel and Cu-bearing steel increased to 5.4 and 2.7 times that in base steel, respectively (Figure 3b). Additionally, the fractions of coarse inclusions (2–3 μm) in Ti- and Cu-bearing steels were 29% and 24%, respectively, which significantly increased in comparison with the base steel (11%) (Figure 3a).

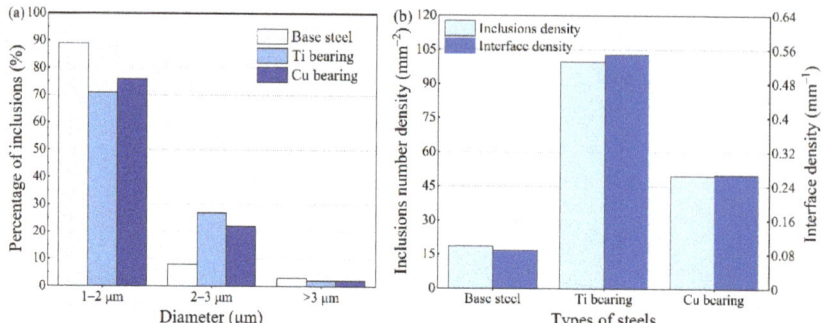

Figure 3. (a) The diameter distribution of inclusions, (b) number density and interface density of inclusions.

Owing to the micro-crevices and high lattice distortion regions around the inclusions, the initiation site of localized corrosion was generally located in the steel matrix/inclusion interface [26,35–38]. Therefore, the length of the interface between inclusions and the steel matrix per unit area was defined as interface density ρ_i, which can be calculated by Equation (2) [39]:

$$\rho_i = \pi \rho R \qquad (2)$$

where ρ is the number density of the inclusions, R is the average diameter of the inclusions. It was noticeable that as Ti and Cu elements content increased, the interface density in Ti-bearing steel and Cu-bearing steel increased to 6.1 and 3.0 times that in base steel (Figure 3b), respectively. The interface density showed a similar trend with the number density, but it was obvious that the interface density of Ti-bearing steels grows more than number density due to the larger average diameter of inclusions, which effectively reflected the influence of inclusions average diameter on corrosion initiation. To gain more insight into the influence of element addition on corrosion behavior, the initiation and propagation process of localized corrosion were investigated by the in-situ immersion tests in the next section.

3.3. Immersion Test

The surface morphology and elements distribution of inclusions in CGHAZ of three types of steels before and after the immersion test were characterized by means of SEM/EDS (Figures 4–6).

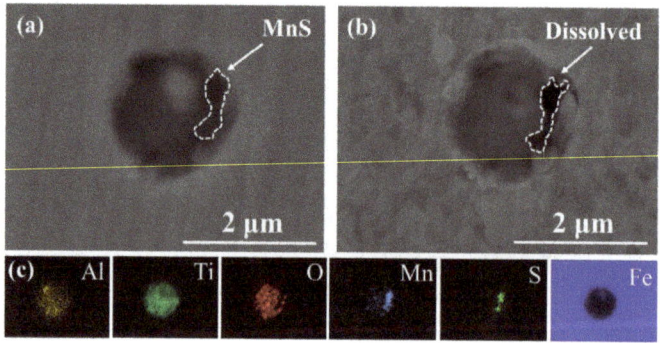

Figure 4. (a) SEM image of a typical complex inclusion in base steel. (b) Corrosion morphology of inclusion in (a). (c) EDS maps of inclusion in (a).

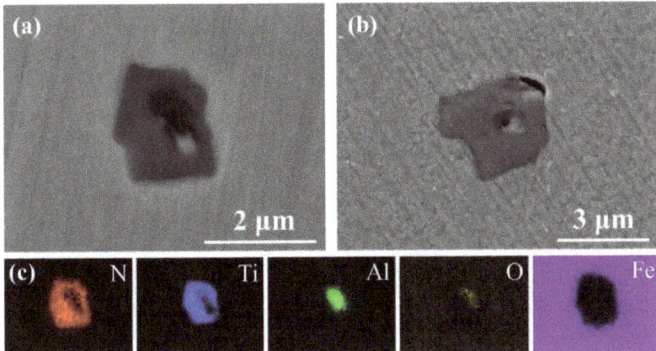

Figure 5. (**a**) SEM image of a typical complex inclusion in Ti-bearing steel. (**b**) Corrosion morphology of a typical inclusion. (**c**) EDS maps of inclusion in (**a**).

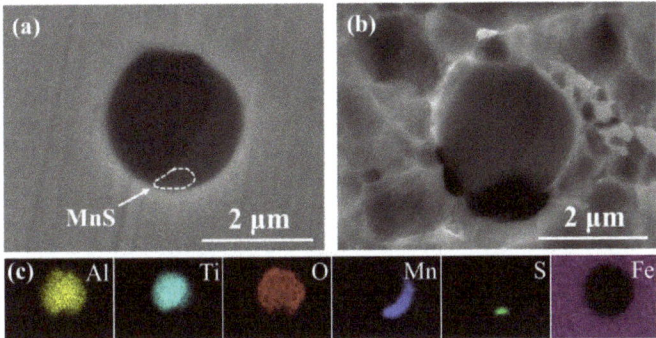

Figure 6. (**a**) SEM image of a typical complex inclusion in Cu-bearing steel. (**b**) Corrosion morphology of inclusion in (**a**). (**c**) EDS maps of inclusion in (**a**).

The complex inclusion in the base steel was composed of two components. The oxide component in complex inclusions was (Al, Ti)-O_x, and the sulfide component is MnS (Figure 4a,c), which is consistent with the literature [26,40,41]. After the immersion tests, the complex inclusions in the base steel partially dissolved (Figure 4b). According to the EDS results, MnS in the complex inclusion completely dissolved and the residual part of the inclusions was (Al, Ti)-O_x (Figure 4). As shown in the potential-pH diagram for MnS in Figure 7a [42], MnS was thermodynamically unstable in near-neutral pH solutions. Moreover, it has been reported [26] that (Al, Ti)-O_x owned higher electrochemical stability in comparison with the steel matrix. This indicates that the dissolution of MnS component in the complex inclusions initiated the localized corrosion [30,31,43,44].

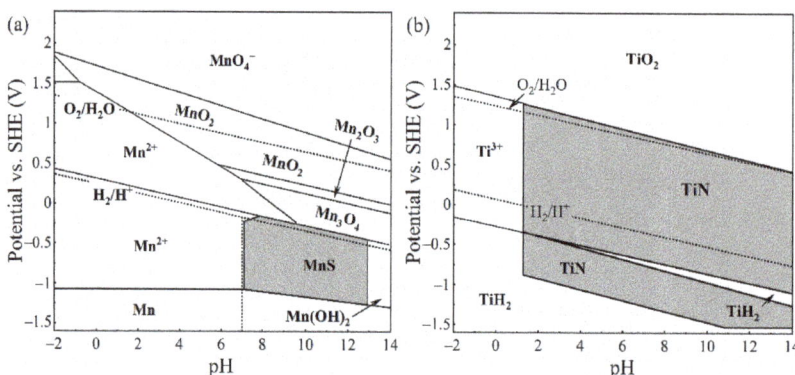

Figure 7. Potential-pH diagrams of MnS-H_2O and TiN-H_2O systems at 25 °C. (**a**) MnS-H_2O system (Adapted with permissions from ref. [42]. 2021 Springer Nature.); (**b**) TiN-H_2O system (Adapted with permissions from ref. [45]. 2021 Elsevier.).

The complex inclusion in the Ti-bearing steel was also composed of two components. The oxide component in complex inclusions was Al_2O_3, and the nitride component was TiN (Figure 5a,c). Besides, the Al_2O_3 component in the complex inclusions was surrounded by the TiN component (Figure 5a,c). After the immersion tests, a micro-crevice generated at the steel matrix/inclusion interface (Figure 5b). As can be seen in the potential-pH diagrams of TiN (Figure 7b) [45], the TiN was thermodynamically stable in the near-neutral pH solutions. Moreover, owing to the difference in thermal coefficient expansions and Young's modulus between the steel matrix and TiN inclusions, micro-crevices and high lattice distortion regions might generate in the steel matrix/inclusion interface [29,36]. Thus, the preferential adsorption of Cl^- in the micro-crevices and high lattice distortion regions accelerated the dissolution of the steel matrix around the TiN inclusions, which initiated the localized corrosion during the immersion test. The composition of inclusions in the Cu-bearing steel was similar to that in the base steel. The oxide component in complex inclusions was (Al, Ti, Mn)-O_x, and the sulfide component was MnS (Figure 6a,c). After the immersion tests, a micro-crevice was detected at the steel matrix/inclusion interface, indicating localized corrosion was induced by the dissolution of MnS component and the adjacent steel matrix (Figure 6b), which was consistent with the localized corrosion behavior of base steel.

Based on the above phenomenon, the localized corrosion behavior induced by inclusions in these three types of steels can be classified into two categories. The first category is MnS-containing complex inclusions. During the immersion test, the localized corrosion was initiated in the complex inclusions rather than the steel matrix (Figures 4 and 6) owing to the lower thermodynamical stability of MnS in near neutral solution. In the second category, the Al_2O_3 and TiN inclusions have higher thermodynamic stability than the steel matrix. Due to the significantly nonuniform deformation between inclusion and the steel matrix during the rolling process, the high lattice distortion regions in the matrix would be the preferential dissolution site at the pit initiation stage [26,29].

3.4. In-Situ SVET Measurements

An SVET measurement was employed to characterize the local electrochemical behavior during the localized corrosion initiation and propagation process, and the local corrosion current density maps are shown in Figure 8. During the SVET test, the local current density in anodic and cathodic sites exhibited considerable disparity (Figure 8). At the initiation stage (15 min), the localized corrosion occurred spontaneously, and the local anodic current density of the three types of steels decreased in the following order (Figure 8): Ti-bearing > Base steel > Cu-bearing. After extended exposure periods (60 min), the relatively high anodic current peak at the center of the maps gradually decreased

(Figure 8). Moreover, in base steel, Cu-bearing steel samples, the anodic current peak almost disappeared at 1 h 30 min (Figure 8), which means localized corrosion gradually transformed to uniform corrosion, whereas a clear anodic current peak was observed in the Ti-bearing sample with exposure time extended to 1 h 30 min (Figure 8).

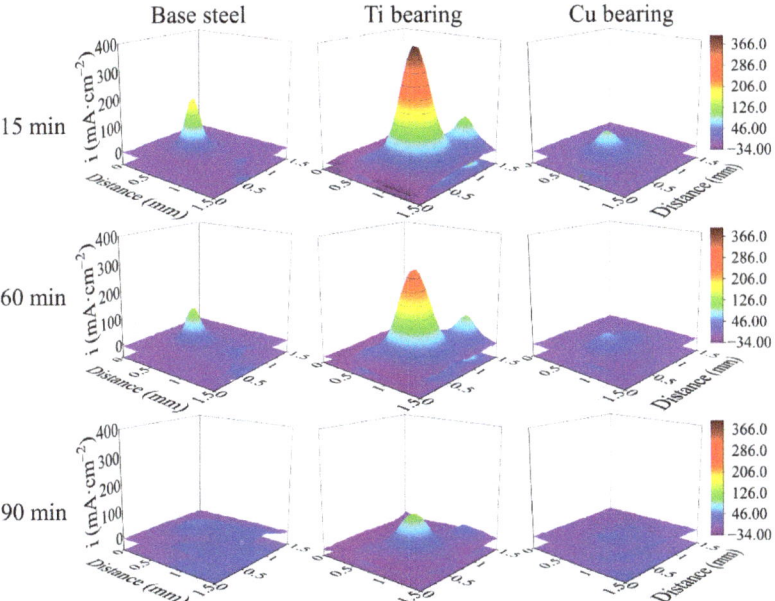

Figure 8. In-situ SVET maps of three types of steels in 0.5% NaCl solution.

To investigate the Cu addition on corrosion resistance of the steels, the corrosion sites after SVET tests were characterized by means of SEM/EDS. Figure 9 shows the morphologies of the corrosion sites and the elemental distribution of the corrosion products. In the anodic corrosion region of these three types of steels, the steel matrix severely dissolved, which contributed to the increase of the anodic current peak (Figures 8 and 9). For these three types of samples, Fe was distributed on the whole surface, O and S were detected to be distributed on the cathodic and anodic regions (Figure 9), respectively. According to the EDS results in Figure 9, the corrosion products in the anodic region of base steel and Ti-bearing steel were iron sulfide, which is consistent with the literature [38,46]. Additionally, the accumulated iron sulfide could promote the uniform dissolution of the steel matrix in Cl^- containing solution [46], whereas the enrichment of Cu was observed in the anodic corrosion region of Cu-bearing sample (Figure 9c). As reported earlier, the solubilized Cu in the steel matrix can deposit on the steel surface during the steel matrix dissolution process and then enhance the compactness of the rust layers and suppress the anodic reactions [18,47]. Thus, the local corrosion current density of Cu-bearing sample was lowered by Cu-enriched rust layer (Figure 8).

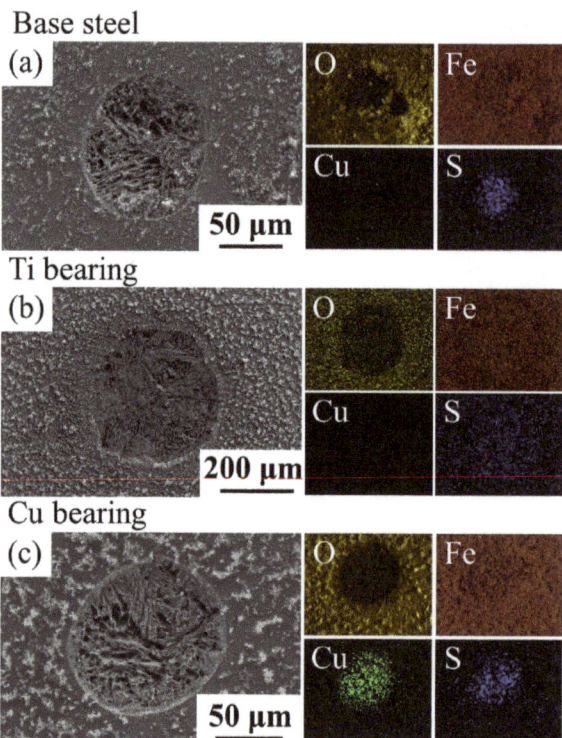

Figure 9. Morphology of the corrosion sites and the element distribution of the corrosion products after SVET tests. (**a**) Base steel. (**b**) Ti-bearing steel. (**c**) Cu-bearing steel.

3.5. Potentiodynamic Polarization Tests

To evaluate the influence of element addition on corrosion resistance in CGHAZ of these three types of steels, the potentiodynamic polarization tests were carried out, and the polarization curves are displayed in Figure 10. As shown in Table 4, the electrochemical parameters were extrapolated by means of the Tafel extrapolation method, which was conducted by stretching the linear parts of the anodic and cathodic curves back to their intersection, the abscissa and ordinate of the intersection was the corrosion current density and corrosion potential, respectively [48]. As shown in Table 4, the corrosion current density increased in the following order: Base steel ≈ Cu-bearing steel < Ti-bearing steel, indicating a decreasing trend of corrosion resistance. With the increased interface density of Ti-bearing steel, the corrosion rate of steels obviously increased compared to base steel. However, Cu-bearing steel was an exception, which had higher interface density and similar corrosion rate compared to that of base steel. According to the EDS results shown in Figure 9, the superior corrosion resistance of Cu-bearing steel was attributed to the Cu-enriched rust layer covered on the surface. Additionally, owing to the different thermal expansion coefficients and Young's modulus between inclusions and the steel matrix, the inclusion/matrix interface with micro-crevices and high lattice distortion regions was the initiation site of localized corrosion. Thus, the higher interface density obviously increased the corrosion rate of Ti-bearing steels [29,36].

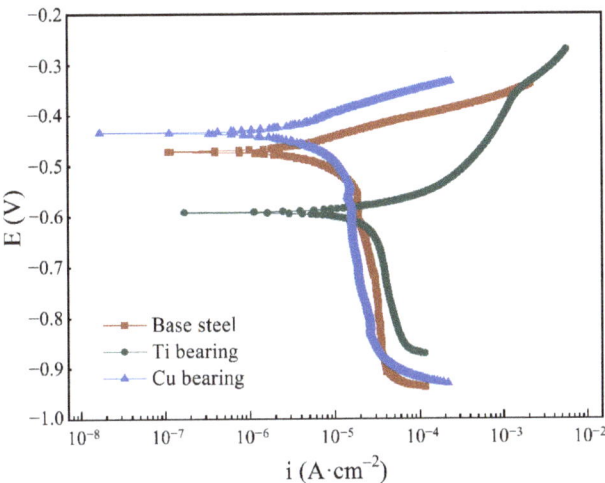

Figure 10. Polarization curves of three types of steels in 0.5% NaCl solution.

Table 4. Fitted electrochemical parameters of polarization curves in Figure 10.

Sample	i ($\times 10^{-6}$ A·cm^{-2})		E (V)		CR * ($\times 10^{-2}$ mm·a^{-1})	
	Mean	Deviation	Mean	Deviation	Mean	Deviation
Base steel	3.33	±0.26	−0.45	±0.017	3.91	±0.30
Ti-bearing	37.70	±3.02	−0.59	±0.037	44.26	±3.58
Cu-bearing	3.12	±0.23	−0.41	±0.016	3.66	±0.37

* CR means corrosion rate.

4. Discussion

Based on the above results, it is evident that the addition of Ti and Cu modified the number density, chemical compositions, and average diameter of inclusions formed in the CGHAZ of low-alloy steels, which had a significant impact on the initiation and propagation of corrosion [49], further leading to the different corrosion resistance of steels.

The schematics of two types of localized corrosion behaviors induced by inclusions in three types of steels are shown in Figure 11. For the first types of inclusions composed of metal oxides (M-O$_x$) and MnS (Figure 11a,b), the localized corrosion was initiated by the dissolution of MnS due to the thermodynamical instability of MnS in near-neutral pH solution [42]. Furthermore, the S$_2$O$_3^{2-}$ and H$^+$ were produced by the electrochemical dissolution of MnS by Equation (3) [46,50]

$$2MnS + 3H_2O \rightarrow 2Mn^{2+} + S_2O_3^{2-} + 6H^+ + 8e^- \quad (3)$$

which decreased the pH of solutions in the micro-crevice located in the steel matrix/inclusion interface (Figure 11b). Subsequently, the M-O$_x$ and the adjacent steel matrix would be dissolved in the aggressive environment.

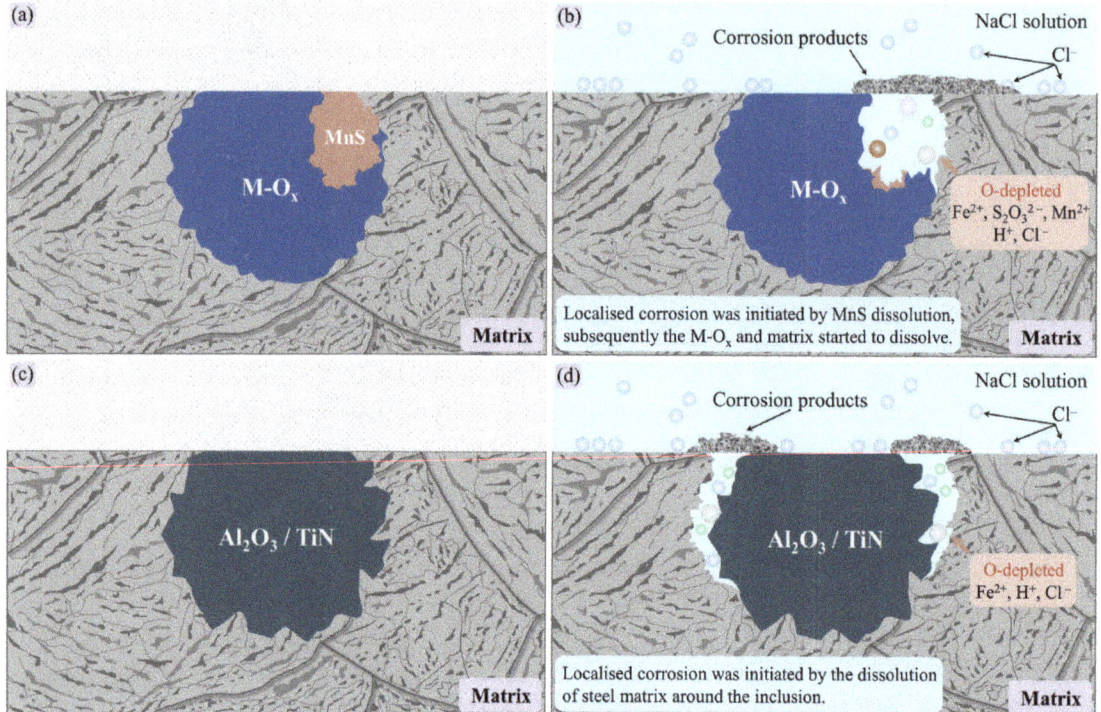

Figure 11. Schematic of the localized corrosion induced by inclusions. (**a**,**b**) Complex inclusions composed of M-O$_x$ and MnS. (**c**,**d**) Inclusions composed of Al$_2$O$_3$/TiN.

For the second type of inclusions, mainly composed of Al$_2$O$_3$/TiN (Figure 11c,d), the dissolution of the steel matrix that was distributed around the inclusions initiated the localized corrosion. Al$_2$O$_3$ and TiN have a much higher Young's modulus in comparison with the steel matrix [51], thus nonuniform deformation appeared between the inclusions and the steel matrix [52]. In addition, the distinctly lower thermal coefficient expansions of Al$_2$O$_3$ and TiN than the steel matrix can also result in the formation of high deformation region in the steel matrix/inclusion interface [36]. It was reported [29,53–56] that the adsorption process of Cl$^-$ can be significantly favored by the local deformation, generally leading to higher electrochemical activity in the deformation matrix. Hence, the steel matrix around inclusions could be easily attacked due to the existence of high deformation region, and it was the preferential dissolution site during the pit initiation process. With the steel matrix at the inclusion/matrix interface dissolved, micro-crevice formed at the interface between inclusion and matrix. The micro-crevice environment would be oxygen-depleted due to the poor convection. Consequently, cathodic reactions in micro-crevice were suppressed [27]. Therefore, the micro-crevice environment became enriched in metal cations, which resulted in the electromigration of Cl-ions into micro-crevice [57]. The pH in the micro-crevice was lowered by the cation hydrolysis

$$Fe^{2+} + 2H_2O \rightarrow Fe(OH)_2 + 2H^+ \tag{4}$$

which would facilitate the propagation of localized corrosion [58].

With the dissolution of MnS or the steel matrix around the Al$_2$O$_3$/TiN inclusions, micro-crevice that was enriched in acid solution was formed at the inclusions/matrix interface. Subsequently, the residual inclusion and adjacent matrix dissolved owing to the existence of acid environment, and the micro-crevice was enlarged and covered by corrosion products (Figure 11b,d). Hence, a catalytic-occluded cell generated in the microcrevice, which would considerably promote the corrosion propagation.

With increased immersion time, numerous corrosion pits generated in the exposed steel surface. Owing to the interaction effects of the nearby corrosion pits, the independent corrosion pits would coalesce into larger corrosion spots [29]. Consequently, the localized corrosion would transform into uniform corrosion in the later stage. As for the uniform corrosion resistance of steel, it was affected by both the interface density of inclusions and the property of rust layers. The schematics of the interface density on corrosion resistance of the steels are displayed in Figure 12. Compared to the base steel, the higher interface density in Ti-bearing steel provided more initiation sites for localized corrosion, induced by the higher number density and the larger average diameter of inclusions. With increased localized corrosion sites in the steel surface, more iron sulfide deposited on the exposure surface around inclusions, which significantly promoted the uniform corrosion of the steel matrix. Hence, the corrosion rate of Ti-bearing steels was obviously higher than that of base steel. Although the interface density in Cu-bearing steel was higher than that in base steel, the corrosion rate of Cu-bearing steel was similar to the base steel (Table 4). According to EDS results shown in Figure 9, the solubilized Cu in the steel matrix deposited on the surface of the anodic corrosion region, which considerably suppressed the anodic reaction [18], resulting in the improvement of corrosion resistance.

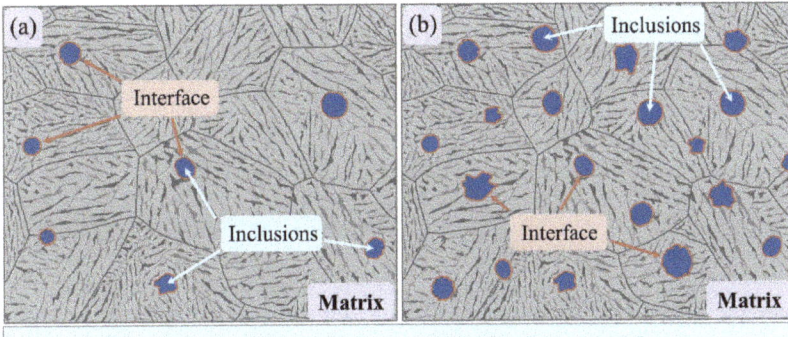

Figure 12. Schematic of element addition on the inclusion modification and corrosion behavior. (**a**) The steel matrix with lower number density and smaller average diameter of inclusions. (**b**) The steel matrix with higher number density and larger average diameter of inclusions.

5. Conclusions

The effects of Ti and Cu addition on the inclusion modification and corrosion behavior in simulated CGHAZ of low-alloy steels were investigated. The conclusions are summarized as follows.

(1) The number density and average diameter of inclusions in Ti-bearing steel (99.77 mm^{-2}, 1.75 µm) and Cu-bearing steel (49.70 mm^{-2}, 1.71 µm) were larger than in base steel (18.53 mm^{-2}, 1.54 µm). The main types of inclusions in Ti-bearing steel were TiN-Al$_2$O$_3$, which was obviously different from that in base steel. In Cu-bearing steel, the main types of inclusions were (Ti, Al, Mn)-O$_x$-MnS, which was similar to that in base steel.

(2) Localized corrosion behaviors induced by inclusions in three types of steels were categorized into two types. For the first type of inclusions, localized corrosion was initiated by the dissolution of MnS. For the second type of inclusions, the dissolution

of the steel matrix around Al_2O_3/TiN initiated the localized corrosion due to the high deformation region in the steel matrix/inclusion interface.

(3) The corrosion rate increases in the following order: Base steel (3.91 × 10^{-2} mm·a^{-1}) ≈ Cu-bearing steel (3.66 × 10^{-2} mm·a^{-1}) < Ti-bearing steel (44.26 × 10^{-2} mm·a^{-1}). With the increased interface density of the steels, the corrosion rate of CGHAZ in the steels obviously increased, whereas the Cu-bearing steel is an exception. The solubilized Cu in the steel matrix can enhance the compactness of the rust layers, then lower the corrosion current density.

Author Contributions: Conceptualization, X.Z. and K.W.; Data curation, Y.W., X.W. and W.W.; Formal analysis, Y.W.; Funding acquisition, X.Z. and K.W.; Investigation, Y.W. and W.W.; Methodology, X.Z.; Supervision, X.Z., J.L. and K.W.; Writing—original draft, Y.W.; Writing—review & editing, X.Z., X.W. and J.L. All authors have read and agreed to the published version of the manuscript.

Funding: This research was funded by the financial support from the National Nature Science Foundation of China (No. 51601138, 51601137, U20A20279, U20A20277), and the 111 Project (No. D18018).

Data Availability Statement: The raw/processed data required to reproduce these findings cannot be shared at this time as the data also forms part of an ongoing study.

Acknowledgments: The financial support from the National Nature Science Foundation of China (No. 51601138, 51601137, U20A20279, U20A20277), and the 111 Project (No. D18018) is highly acknowledged. We also would like to thank Guohong Zhang at the Analytical & Testing Center of Wuhan University of Science and Technology for the help in the on SEM/EDS analysis.

Conflicts of Interest: The authors declare no conflict of interest.

References

1. Huang, Y.; Jin, X.; Cai, G. Evolution of microstructure and mechanical properties of a new high strength steel containing Ce element. *J. Mater. Res.* **2017**, *32*, 3894–3903. [CrossRef]
2. Chen, A.-H.; Xu, J.-Q.; Li, R.; Li, H.-L. Corrosion resistance of high performance weathering steel for bridge building applications. *J. Iron Steel Res. Int.* **2012**, *19*, 59–63. [CrossRef]
3. Angelini, E.; Grassini, S.; Parvis, M.; Zucchi, F. An in situ investigation of the corrosion behaviour of a weathering steel work of art. *Surf. Interface Anal.* **2012**, *44*, 942–946. [CrossRef]
4. Atkinson, H.V.; Shi, G. Characterization of inclusions in clean steels: A review including the statistics of extremes methods. *Prog. Mater. Sci.* **2003**, *48*, 457–520. [CrossRef]
5. Meng, Q.; Frankel, G.S.; Colijn, H.O.; Goss, S.H. Stainless-steel corrosion and MnS inclusions. *Nature* **2003**, *424*, 389–390. [CrossRef] [PubMed]
6. Mohammadi, F.; Eliyan, F.F.; Alfantazi, A. Corrosion of simulated weld HAZ of API X-80 pipeline steel. *Corros. Sci.* **2012**, *63*, 323–333. [CrossRef]
7. Hou, B.; Li, X.; Ma, X.; Du, C.; Zhang, D.; Zheng, M.; Xu, W.; Lu, D.; Ma, F. The cost of corrosion in China. *NPJ Mater. Degrad.* **2017**, *1*, 4. [CrossRef]
8. Yang, Y.; Yan, B.; Li, J.; Wang, J. The effect of large heat input on the microstructure and corrosion behaviour of simulated heat affected zone in 2205 duplex stainless steel. *Corros. Sci.* **2011**, *53*, 3756–3763. [CrossRef]
9. Tomków, J.; Fydrych, D.; Wilk, K. Effect of electrode waterproof coating on quality of underwater wet welded joints. *Materials* **2020**, *13*, 2947. [CrossRef] [PubMed]
10. Missori, S.; Costanza, G.; Sili, A.; Tata, M.E. Metallurgical modifications and residual stress in welded steel with average carbon content. *Weld. Int.* **2015**, *29*, 124–130. [CrossRef]
11. Wan, X.L.; Wu, K.M.; Huang, G.; Wei, R.; Cheng, L. In situ observation of austenite grain growth behavior in the simulated coarse-grained heat-affected zone of ti-microalloyed steels. *Int. J. Miner. Metall. Mater.* **2014**, *21*, 878–885. [CrossRef]
12. Shi, J.; Hu, J.; Wang, C.; Wang, C.-Y.; Dong, H.; Cao, W. quan Ultrafine grained duplex structure developed by ART-annealing in cold rolled medium-mn steels. *J. Iron Steel Res. Int.* **2014**, *21*, 208–214. [CrossRef]
13. Shi, J.; Sun, X.; Wang, M.; Hui, W.; Dong, H.; Cao, W. Enhanced work-hardening behavior and mechanical properties in ultrafine-grained steels with large-fractioned metastable austenite. *Scr. Mater.* **2010**, *63*, 815–818. [CrossRef]
14. Yan, W.; Shan, Y.Y.; Yang, K. Effect of TiN inclusions on the impact toughness of low-carbon microalloyed steels. *Metall. Mater. Trans. A Phys. Metall. Mater. Sci.* **2006**, *37*, 2147–2158. [CrossRef]
15. Shen, Y.; Wan, X.; Liu, Y.; Li, G.; Xue, Z.; Wu, K. The significant impact of Ti content on microstructure–toughness relationship in the simulated coarse-grained heated-affected zone of high-strength low-alloy steels. *Ironmak. Steelmak.* **2019**, *46*, 584–596. [CrossRef]

16. Liu, Y.; Li, G.; Wan, X.; Wang, H.; Wu, K.; Misra, R.D.K. The role of Cu and Al addition on the microstructure and fracture characteristics in the simulated coarse-grained heat-affected zone of high-strength low-alloy steels with superior toughness. *Mater. Sci. Technol.* **2017**, *33*, 1750–1764. [CrossRef]
17. Wan, X.L.; Wu, K.M.; Huang, G.; Nune, K.C.; Li, Y.; Cheng, L. Toughness improvement by Cu addition in the simulated coarse-grained heat-affected zone of high-strength low-alloy steels. *Sci. Technol. Weld. Join.* **2016**, *21*, 295–302. [CrossRef]
18. Samusawa, I.; Nakayama, S. Influence of plastic deformation and Cu addition on corrosion of carbon steel in acidic aqueous solution. *Corros. Sci.* **2019**, *159*, 108122. [CrossRef]
19. Kamimura, T.; Nasu, S.; Segi, T.; Tazaki, T.; Morimoto, S.; Miyuki, H. Corrosion behavior of steel under wet and dry cycles containing Cr^{3+} ion. *Corros. Sci.* **2003**, *45*, 1863–1879. [CrossRef]
20. Jeon, S.-H.; Kim, S.-T.; Lee, J.-S.; Lee, I.-S.; Park, Y.-S. Effects of sulfur addition on the formation of inclusions and the corrosion behavior of super duplex stainless steels in chloride solutions of different pH. *Mater. Trans.* **2012**, *53*, 1617–1626. [CrossRef]
21. Xu, Q.; Gao, K.; Lv, W.; Pang, X. Effects of alloyed Cr and Cu on the corrosion behavior of low-alloy steel in a simulated groundwater solution. *Corros. Sci.* **2016**, *102*, 114–124. [CrossRef]
22. Zhu, T.; Huang, F.; Liu, J.; Hu, Q.; Li, W. Effects of inclusion on corrosion resistance of weathering steel in simulated industrial atmosphere. *Anti-Corros. Methods Mater.* **2016**, *63*, 490–498. [CrossRef]
23. Zhang, X.; Wei, W.; Cheng, L.; Liu, J.; Wu, K.; Liu, M. Effects of niobium and rare earth elements on microstructure and initial marine corrosion behavior of low-alloy steels. *Appl. Surf. Sci.* **2019**, *475*, 83–93. [CrossRef]
24. Wei, W.; Wu, K.; Zhang, X.; Liu, J.; Qiu, P.; Cheng, L. In-situ characterization of initial marine corrosion induced by rare-earth elements modified inclusions in Zr-Ti deoxidized low-alloy steels. *J. Mater. Res. Technol.* **2019**, *9*, 1412–1424. [CrossRef]
25. Tang, M.; Wu, K.; Liu, J.; Cheng, L.; Zhang, X.; Chen, Y. Mechanism understanding of the role of rare earth inclusions in the initial marine corrosion process of microalloyed steels. *Materials* **2019**, *12*, 3359. [CrossRef]
26. Liu, C.; Jiang, Z.; Zhao, J.; Cheng, X.; Liu, Z.; Zhang, D.; Li, X. Influence of rare earth metals on mechanisms of localised corrosion induced by inclusions in Zr-Ti deoxidised low alloy steel. *Corros. Sci.* **2020**, *166*, 108463. [CrossRef]
27. Liu, C.; Revilla, R.I.; Liu, Z.; Zhang, D.; Li, X.; Terryn, H. Effect of inclusions modified by rare earth elements (Ce, La) on localized marine corrosion in Q460NH weathering steel. *Corros. Sci.* **2017**, *129*, 82–90. [CrossRef]
28. Wei, W.; Wu, K.; Liu, J.; Cheng, L.; Zhang, X. Initiation and propagation of localized corrosion induced by (Zr, Ti, Al)-Ox inclusions in low-alloy steels in marine environment. *J. Iron Steel Res. Int.* **2020**, *20*, 492–502. [CrossRef]
29. Liu, C.; Revilla, R.I.; Zhang, D.; Liu, Z.; Lutz, A.; Zhang, F.; Zhao, T.; Ma, H.; Li, X.; Terryn, H. Role of Al2O3 inclusions on the localized corrosion of Q460NH weathering steel in marine environment. *Corros. Sci.* **2018**, *138*, 96–104. [CrossRef]
30. Zheng, S.J.; Wang, Y.J.; Zhang, B.; Zhu, Y.L.; Liu, C.; Hu, P.; Ma, X.L. Identification of MnCr2O4 nano-octahedron in catalysing pitting corrosion of austenitic stainless steels. *Acta Mater.* **2010**, *58*, 5070–5085. [CrossRef]
31. Avci, R.; Davis, B.H.; Wolfenden, M.L.; Beech, I.B.; Lucas, K.; Paul, D. Mechanism of MnS-mediated pit initiation and propagation in carbon steel in an anaerobic sulfidogenic media. *Corros. Sci.* **2013**, *76*, 267–274. [CrossRef]
32. Bastos, A.C.; Simões, A.M.; Ferreira, M.G. Corrosion of electrogalvanized steel in 0.1 M NaCl studied by SVET. *Port. Electrochim. Acta* **2003**, *21*, 371–387. [CrossRef]
33. Zheng, J.; Hu, X.; Pan, C.; Fu, S.; Lin, P.; Chou, K. Effects of inclusions on the resistance to pitting corrosion of S32205 duplex stainless steel. *Mater. Corros.* **2018**, *69*, 572–579. [CrossRef]
34. Ha, H.Y.; Park, C.J.; Kwon, H.S. Effects of misch metal on the formation of non-metallic inclusions and the associated resistance to pitting corrosion in 25% Cr duplex stainless steels. *Scr. Mater.* **2006**, *55*, 991–994. [CrossRef]
35. Zheng, S.; Li, C.; Qi, Y.; Chen, L.; Chen, C. Mechanism of (Mg,Al,Ca)-oxide inclusion-induced pitting corrosion in 316L stainless steel exposed to sulphur environments containing chloride ion. *Corros. Sci.* **2013**, *67*, 20–31. [CrossRef]
36. Wang, L.; Xin, J.; Cheng, L.; Zhao, K.; Sun, B.; Li, J.; Wang, X.; Cui, Z. Influence of inclusions on initiation of pitting corrosion and stress corrosion cracking of X70 steel in near-neutral pH environment. *Corros. Sci.* **2019**, *147*, 108–127. [CrossRef]
37. Nishimoto, M.; Muto, I.; Sugawara, Y.; Hara, N. Passivity of (Mn,Cr)S inclusions in type 304 stainless steel: The role of Cr and the critical concentration for preventing inclusion dissolution in NaCl solution. *Corros. Sci.* **2020**, *176*, 109060. [CrossRef]
38. Tokuda, S.; Muto, I.; Sugawara, Y.; Hara, N. Pit initiation on sensitized Type 304 stainless steel under applied stress: Correlation of stress, Cr-depletion, and inclusion dissolution. *Corros. Sci.* **2020**, *167*, 108506. [CrossRef]
39. Jeon, S.H.; Kim, S.T.; Lee, I.S.; Park, J.H.; Kim, K.T.; Kim, J.S.; Park, Y.S. Effects of copper addition on the formation of inclusions and the resistance to pitting corrosion of high performance duplex stainless steels. *Corros. Sci.* **2011**, *53*, 1408–1416. [CrossRef]
40. Pu, J.; Yu, S.F.; Li, Y.Y. Effects of Zr-Ti on the microstructure and properties of flux aided backing submerged arc weld metals. *J. Alloys Compd.* **2017**, *692*, 351–358. [CrossRef]
41. Chai, F.; Yang, C.-F.; Su, H.; Zhang, Y.-Q.; Xu, Z. Effect of Zr addition to Ti-killed steel on inclusion formation and microstructural evolution in welding induced coarse-grained heat affected zone. *Acta Metall. Sin.* **2008**, *21*, 220–226. [CrossRef]
42. Tyurin, A.G.; Pyshmintsev, I.Y.; Kostitsyna, I.V.; Zubkova, I.M. Thermodynamics of chemical and electrochemical stability of corrosion active nonmetal inclusions. *Prot. Met.* **2007**, *43*, 34–44. [CrossRef]
43. Baker, M.A.; Castle, J.E. The initiation of pitting corrosion at MnS inclusions. *Corros. Sci.* **1993**, *34*, 667–682. [CrossRef]
44. Lin, B.; Hu, R.; Ye, C.; Li, Y.; Lin, C. A study on the initiation of pitting corrosion in carbon steel in chloride-containing media using scanning electrochemical probes. *Electrochim. Acta* **2010**, *55*, 6542–6545. [CrossRef]

45. Heide, N.; Schultze, J.W. Corrosion stability of TiN prepared by ion implantation and PVD. *Nucl. Inst. Methods Phys. Res. B* **1993**, *80*, 467–471. [CrossRef]
46. Chiba, A.; Muto, I.; Sugawara, Y.; Hara, N. Pit initiation mechanism at MnS inclusions in stainless steel: Synergistic effect of elemental sulfur and chloride Ions. *J. Electrochem. Soc.* **2013**, *160*, C511–C520. [CrossRef]
47. Chen, Y.Y.; Tzeng, H.J.; Wei, L.I.; Wang, L.H.; Oung, J.C.; Shih, H.C. Corrosion resistance and mechanical properties of low-alloy steels under atmospheric conditions. *Corros. Sci.* **2005**, *47*, 1001–1021. [CrossRef]
48. McCafferty, E. Validation of corrosion rates measured by the Tafel extrapolation method. *Corros. Sci.* **2005**, *47*, 3202–3215. [CrossRef]
49. Shibaeva, T.V.; Laurinavichyute, V.K.; Tsirlina, G.A.; Arsenkin, A.M.; Grigorovich, K.V. The effect of microstructure and non-metallic inclusions on corrosion behavior of low carbon steel in chloride containing solutions. *Corros. Sci.* **2014**, *80*, 299–308. [CrossRef]
50. Alkire, R.C.; Lott, S.E. The Role of Inclusions on Initiation of Crevice Corrosion of Stainless Steel: II. Theoretical Studies. *J. Electrochem. Soc.* **1989**, *136*, 3256–3262. [CrossRef]
51. Juvonen, P. Effects of Non-Metallic Inclusions on Fatigue Properties of Calcium Treated Steels. Ph.D. Thesis, Helsinki University of Technology, Espoo, Finland, 10 December 2004.
52. Sahal, M.; Creus, J.; Sabot, R.; Feaugas, X. Consequences of plastic strain on the dissolution process of polycrystalline nickel in H2SO4 solution. *Scr. Mater.* **2004**, *51*, 869–873. [CrossRef]
53. Large, D.; Sabot, R.; Feaugas, X. Influence of stress-strain field on the dissolution process of polycrystalline nickel in H2SO4 solution: An original in situ method. *Electrochim. Acta* **2007**, *52*, 7746–7753. [CrossRef]
54. Li, W.; Li, D.Y. Variations of work function and corrosion behaviors of deformed copper surfaces. *Appl. Surf. Sci.* **2005**, *240*, 388–395. [CrossRef]
55. Sahal, M.; Creus, J.; Sabot, R.; Feaugas, X. The effects of dislocation patterns on the dissolution process of polycrystalline nickel. *Acta Mater.* **2006**, *54*, 2157–2167. [CrossRef]
56. Gutman, E.M.; Solovioff, G.; Eliezer, D. The mechanochemical behavior of type of 316L stainless steel. *Corros. Sci.* **1999**, *24*, 12–15. [CrossRef]
57. Brossia, C.S.; Kelly, R.G. Influence of alloy sulfur content and bulk electrolyte composition on crevice corrosion initiation of austenitic stainless steel. *Corrosion* **1998**, *54*, 145–154. [CrossRef]
58. Frankel, G.S. Pitting corrosion of metals. *J. Electrochem. Soc.* **1998**, *145*, 2186–2198. [CrossRef]

Article

Optimizing the Required Cathodic Protection Current for Pre-Buried Pipelines Using Electrochemical Acceleration Methods

Nguyen-Thuy Chung †, Min-Sung Hong † and Jung-Gu Kim *

School of Advanced Materials Engineering, Sungkyunkwan University, 300 Chunchun-Dong, Jangan-Gu, Suwon 440-746, Korea; chung.ngthuy@gmail.com (N.-T.C.); smith803@skku.edu (M.-S.H.)
* Correspondence: kimjg@skku.edu
† Both authors contributed equally in this work.

Abstract: Several corrosion mitigation methods are generally applied to pipelines exposed to corrosive environments. However, in the case of pre-buried pipelines, the only option for corrosion inhibition is cathodic protection (CP). To apply CP, the required current should be defined even though the pipeline is covered with various oxide layers. In this study, an electrochemical acceleration test was used to investigate the synthetic soil corrosion of a pre-buried pipeline. Potentiodynamic polarization experiments were first conducted to ascertain the corrosion current density in the environment, and galvanostatic measurements were performed to accelerate corrosion according to the operating time. In addition, corrosion current density and the properties of the rust layer were investigated via potentiodynamic polarization tests and electrochemical impedance spectroscopy (EIS) tests. The variation in surface corrosion was subsequently analyzed via optical microscopy (OM) and X-ray diffraction (XRD) measurements. Finally, an empirical equation for the optimized CP current requirement, according to the pipeline service time, was derived. This equation can be applied to any corroded pipeline.

Keywords: cathodic protection; buried pipelines; rust layer; electrochemical acceleration test; applied current density; required current

1. Introduction

Buried steel pipelines commonly play an important role worldwide as a means of transporting gases and liquids over long distances to meet several demands [1,2]. These pipelines are utilized in various soil environments, where they are exposed to chloride ions, humidity, and microbiological factors throughout their life, which could result in corrosion [2]. Corrosion is the main degradation mechanism that can reduce the structural integrity of buried pipelines [3]. The most basic method to protect a pipeline from corrosion is the application of an external coating to block corrosive ions, oxygen, and water [4]. However, the process of applying a coating on a pipeline should be evaluated at the initial stage, since it is nearly impossible to coat a pipeline that has already been buried. Therefore, additional methods should be considered to protect pre-buried pipelines.

Cathodic protection (CP) has been used in conjunction with organic coatings as the first method to control metal corrosion. In addition, CP can be utilized for pipelines that have already been buried [5]. In order for this approach to function adequately, it is important to calculate the required CP current according to the corrosion rate, operating time, environment, and other factors [6,7]. While it is quite simple to calculate the required current according to international standards when CP is applied at the initial stage, more thorough considerations are needed to implement the CP approach on corroded pipelines, since the rust layer could affect the surface properties of the structure. However, studies regarding the proper CP current required in pre-buried pipelines are still lacking [8]. The

NACE Standard SP0408-2019, i.e., Cathodic Protection of Reinforcing Steel in Buried or Submerged Concrete Structures, indicates how the CP method should be used for buried and submerged pipelines [9]. However, the standard does not specify the CP current required for corroded pipelines according to the corrosion time. For this reason, the optimized CP current required for a buried pipeline is derived in this work according to the corrosion time in a soil environment.

In this study, electrochemical tests were conducted to calculate the optimized CP current required according to the corrosion time. Galvanostatic tests were also carried out to accelerate the corrosion for 0.25, 1.2, 2.5, 5, and 7.4 years; electrochemical impedance spectroscopy (EIS) and potentiodynamic polarization experiments were performed after each test. After the electrochemical investigations, the surface of the corroded specimens was analyzed via optical microscopy (OM) and X-ray diffraction (XRD). Finally, an equation for the optimized CP current requirement concerning the pipeline service time was derived from the test results.

2. Materials and Methods

2.1. Specimen and Solution Preparation

Carbon steel (SPW400) with 8.7 mm thickness, a general material utilized for buried pipelines, was employed as a test specimen; the specific composition is listed in Table 1 (Korean Standards). Before conducting the experiments, the surfaces of the specimens were polished with 600-grit silicon carbide (SiC) paper, rinsed in ethanol via ultrasonication, and finally dried with air. The corrosion environment consisted of a synthetic soil solution with pH value is 6.8 which the pH was conditioned by diluted HNO_3 and NaOH., the chemical composition of which is listed in Table 2.

Table 1. Chemical composition of SPW400 (wt.%).

Fe	C	P	S
Bal.	0.25 max	0.04 max.	0.004 max.

Table 2. Chemical composition of the synthetic soil solution (ppm).

$CaCl_2$ (ppm)	$MgSO_4 \cdot 7H_2O$ (ppm)	$NaHCO_3$ (ppm)	H_2SO_4 (ppm)	HNO_3 (ppm)
133.2	59.0	208.0	48.0	21.8

2.2. Electrochemical Tests

All electrochemical tests were based on a three-electrode system configuration. (VSP 300, Bio-Logic SAS, Seyssinet-Pariset, France). The test specimen consisted of a working electrode; a graphite rod and saturated calomel electrode (SCE) were used as the counter and reference electrodes, respectively. The exposed area of the specimen was 0.25 cm^2. An open-circuit potential (OCP) was established within 30 min before the electrochemical tests. The corrosion properties of the steel pipeline were evaluated via potentiodynamic polarization tests performed from an initial potential of -0.4 VOCP to a final potential of 1.2 VSCE with a sweep rate of 1 mV/s. Based on the potentiodynamic polarization results, the acceleration time for galvanostatic testing was calculated according to Faraday's law. EIS experiments were subsequently performed with a 20 mV amplitude in the frequency range from 100 kHz to 10 mHz, and potentiodynamic polarization tests were conducted to obtain the corrosion current density after the corrosion acceleration.

2.3. Surface Analysis

Optical microscopy (SZ61, Olympus, Tokyo, Japan) was used to investigate the relationship between thickness loss and corrosion time to determine the corrosion rate through thickness loss. The corrosion products were analyzed via XRD (D8 Advance, Bruker Co., Karlsruhe, Germany) The XRD patterns were obtained using a Cu Kα target at 50 kV and 250 mA over a 2θ range of 20–70° with a scan rate of 2° min^{-1}.

3. Results and Discussion

3.1. Potentiodynamic Polarization and Galvanostatic Tests

To simulate testing over a period of years, the acceleration time should be calculated according to Faraday's law [10]:

$$Q = It \quad (1)$$

where Q is the total electric charge passed through the substance, t is the total time over which a constant current is applied, and I is the constant current for corrosion, i.e., the corrosion current density. According to Equation (1), as the total amount of electricity transferred through the specimen is kept constant, the time is reduced but the current must increase proportionally [11]. Therefore, in order to determine the experimental time, it is necessary to know the corrosion current density for a material in a given environment. The corrosion current density can be measured via potentiodynamic polarization testing. Figure 1 shows the potentiodynamic polarization curve of carbon steel in a synthetic soil solution, whereas the results of the potentiodynamic polarization test are displayed in Table 3. According to the test results, the specimen exhibits the corrosion current density of 4.2 μA/cm^2, which leads to a corrosion rate of 0.098 mm/year. From this value, the acceleration time can be calculated using Equation (1). The values for the acceleration of the corrosion time depend on the purpose of the study, the applied current density, i.e., the acceleration current from the system, was set to 5000 μA/cm^2, which is about 1,200 times larger than the corrosion current density. Corrosion times were set to be 0.25, 1.2, 2.5, 5, and 7.4 years in this study; all calculated values are listed in Table 4. These findings were applied to the galvanostatic tests in order to accelerate the corrosion of steel.

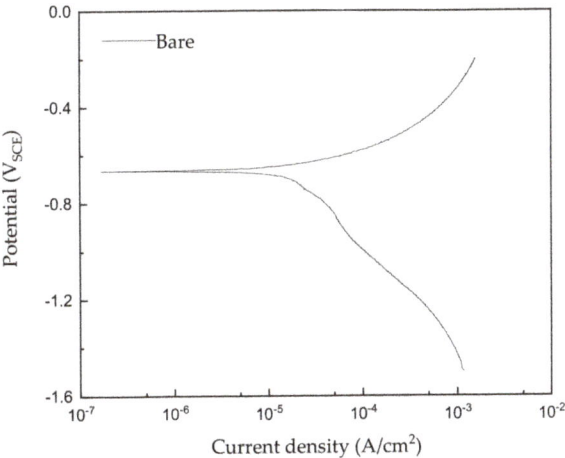

Figure 1. Potentiodynamic polarization curves for pipeline steel in a synthetic soil solution at room temperature.

Table 3. Potentiodynamic polarization results for pipeline steel.

E_{corr} (mV_{SCE})	i_{corr} (μA/cm^2)	Corrosion Rate (mm/year)
−676.8	4.2	0.098

Table 4. Conditions for accelerated corrosion tests as a function of the service time.

Un-Accelerated Condition			Accelerated Condition	
Corrosion Current Density ($\mu A/cm^2$)	4.2	→	Accelerating Current Density ($\mu A/cm^2$)	5000
Real time for corrosion	0.25 years	→	Accelerated time for corrosion	1.8 h
	1.2 years	→		9 h
	2.5 years	→		18 h
	5 years	→		36 h
	7.4 years	→		54 h

3.2. Surface Analysis

After performing the acceleration tests using galvanostatic methods, surface analysis investigations were performed via OM and XRD to measure the thickness loss and rust chemical composition. Through examining the cross-section of the specimens, the thickness loss during the corrosion process can be determined. The cross-sections of the accelerated specimens are shown in Figure 2. It can be seen that, as the corrosion time increases, more rust accumulates, and the thickness loss gradually becomes greater. As shown in Figure 3a, the thickness loss for pipeline steel as a function of the corrosion time continuously increases. The thickness loss increases with the corrosion time regardless of whether the corrosion rate is fast or slow. Therefore, in order to describe the corrosion rate, the thickness loss should be divided by the total corrosion time [8], as shown in Figure 3b. From this figure, it can be seen that the corrosion rate is accelerated according to the corrosion time. However, the corrosion rate does not increase linearly, since its slope decreases as the corrosion time increases. It can be predicted that it will remain stable after a certain corrosion time, as an effect of the corrosive product on the corrosion process. This result will be more clearly demonstrated in the following section.

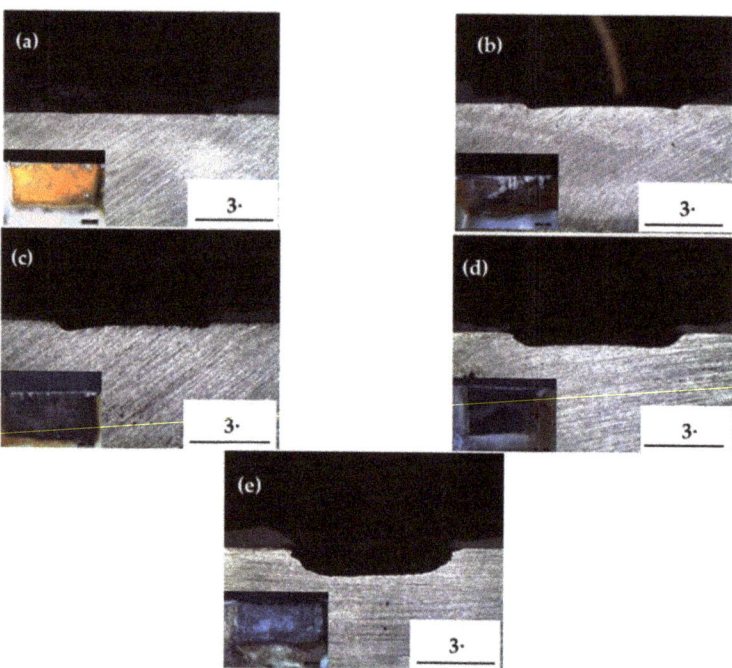

Figure 2. Cross-section images with magnification are x10 and x45 of the specimens after acceleration tests of (**a**) 0.25 years, (**b**) 1.2 years, (**c**) 2.5 years, (**d**) 5 years, and (**e**) 7.4 years.

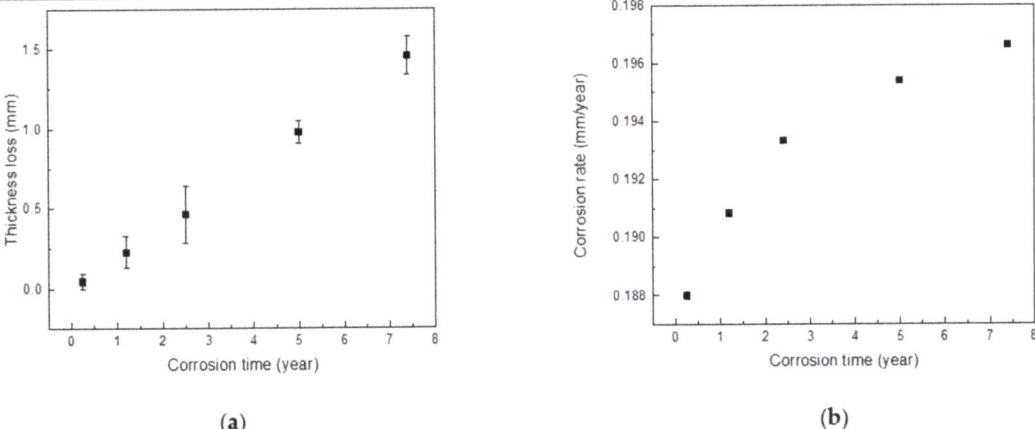

Figure 3. (a) Thickness loss of pipeline steel as a function of the corrosion time and (b) instantaneous corrosion rate of carbon steel as a function of the corrosion time.

To explore the influence of the corrosion products on the corrosion process, it is necessary to first determine the components of the rust layer. XRD measurements of the corrosion products were performed after completing the experiments. The corrosion rate at different exposure periods is related to the composition and structure of the rust layer. Figure 4 shows the XRD patterns of powdered rust on pipeline steel. Upon comparison to the reference patterns [12,13], it can be seen that the rust layers that form for different corrosion time durations exhibit the constituents that are mainly assigned to lepidocrocite ($\gamma-FeOOH$) and magnetite (Fe_3O_4) phases. In particular, the amount of lepidocrocite (brown rust) observed for the 0.25 and 1.2 years experiments accounts for a larger percent, it can be visible to the naked eye. However, as the corrosion time increases, the amount of magnetite gradually increases. By contrast, Fe_3O_4 becomes the main component of the rust layer upon increasing the immersion time. The rapid Fe_3O_4 generation indicates that this material is directly converted from other corrosion products rather than being generated in the oxygen-depleted environment. Additionally, other studies have reported that Goethite ($\alpha-FeOOH$) is located in the inner rust layer at the initial stage [14,15].

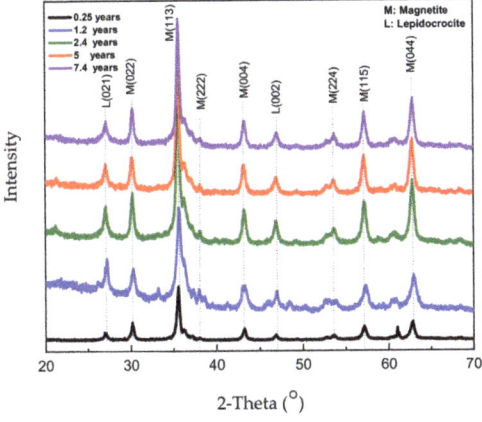

Figure 4. XRD patterns obtained for powdered rust formed after the galvanostatic tests.

3.3. Electrochemical Acceleration Test Results

Figure 5 shows the potentiodynamic polarization curves obtained for corroded steel in a synthetic soil solution as a function of the acceleration time. The electrochemical parameters obtained via the polarization test are listed in Table 5. It can be seen that the potentiodynamic polarization curve of the non-rust steel is quite different from those of the corroded steels, implying differences in the reaction process and corrosion mechanism. Normally, when there is no rust layer, the corrosion process of the carbon steel electrode is controlled by oxygen diffusion: if the rust layer hinders the oxygen diffusion process, the corrosion process will be inhibited, results will increase the Tafel slope of the cathodic polarization curve. However, the experimental results show that, when the carbon steel surface is covered with a rust layer, the Tafel slope of the cathodic curve is reduced, indicating that the cathodic process has been promoted. Therefore, it is believed that the rust layer formed by carbon steel in a synthetic soil solution cannot effectively hinder the diffusion of oxygen, thus leading to an acceleration of the corrosion process. As the corrosion time increases, the corrosion current density (i_{corr} mA/cm^2) increases gradually, and E_{corr} moves toward positive values. The amount of rust increases even upon increasing the corrosion time; this rust layer not only cannot protect the carbon steel but also accelerates its corrosion further. The corrosion current density values listed in Table 5 are plotted in Figure 6, and a variation in the corrosion current density can be noted according to the corrosion time (y/year). In particular, the corrosion current density initially exhibits a continuous increase, but the rate of increase becomes less pronounced after 5 years, i.e., the expected saturation is observed.

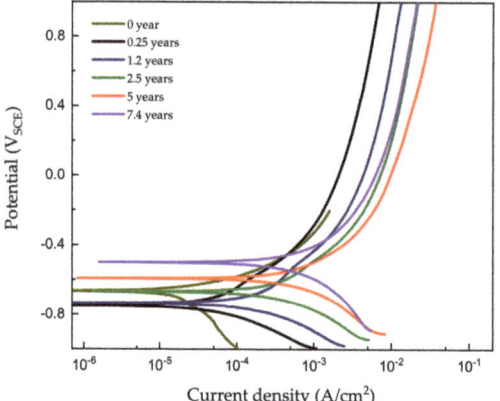

Figure 5. Potentiodynamic polarization curves of pipeline steel in a synthetic soil solution as a function of the corrosion time after galvanostatic tests.

Table 5. Potentiodynamic polarization results of pipeline steel as a function of the acceleration time.

Acceleration Time. (year)	Potential (mV_{SCE})	i_{corr} (μA/cm^2)	β_a (mV_{SCE})	β_c (mV_{SCE})
0	−676.8	4.2	127.9	365.5
0.25	−732.4	47.7	282.6	200.2
1.2	−722.5	173.2	386.2	227.1
2.5	−644.6	324.6	317.5	257.7
5	−579.3	487.6	258.4	261.5
7.4	−499.2	547.6	316.9	323.2

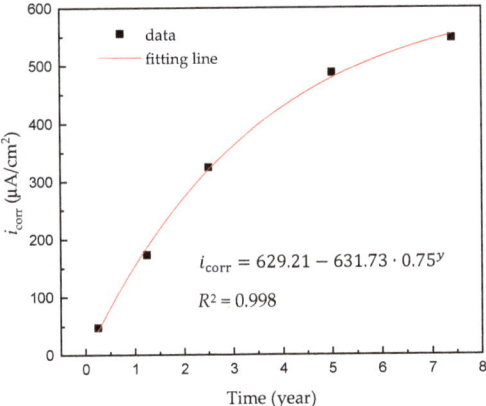

Figure 6. Variation of corrosion current density for carbon steel pipeline in a soil solution according to the acceleration time.

The corrosion process shown in Figure 7 explains the corrosion mechanism of pipeline steel in the synthetic soil solution. The rusting of ferrous is an electrochemical process that begins with the transfer of electrons from iron to oxygen. Ferrous is the reducing agent, whereas oxygen is the oxidizing agent. The rate of corrosion is affected by electrolytes. The reaction is the reduction of oxygen [16]:

$$O_2 + 2H_2O + 4e \rightarrow 4OH^- \qquad (2)$$

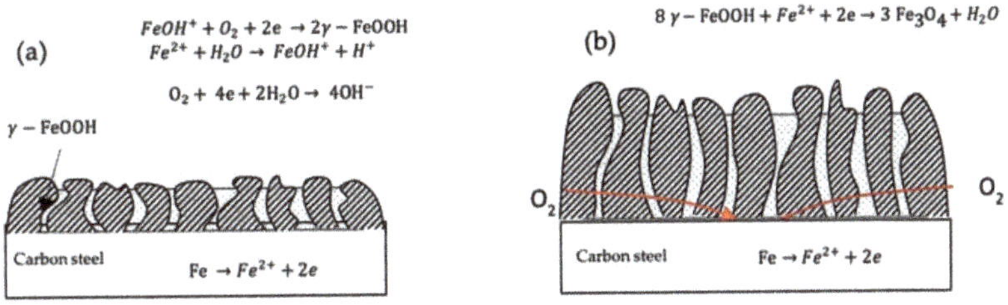

Figure 7. Corrosion mechanism for a buried pipeline according to time: (**a**) initial stage and (**b**) later stage.

The electrons for the above reaction are provided by the oxidation of iron, which may be described as follows [16]:

$$Fe \rightarrow Fe^{2+} + 2e \qquad (3)$$

The ferrous ions (Fe^{2+}) will form hydrated ions in solution [16]:

$$Fe^{2+} + H_2O \rightarrow FeOH^+ + H^+ \qquad (4)$$

The intermediate corrosion product $FeOH^+$ can be oxidized quickly by O_2 and converted into lepidocrocite ($\gamma-FeOOH$), the brown rust $\gamma-FeOOH$ accumulates on the metal surface and forms the early rust layer (the outer rust layer) according to [17]:

$$FeOH^+ + O_2 + 2e \rightarrow 2\gamma - FeOOH \qquad (5)$$

Hence, γ-FeOOH accumulates on the surface of the metal and forms the early rust layer. The γ-FeOOH layer, which is in electrical contact with the metal, can be reduced according to [16–18]:

$$8\,\gamma-\text{FeOOH} + \text{Fe}^{2+} + 2e \rightarrow 3\text{Fe}_3\text{O}_4 + 4\text{H}_2\text{O} \tag{6}$$

When the reduction potential of γ-FeOOH is higher than the corrosion potential of carbon steel in the solution, γ-FeOOH can be reduced and works as an oxidant. Fe_3O_4 can accumulate at the interface between the γ-FeOOH and the surface metal, forming the inner rust layer. In previous works, it has been reported that ferrous ions and electrons can pass through Fe_3O_4, and hence the cathodic reaction can occur on the surface of the Fe_3O_4 [16,19]. Additionally, the value of the oxygen diffusion coefficient (D) it has been proved that oxygen is reduced on the surface of the rust layer. Therefore, it can be concluded that the Fe_3O_4 layer is a large cathode area, and oxygen is reduced on its surface therefore the corrosion rate of carbon steel can be promoted significantly. Other studies have demonstrated that, when the metal corrosion rate is high, this rate can be controlled by limiting the diffusion of the oxidant. This is because when the outer rust layer of γ-FeOOH loosens and drifts out of the soil solution, this layer cannot hinder the diffusion process of oxygen effectively [20]. Corrosion at the surface is determined by the rate of oxygen diffusion into the pores of the Fe_3O_4 layer; once equilibrium is reached, the corrosion current density is saturated [18]. The limiting diffusion current density can be described as follows [16,21]:

$$i_L = -nFD\frac{C_0}{\delta} \tag{7}$$

where D is the diffusion coefficient of oxygen, C_0 is the concentration of oxygen in the solution and δ is the thickness of the diffusion layer. All of the parameters in the above equation were kept constant with all experimental condition unchanged, therefore i_L has a specific value and the corrosion rate of carbon steel stabilizes at a certain value. It was found that, while such scales exhibit high electrochemical activity, difficulties in using in-situ spectroscopic techniques for electrochemical experiments have led to a lack of agreement in the reaction schemes proposed by different authors. Consequently, the importance of reaction (b.1) during the corrosion process of iron or carbon steel in soil could not be clarified until very recently, and this reaction is extracted from articles on atmospheric corrosion, not exactly in articles on soil corrosion. However, corrosion in soil is aqueous, and the mechanism is electrochemical, but the conditions in the soil can range from 'atmospheric' to 'completely immersed' [22]. Likewise, reaction (b.1) can explain why the corrosion current density is saturated in this study when corrosion occurs in soil over a long time exposure; the main component of corrosion products steel pipelines in soil are γ-FeOOH and Fe_3O_4.

The properties of the rust layer were further investigated via in-situ EIS measurements to obtain real-time corrosion data in a continuous process. Figure 8 illustrates the equivalent circuit used for a short immersion time, which was used to fit the data [19,20]. At very high-frequency (HF), the imaginary component, Z'', disappears, leaving only the electrolyte resistance R_e. For the 0.25 years case, the outer rust layer of γ-FeOOH remains on the surface of the electrode; this data can be fitted using the circuit in Figure 8. However, the results of the Nyquist plot, as shown in Figure 9, describe an increase of the HF resistance with increasing corrosion time. This is because, after a certain period of corrosion time, the amount of rust created is quite large and the exposure area is small (0.25 cm^2). Thus, not all the rust (especially γ-FeOOH) can cling to the metal surface; it gradually loosens and drifts out of the solution. Hence, the fitting value of the HF resistance corresponds to not only the R_e value but it is replaced the sum of R_e and R_0, the latter being the resistance of the layer which grows with the immersion time (i.e., $R_{HF} = R_e + R_0$) [21].

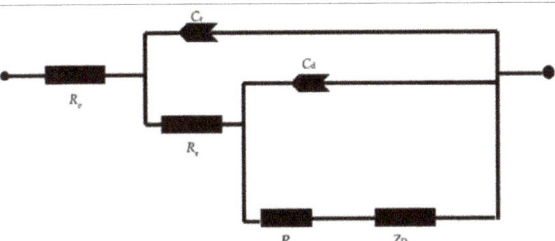

Figure 8. The equivalent electrical circuit for carbon steel/solution products/water interphase is used to describe how rust adheres to metal surfaces. R_e: electrolyte resistance, C_f: film capacity, R_f: electrolyte resistance across the film, C_d: double layer capacity, R_t: charge transfer resistance, Z_D: diffusional impedance.

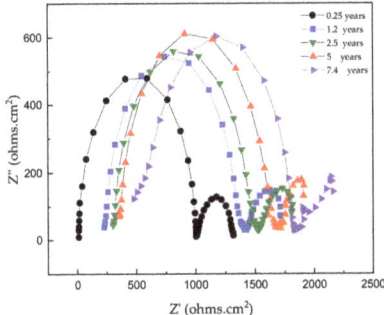

Figure 9. Impedance spectra are presented in the form of a Nyquist plot.

The equivalent circuit was then modified with R_0 and C_0, where C_0 is the layer capacitance, whose low value is out of the range of the frequency domain used for the measurements. This new equivalent circuit is shown in Figure 10a. This modification of the equivalent circuit means that the interface is coated by two layers characterized by the pairs of parameters R_0, C_0, and R_1, C_1. The increase of the R_0 value with time corresponds to the increase of the thickness of the porous brown γ−FeOOH external layer. The equivalent circuit in Figure 10a can be fitted for the 1.2, 2.5, and 5 years cases [20]. The R_1 value is almost constant, with a mean value of 1000 $\Omega \cdot cm^2$. Furthermore, the experiments with 1.2, 2.5, and 5 years of corrosion times at low frequencies are complicated by the influence of the γ−FeOOH rust on the metal surface, with a proportion of it being drifted out of the solution. The equivalent circuit, shown in Figure 10b, is only adequate for the 7.4 years case. After this corrosion time, almost all brown rust scale loosens and leaches out of the solution, and the main component of rust which remains on the metal surface is black rust (Fe_3O_4) [16]. Since the Fe_3O_4 layer cannot be distinguished from the remaining metal via electrical measurements, owing to its good electronic conductivity [20], it was believed that the thick inner rust Fe_3O_4 layer could not be detected via EIS measurements. Consequently, it was inferred that the corrosion rate of carbon steel in a synthetic soil solution was determined by the limiting diffusion rate of oxygen. Hence, the equivalent circuit is shown in Figure 10b is proposed for the rusted electrodes. At low frequency, determining the resistance under the resistors of the surface rust is a very complex process, probably due to the oxygen capability to diffuse into the pores. As long as oxygen is saturated inside the black rust, the corrosion rate of carbon steel remains stable at a certain value. This is still reasonable and uniform in the case of actual underground environments. When the black rust increases rapidly, the scale is too thick, the outer layer will break the structure that clings to the metal surface, it will also gradually loosen and expose the inner rust layer in direct contact with the soil environment.

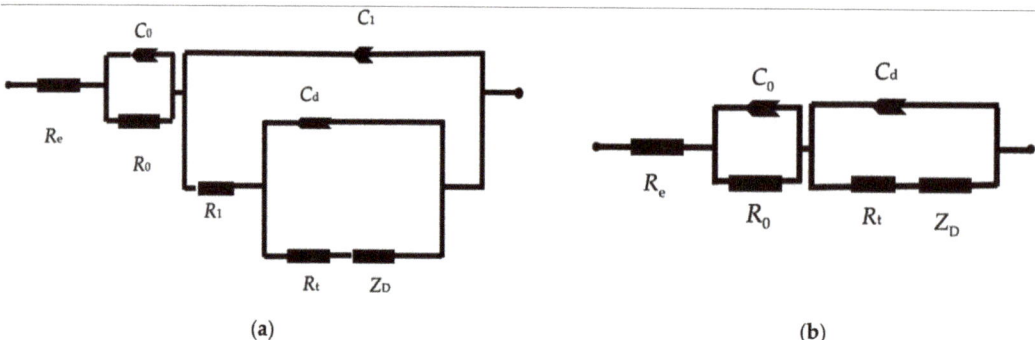

Figure 10. The equivalent electrical circuits for the carbon steel/corrosion products/soil solution interface: (**a**) a portion of the rust loosens out of the solution; (**b**) a portion of the rust (Fe_3O_4) adheres to the metal surfaces. R_e: electrolyte resistance, C_0, and C_1: film capacitance, R_0, and R_1: film resistance, C_d: double layer capacity, R_t: charge-transfer resistance, Z_D: diffusion impedance.

3.4. Required Current

A series of studies have been performed to examine the influence of soil conductivity on cathodic protection systems applied to the protection of buried pipelines [23]. As can be seen in the following two equations, the total current increases due to the increase in soil conductivity:

$$I = 237k \tag{8}$$

where $2 \leq k \leq 60$ and:

$$I = 1895 \cdot e^{0.0373 \cdot k} \tag{9}$$

where $61 \leq k \leq 120$.

Here, I is the total required current in mA, and k is the soil conductivity in mS/m. These two equations govern the relationship between the total current required for protection and the conductivity and are valid within two separate regimes. Indeed, Equation (8) is a linear expression, whereas Equation (9) is an exponential relationship. However, the above two formulas apply to newly buried pipelines and not to pipelines that have been buried for a long time. Because of the corrosion mechanism for buried pipelines, the required current depends not only on the conductivity of the soil but also on the rust covering the cathode. Therefore, the specific required current to be applied for the CP approach depends on the service time of the buried pipeline. To devise a CP strategy, the applied CP current density (i_{app}) should be calculated using the Evan's diagram, as shown in Figure 11 [7,24,25].

As in previous studies, the applied current density was calculated as the difference between the anodic polarization curve and the cathodic polarization curve at -800 mV$_{SCE}$ (the protective potential for C-steel) [26]. The corresponding calculation results are shown in Table 6, and the values of the applied current density as a function of the corrosion time are plotted in Figure 12. The obtained R^2 value of 0.96 suggests that the fitting is reliable, and the empirical equation attained after an analysis of the fitting is:

$$i_{app} = 6.2 - 6.6 \times 0.9^y \tag{10}$$

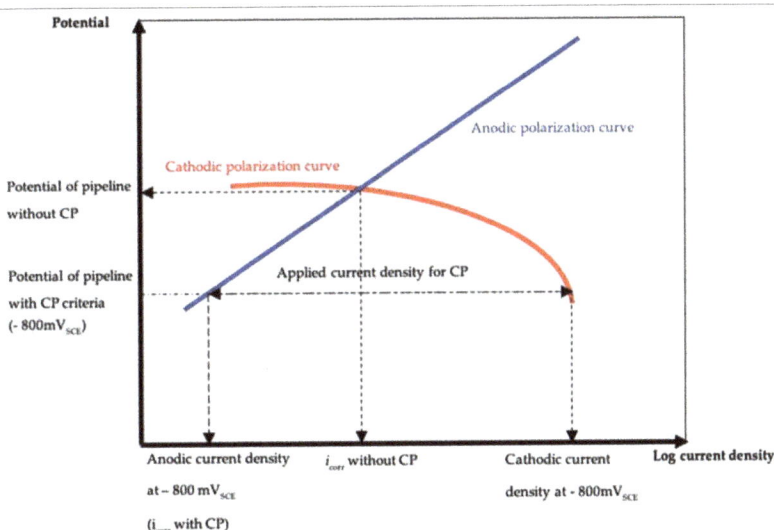

Figure 11. Evan's diagram, indicating the relationship between the applied current density and protective potential [7,24,25].

Table 6. Results for the calculation of parameters: the anodic current density and applied current density are based on the experiment results.

Acceleration Time (year)	Anodic Current Density at -800 mV$_{SCE}$ (μA/cm^2)	Cathodic Current Density at -800 mV$_{SCE}$ (μA/cm^2)	Applied Current Density (mA/cm^2)
0.25	47.5	79.8	0.0323
1.2	173.0	317.8	0.1448
2.5	324.1	1355.4	1.0313
5	486.7	3177.3	2.6906
7.4	546.7	3589.7	3.0430

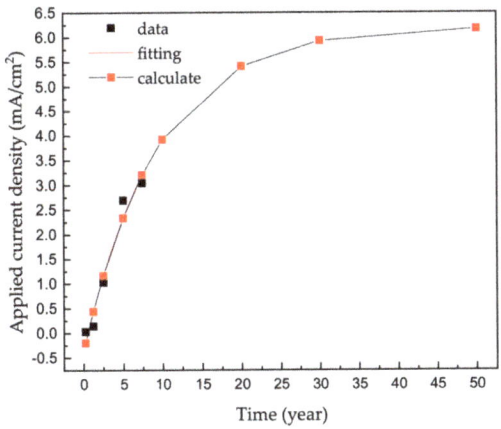

Figure 12. Applied current density as a function of the corrosion time for steel pipelines in a soil solution.

As can be seen from Figure 12, according to the equation corresponding to a corrosion time exceeding 50 years, i_{app} approaches the maximum value of 6.2 mA/cm^2. The purpose of this study is to assess the electrical current applied to an underground pipe. Under standard DNV-RP-B401, the required current, I_{req}, needs to provide an adequate polarizing capacity and maintain cathodic protection during the design life. I_{req} can be calculated from the individual surface area, A, coating breakdown factor, f_c, and relevant design current density, i_c, if applicable [26]:

$$I_{req} = A \cdot i_c \cdot f_c \tag{11}$$

The coating breakdown factor describes the anticipated reduction in the cathodic current due to the application of an electrically insulating coating. When $f_c = 0$, the coating is 100% electrically insulating, and the cathodic current density thus decreases to zero. In contrast, $f_c = 1$ means that the coating has no current-reducing properties. In general, the relevant design current density can be considered to be the applied current density value that has been measured above. Equation (11) may be written as:

$$I_{req} = A \cdot (6.2 - 6.6 \times 0.9^y) \tag{12}$$

Equation (12) should be applied to design a CP system for buried pipeline steel, according to the pipeline service time and surface area.

4. Conclusions

In this study, electrochemical tests were performed on pre-buried pipelines to optimize the required current needed for the CP strategy according to the pipeline service time. This research is of great interest in engineering applications by a novel conducted study. According to the results, the following conclusions can be pointed out:

- The corrosion current density of pipeline steel in a soil environment reaches a stable and saturated state after a long time, due to the oxygen limited diffusion rate in the oxide layer at equilibrium.
- When designing a CP system for buried pipeline steel, the required CP current should be determined according to the logarithmic equation derived in this study.

Author Contributions: Conceptualization, M.-S.H.; methodology, N.-T.C.; validation, M.-S.H. and J.-G.K.; formal analysis, N.-T.C.; investigation, N.-T.C.; resources, N.-T.C.; data curation, M.-S.H. and N.-T.C.; writing—original draft preparation, N.-T.C.; writing—review and editing, M.-S.H. and J.-G.K.; visualization, M.-S.H.; supervision, J.-G.K.; project administration, M.-S.H. and J.-G.K. All authors have read and agreed to the published version of the manuscript.

Funding: This research was supported by the program for fostering next-generation researchers in engineering of National Research Foundation of Korea (NRF) funded by the Ministry of Science and ICT (2017H1D8A2031628). This work was also supported by an NRF grant funded by the Korean Government (NRF-2020-Research Staff Program) (NRF-2020R1I1A01074866).

Institutional Review Board Statement: Not applicable.

Informed Consent Statement: Not applicable.

Data Availability Statement: Data sharing is not applicable to this article.

Conflicts of Interest: The authors declare no conflict of interest.

References

1. El-Lateef, H.A.; Abbasov, V.M.; Aliyeva, L.I.; Ismayilov, T.A. Corrosion Protection of Steel Pipelines Against CO_2 Corrosion—A Review. *Chem. J.* **2012**, *2*, 52–63.
2. Noor, N.M.; Lim, K.S.; Yahaya, N.; Abdullah, A. Corrosion study on x70-carbon steel material influenced by soil engineering properties. *Adv. Mater. Res.* **2011**, *311*, 875–880. [CrossRef]
3. Jack, T.R.; Wilmott, M.J.; Sutherby, R.L.; Worthingham, R.G. Evaluating performance of coatings exposed to biologically active soils. *Mater. Perform.* **1996**, *35*, 39–45.
4. Stewart, M.; Arnold, K. Surface production operations. *Pump Compress. Syst. Mech. Des.* **2008**. [CrossRef]
5. Kennelley, K.; Bone, L.; Orazem, M. Current and potential distribution on a coated pipeline with holidays part I—Model and experimental verification. *Corrosion* **1993**, *49*, 199–210. [CrossRef]

6. Ameh, E.; Ikpeseni, S. Pipelines cathodic protection design methodologies for impressed current and sacrificial anode systems. *Niger. J. Technol.* **2017**, *36*, 1072–1077. [CrossRef]
7. Hong, M.-S.; So, Y.-S.; Kim, J.-G. Optimization of cathodic protection design for pre-insulated pipeline in district heating system using computational simulation. *Materials* **2019**, *12*, 1761. [CrossRef]
8. Liu, Y.; Wang, Z.; Wei, Y. Influence of Seawater on the Carbon Steel Initial Corrosion Behavior. *Int. J. Electrochem. Sci* **2019**, *14*, 1147–1162. [CrossRef]
9. NACE International. *Cathodic Protection of Reinforcing Steel in Buried or Submerged Concrete Structrures*; SP0408; NACE International: Houston, TX, USA, 2008.
10. Michael, F. VI. Experimental Researches in Electricity–Seventh Series. *Phil. Trans. R. Soc.* **1834**, *124*, 77–122.
11. Hong, M.-S.; So, Y.-S.; Lim, J.-M.; Kim, J.-G. Evaluation of internal corrosion property in district heating pipeline using fracture mechanics and electrochemical acceleration kinetics. *J. Ind. Eng. Chem.* **2020**, *94*, 253–263. [CrossRef]
12. Lapina, A.; Holtappels, P.; Mogensen, M. Conductivity at Low Humidity of Materials Derived from Ferroxane Particles. *Int. J. Electrochem.* **2012**, *2012*. [CrossRef]
13. Zhao, X.; Guo, X.; Yang, Z.; Liu, H.; Qian, Q. Phase-controlled preparation of iron (oxyhydr) oxide nanocrystallines for heavy metal removal. *J. Nanopart. Res.* **2011**, *13*, 2853–2864. [CrossRef]
14. Leban, M.B.; Kosec, T. Characterization of corrosion products formed on mild steel in deoxygenated water by Raman spectroscopy and energy dispersive X-ray spectrometry. *Eng. Fail. Anal.* **2017**, *79*, 940–950. [CrossRef]
15. Wei, B.; Qin, Q.; Bai, Y.; Yu, C.; Xu, J.; Sun, C.; Ke, W. Short-period corrosion of X80 pipeline steel induced by AC current in acidic red soil. *Eng. Fail. Anal.* **2019**, *105*, 156–175. [CrossRef]
16. Hu, J.; Cao, S.A.; Xie, J. EIS study on the corrosion behavior of rusted carbon steel in 3% NaCl solution. *Anti-Corros. Method. Mater.* **2013**, *60*, 100–105. [CrossRef]
17. Hœrlé, S.; Mazaudier, F.; Dillmann, P.; Santarini, G. Advances in understanding atmospheric corrosion of iron. II. Mechanistic modelling of wet–dry cycles. *Corros. Sci.* **2004**, *46*, 1431–1465. [CrossRef]
18. Stratmann, M. The Atmospheric Corrosion of Iron—A Discussion of the Physico-Chemical Fundamentals of this Omnipresent Corrosion Process Invited Review. *Ber. Bunsenges. Phys. Chem.* **1990**, *94*, 626–639. [CrossRef]
19. Bousselmi, L.; Fiaud, C.; Tribollet, B.; Triki, E. The characterisation of the coated layer at the interface carbon steel-natural salt water by impedance spectroscopy. *Corros. Sci.* **1997**, *39*, 1711–1724. [CrossRef]
20. Bousselmi, L.; Fiaud, C.; Tribollet, B.; Triki, E. Impedance spectroscopic study of a steel electrode in condition of scaling and corrosion: Interphase model. *Electrochim. Acta* **1999**, *44*, 4357–4363. [CrossRef]
21. Stansbury, E.E.; Buchanan, R.A. *Fundamentals of Electrochemical Corrosion*; ASM International, Material Information Society: Materials Park, OH, USA, 2000.
22. Sheir, L.; Jarman, R.; Burstein, G. Corrosion: Metal/Environment Reactions. *Newnes-Butterworths Lond.* **1994**, *8*, 3–8.
23. Al-Hazzaa, M.; Al-Abdullatif, M. Effect of Soil Conductivity on the Design of Cathodic Protection Systems Used in the Prevention of Pipeline Corrosion. *J. King Saud Univ.-Eng. Sci.* **2010**, *22*, 111–116. [CrossRef]
24. Denny, A.; Jones, D.A. *Principles and Prevention of Corrosion*, 2nd ed.; Macmillan Pulishing: New York, NY, USA, 1992.
25. Roberge, P.R. *Corrosion Inspection and Monitoring*; John Wiley & Sons: Hoboken, NJ, USA, 2007; Volume 2.
26. Det Norske Veritas. *Recommended Practice Dnv-Rp-B401 Cathodic Protection Design*; Det Norske Veritas: Oslo, Norway, 2010.

Article

A Novel Testing Method for Examining Corrosion Behavior of Reinforcing Steel in Simulated Concrete Pore Solutions

Yanru Li [1], Jiazhao Liu [2], Zhijun Dong [3], Shaobang Xing [4], Yajun Lv [1] and Dawang Li [1,*]

1. Guangdong Provincial Key Laboratory of Durability for Marine Civil Engineering, Shenzhen University, Shenzhen 518060, China; liyanru6885@126.com (Y.L.); darkdanking@126.com (Y.L.)
2. Shenzhen Luohu District Bureau of Construction Works, Shenzhen 518060, China; Liujz9818@126.com
3. Shenzhen Institute of Information Technology, Shenzhen 518172, China; dongzj@sziit.edu.cn
4. Hong Kong Huayi Design Consultants (S.Z) Ltd., Shenzhen 518060, China; shaobangxing@126.com
* Correspondence: lidw@szu.edu.cn

Received: 29 October 2020; Accepted: 22 November 2020; Published: 24 November 2020

Abstract: In this paper, a new mechanical-based experimental method is proposed to determine the corrosion initiation and subsequent corrosion behavior of steel in simulated concrete pore solutions. The proposed experiment is used to investigate the corrosion of the steel wire under various different conditions and to examine the effects of pre-stress level in steel wire, passivation time of steel wire, composition and concentration of simulated concrete pore solution on the corrosion initiation, and subsequent corrosion development in the steel wire. The experimental results show that the reduction rate of the cross-section area of the steel wire increases with the increase of chloride concentration or decrease of pH value in the solution. However, for the case where the chloride concentration is high and the pH value is low, there is a slight decrease in the corrosion rate due to the coating function of the corrosion products surrounding the wire.

Keywords: reinforcing steel; corrosion; chloride; experiment; bow shaped device; concrete

1. Introduction

Metal corrosion is a common phenomenon which can cause serious deterioration of the service performance of engineering structures. Reinforcing steel corrosion, frequently found in concrete structures exposed to chloride environments, is a typical example [1,2]. Reinforcing steel corrosion can be defined as an electrochemical process of a metal in relation to its surrounding environment, which can be represented by two electrochemical reactions of the dissolution of iron at anodic sites and the corresponding oxygen reduction at cathodic sites. The chemical reactions of corrosion are the same no matter the passive layer around steel breaks down by chloride attack or by carbonation. When depassivation occurs, areas of rust will start appearing on the affected area of the steel surface in the presence of water and oxygen. The affected steel areas become anodic whilst the passive areas are cathodic. A corrosion cell develops on steel surface because of the difference in electrical potential between the anodic and cathodic areas, while a current flows from the anode to the cathode transported by the ions in the electrolyte. The faster the solid iron is converted to ions, the greater the corrosion and the larger the current flowing in the corrosion cell. The detailed description of the electrochemical process of the corrosion of steel bars in concrete can be found in technical books, for example, [3–5].

Reinforcing steel embedded in concrete is usually in a passive state due to the high alkalinity of concrete pore solution, which protects the steel bars from corrosion in some extent. However, the passive state could disappear if the concrete is exposed to a marine environment since chloride ions can penetrate into concrete and cause rupture of the passive layer [6,7]. Numerous studies exist in literature on chloride-induced corrosion of reinforcement in concrete under different conditions. Angst and Vennesland [8] and Angst et al. [9] presented excellent review articles, which not only described the concept of the chloride threshold value (CTV) and discussed the factors that affect the CTV, but also assessed various measurement techniques for reinforcing corrosion. Recently, Green [10] also presented a review paper, which described the fundamental and mechanistic aspects of the protection afforded to steel reinforcement by concrete, corrosion of steel reinforcement, corrosion products composition and development, chloride- and carbonation-induced corrosion mechanisms, leaching induced corrosion of reinforcement and reinforcing steel stray current corrosion and interference.

In the literature, two types of methods have been reported to determine the CTV. One is from a scientific point of view, in which the CTV is defined as the chloride content required for depassivation of the steel [11–14]. The other is from an engineering point of view, in which the CTV is taken as the chloride content associated with visible corrosion of steel or acceptable deterioration of a RC structure [3,4,7,11]. Note that, as far as the durability is concerned, the first definition is more related to materials which is considered to be in the initiation stage of corrosion; whereas the second one is more associated with structures which is considered to be in the propagation stage of corrosion. These two definitions could lead to different CTVs. In general, the engineering definition leads to a higher CTV and results in a large scatter of CTVs [12]. Simulated concrete pore solutions have been used to study the corrosion initiation of reinforcing steel by using laboratory testing methods [15,16]. Many researchers have examined the effect of surface finishing of steel bars on their corrosion initiation in simulated concrete pore solutions with different pH values [17–20]. Testing methods for on-site corrosion monitoring of reinforced concrete structures have been reported [21–23].

There are many electrochemical techniques that could be used to determine the CTV associated with the depassivation and subsequent corrosion behavior of reinforcing steel bars in concrete, such as monitoring the macrocell current between an anode and a cathode [24], measuring the half-cell potential [25], measuring the polarisation resistance [26], and measuring the electrochemical impedance spectroscopy [27]. However, most of these methods do not have good repeatability. Also, the CTVs obtained using different methods are found to be quite different. For example, according to Angst and Vennesland [8] the CTV varies from 0.02% to 3.08% total chloride by binder weight. Recently, Meira et al. [28] reported that average CTV ranges from 1.82% to 2.45% of cement weight for laboratory measurements and between 0.88% and 1.58% of cement weight for field exposure experiments. They concluded that the large difference could be attributed to the environmental interaction.

The wide spread of results on CTV reported in literature may be attributed to the different definitions for the corrosion initiation, the techniques used to determine the CTV [8], the surface finishing of the steel bars [17], the concrete properties and the aggressiveness of the environment [29]. In this paper a new experimental method of using a novel bow shaped device is proposed. The experimental device is then used to investigate the corrosion of the steel wire under various different conditions. The effects of pre-stress level in steel wire, passivation time of steel wire, composition and concentration of simulated concrete pore solution on the corrosion initiation and subsequent corrosion development in the steel wire are examined and a detailed discussion is provided.

2. Method and Device

A new bow shaped device is described herein, which is used for examining steel wire corrosion in a simulated concrete pore solution along with the corresponding cross-section area reduction rate of the wire induced by steel corrosion.

Consider the corrosion problem of a steel wire of uniform cross-section, subjected to a tensile force F when it is immersed in a corrosive solution. Assume that the steel wire has had a passivation treatment and there is a thin passivation film surrounding its surface. Let A_0 be the initial cross-section area of the steel wire and σ_{ult} be the ultimate tensile strength of the steel wire.

In general, the corrosion process of the steel wire after it is immersed in the corrosive solution can be divided into two stages. One is the corrosion initiation stage during which the passivation film is destroyed. At this stage, there is no actual corrosion taking place in the steel and thus the cross-section area of the steel wire remains unchanged. This stage ends when the passivation film is destroyed and the steel is about to start to corrode. The other is the corrosion stage during which the steel wire corrodes continuously. At this stage, the cross-section area of the steel wire decreases gradually with increasing time. This stage ends when the steel wire breaks due to the reduction of its cross-section area caused by the corrosion and the corroded steel wire is no longer able to sustain the applied force F.

Let t_s be the time at the end of the corrosion initiation stage and t_r be the time at the end of the corrosion stage. Thus, the average reduction rate of the cross-section area of the steel wire can be expressed as follows,

$$v = \frac{A_0 - A(t_r)}{t_r - t_s} \tag{1}$$

where v is the corrosion reduction rate of the cross-section area of the steel wire and $A(t_r)$ is the residual cross-section area of the corroded steel wire at the break time. Note that $A(t_r)$ can be expressed in terms of the applied force, that is $A(t_r) = F/\sigma_{ult} = A_0 F/F_{max}$, where $F_{max} = \sigma_{ult} A_0$ is the maximum force that the steel wire can sustain. Thus, Equation (1) can be rewritten as follows:

$$v = \frac{A_0 - \frac{F}{\sigma_{ult}}}{t_r - t_s} = \frac{A_0}{t_r - t_s} \cdot \left(\frac{F_{max} - F}{F_{max}}\right) \tag{2}$$

From the above equation, the following equation can be obtained,

$$t_r - t_s = \frac{A_0}{v} \cdot \left(\frac{F_{max} - F}{F_{max}}\right) \tag{3}$$

Note that, for given environment conditions the average corrosion reduction rate of the cross-section area of the steel wire should be a finite number. Thus, it can be found from Equation (3) that when $A_0 \to 0$ or $F \to F_{max}$, $t_r \to t_s$. This indicates that the break time of the steel wire could be used to represent the corrosion initiation time if the diameter of the steel wire were very small or the applied force were very close to the maximum force of the steel wire.

The concept described above can be fulfilled by using the experiment of a bow shaped device as shown in Figure 1a, in which the frame of the bow can be obtained by bending a straight plexiglass ruler and the wire is a steel wire that is to be tested for corrosion. The dimensions of the bow and the length of the wire can be adjusted based on the pre-tensile stress of the wire required in the test. After the plexiglass ruler is bent to a pre-calculated curvature, the steel wire is fixed to the small holes predrilled on the plexiglass ruler near its two ends. Copper conductive wires are connected to the ends of the steel wire at the fixing points. The other ends of the copper wires are connected to a data recorder or electrochemical workstation, as shown in Figure 1, to monitor the change of signals and/or record the times at which the steel wire starts corroding and/or breaks after it is immersed in the corrosive solution.

In order to make sure the corrosion of the steel wire occurs in the middle part of the wire after it is immersed in the corrosive solution, the two ends of the steel wire near the fixing points are coated with epoxy resin. The time is recorded immediately after the bow is immersed in the solution. After the steel wire is submerged in the corrosive solution, its passivation film will be destroyed first and then corrosion starts. How fast this process takes place is dependent on the corrosive solution used in the test. Nevertheless, when a visible corrosion is observed in the pre-stressed wire, the corresponding time is recorded as t_s, whereas the time when the wire breaks is recorded as t_r.

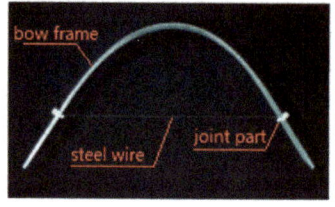

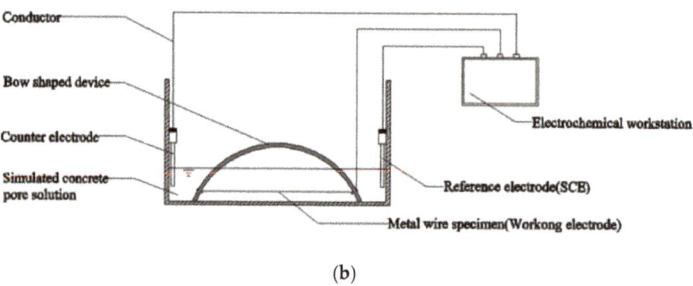

Figure 1. Illustration of corrosion test of steel wire in a corrosive solution. (**a**) Image and (**b**) schematic of corrosion test of steel wire in a corrosive solution.

3. Experimental Programme

The present experimental method is developed based on the mechanical principle in which the break of a pre-stressed steel wire is purely due to the corrosion-induced increase of the stress, which reaches to the ultimate tensile strength of the wire. In the present experiment, the frame of the bow is made from plexiglass fragment of rectangular cross-section (3 mm × 40 mm). The steel wire was obtained from a Chinese company (Mingjunchangrong, Foshan, China), has a diameter of 0.1413 mm, and is made from high strength steel. The tensile test of the steel wire alone shows that the ultimate tensile strength of the steel wire can reach to 2931 MPa. The chemical compositions of the steel wire obtained from five randomly taken sampling points on the wire are shown in Table 1. Before the bow is immersed in the solution, the steel wire is first cleaned using absolute ethanol and then washed using deionized water. After it is dried, the steel wire is subjected to a passivation treatment by immersing it into a saturated $Ca(OH)_2$ solution for a period of 1–31 days depending on the purpose of tests. All of the tests are conducted at room temperature (about 25 °C) in a controlled laboratory condition.

Table 1. Chemical compositions of steel wire (% in mass).

Sample Points	C	O	Na	P	Mn	Co	Zn	Si	Fe
1	3.91	7.85	0.04	2.84	1.03	1.47	5.14	0.29	77.43
2	4.31	4.32	0.05	1.5	1.18	1.39	2.82	0.87	83.56
3	4.62	8.48	0.06	3.07	1.67	1.28	5.89	0.55	71.55
4	6.16	1	0.05	0.53	2.61	2.82	3.53	0.86	82.43
5	6.72	5.82	0.05	2.48	2.17	2.19	7.42	0.34	72.81

The pre-stress applied on the steel wire can be calculated based on the geometry of the deformed bow. Alternatively, it can be determined using a vibration method [30] from which the pre-force can be calculated as follows,

$$F = 4ml^2 f^2 \qquad (4)$$

where m is the mass of the steel wire per-unit length, l is the length of the steel wire between the two fixed points, and f in Hz is the fundamental frequency of the steel wire in the bow shaped device, which can be obtained by using the laser doppler vibrometer (LDV) instrument (see Figure 2).

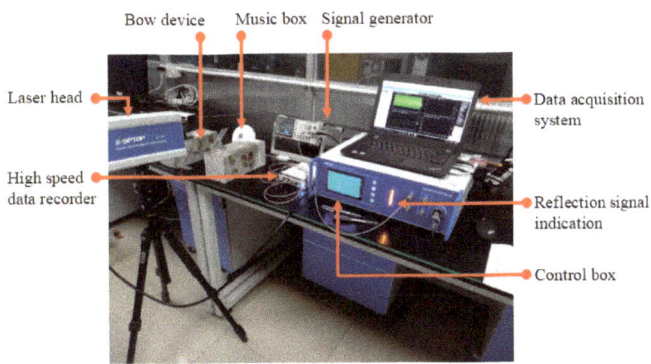

Figure 2. Frequency measurement of steel wire in bow shaped device using LDV.

The polarization resistance R_p is obtained from the electrochemical workstation via a three-electrode system (Figure 1b) by using linear polarization measurement, in which the working electrode is the steel wire immersed in the solution, the auxiliary electrode (counter) is the platinum electrode, and the reference electrode is the saturated Calomel/KCl electrode (SCE). The parametric values used for the linear polarization measurement are as follows. The electro-potential range for scanning is taken as ±10 mV relative to that of the working electrode. The scanning speed is 0.1667 mV/s. A total of 201 points are scanned. The data obtained are recorded and processed using Versastudio software (Princeton Applied Research, Oak Ridge, TN, USA). The polarization resistance R_p is then calculated by the ratio of the slope of E-I curve at zero-potential point to the exposed area of the wire.

The visual inspection is carried out by using 3R Anyty high-resolution portable microscope (Anyty, Tokyo, Japan), which has camera and video functions on the amplified object. The use of the portable microscope not only overcomes the inaccuracy of the traditional artificial visual inspection on the rust product of the wire, but also enables the evolution of the wire from corrosion initiation to final break to be recorded. In the present experiment the magnification rate of 30 times is used. As an example, Figure 3 shows the typical images of the steel wire at the four different stages: (a) immediately after it is immersed into the solution, (b) corrosion is just initiated, (c) corrosion is in development, and (d) wire breaks. The details about when and under what experimental conditions these photos were taken are given in Figure 4, respectively.

As the objective of the present experimental study is to investigate the corrosion initiation and corresponding corrosion behavior of the reinforcing steel in concrete, various simulated concrete pore solutions with different Cl^- and/or OH^- concentrations are employed as the corrosive solution used in the experiments. Table 2 gives the details of the solutions used, in which the three solutions represent the control solution (pH = 7), completely carbonated solution (pH = 8.3), and partially carbonated solution (pH = 10.3), respectively [31,32]. In addition, different Cl^- concentrations are also used in the solutions of the same pH value in order to examine the effect of Cl^- on the corrosion process of the steel wire.

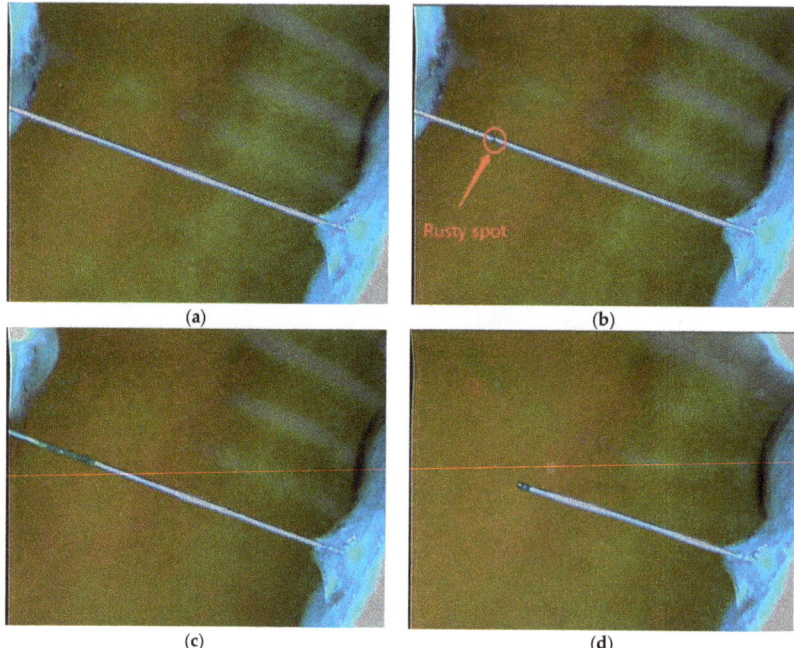

Figure 3. Images of steel wire at different stages. (**a**) Immediately after immersed into solution, (**b**) corrosion just initiated where a rust spot was viewed, (**c**) corrosion in development where corrosion region has been developed, and (**d**) after wire breaks (The details about when and under what experimental conditions these photos were taken are given in Figure 4, respectively).

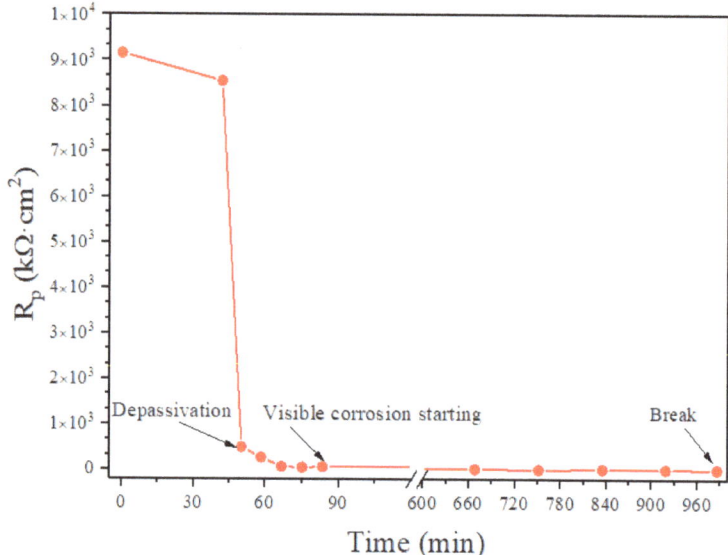

Figure 4. Time-history of R_p value and key time points for depassivation, corrosion initiation and final break (F = 42 N, 7-day passivation, Cl$^-$ = 0.025 mol/L, pH = 10.3).

Table 2. Details of simulated concrete pore solutions used in experiments.

Type of Solution	pH	OH$^-$ mol/L	CL$^-$ mol/L
Deionized H$_2$O	7	1×10^{-7}	0, 0.00001, 0.0001, 0.005, 0.0025, 0.0125, 0.025, 0.05, 0.1, 0.5, 1.0
Saturated CaCO$_3$	8.3	$1 \times 10^{-5.7}$	0.0025, 0.0125, 0.025, 0.05, 0.1, 0.5, 1.0
0.53 g/L NaHCO$_3$ + 1.26 g/L Na$_2$CO$_3$	10.3	$1 \times 10^{-4.3}$	0.0125, 0.025, 0.05, 0.1, 0.5, 1.0, 1.25

4. Experimental Results

4.1. Polarization Resistance Measurement

In order to make a comparison between the present experimental method and other methods, polarization resistance is measured during the immersion tests of the bow shaped device in the simulated concrete pore solution of pH = 10.3 and Cl$^-$ = 0.025 mol/L. Before the test the steel wire has undergone a passivation treatment for seven days. The pre-tensile force used for the test is F = 42 N. The polarization resistance is measured between the steel wire and a counter electrode (see Figure 1b). Figure 4 shows the variation of the R_p with the time obtained from one of the conducted 15 tests. It can be seen from the figure that there is an abrupt change in R_p response curve around the time t = 45 min. This time represents the breakdown time of the passivation film surrounding the steel wire. Before that time the steel wire is protected by the passivation film, which makes the corrosive reactions difficult to take place and thus results in a high polarization resistance. However, after the passivation film is destroyed, the polarization resistance decreases rapidly, and thus the corrosion starts.

A careful examination on the steel wire surface during the polarization resistance test shows that, after the breakdown of the passivation film the corrosion in the steel wire starts. The visible corrosion of the steel wire is observed at the time of about t = 80 min (Figure 3b). The corrosion continues until to the time t = 990 min at which the steel wire breaks (Figure 3d) due to the corrosion-induced reduction of the cross-section area of the steel wire

Figure 5 shows the polarization curves obtained experimentally at four different stages, which represent the corrosion level of the wire at the initial stage of immersion, the stage just after depassivation, the stage when the rust is visually detected, and the stage after final break. The polarization curves were obtained from the electrochemical workstation via a three-electrode system (Figure 1b) by using linear polarization measurement, in which the working electrode is the steel wire immersed in the solution, the counter electrode is the platinum electrode, and the reference electrode is the saturated Calomel/KCl electrode (SCE). The corrosion potential values of these four stages are found to be −0.162 V, −0.322 V, −0.361 V, and −0.418 V, respectively. It can be seen from the figure that the corrosion current has a jump from the initial stage of immersion to the depassivation stage, whereas after the depassivation, the increase of the corrosion current becomes rather slow, representing the gradual development of the corrosion.

Figure 6 shows the breakdown time of the passivation film obtained from the polarization resistance measurement, the corrosion starting time visibly observed during the test, and the final break time of the steel wire obtained from the 15 immersion tests under identical conditions. It can be seen from the figure that, overall, the passivation breakdown time obtained from the polarization resistance method is close to the corrosion starting time visibly observed in the test and they also have the similar trend; whereas the break time of the steel wire is found to vary between specimens. The correlation analyses between these three times are shown in Figure 7. It can be seen from the figure that there is a good correlation between the breakdown time of the passivation film and the corrosion starting time visibly observed with the correlation coefficient of 0.96. In contrast, the correlations between the breakdown time of the passivation film and the break time and between the corrosion starting time visibly observed and the break time are not very good, with the correlation coefficients of 0.66 and 0.67, respectively.

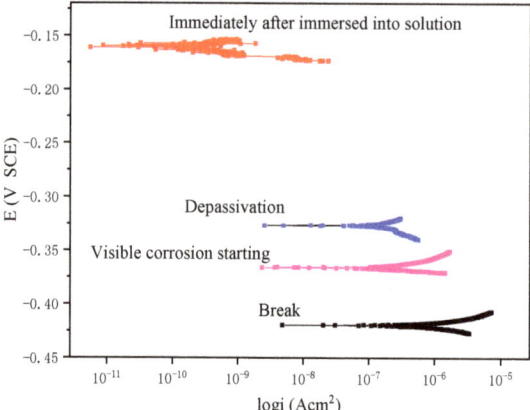

Figure 5. Polarization curves at different stages.

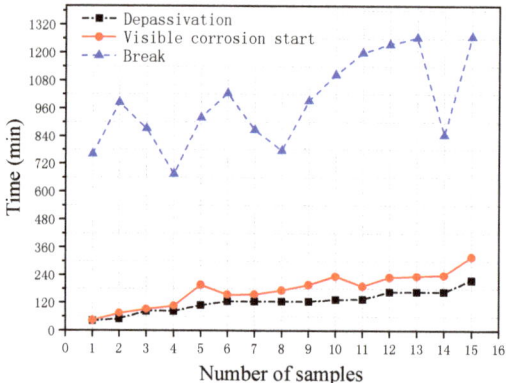

Figure 6. Depassivation time, corrosion starting time, and wire break time obtained from 15 corrosion tests (F = 42 N, 7-day passivation, Cl^- = 0.025 mol/L, pH = 10.3).

4.2. Effect of Prestress of Steel Wire

The effect of the prestress of the steel wire on its corrosion is investigated by using the immersion tests of the bow shaped device in the simulated concrete pore solution of pH = 7.0 and Cl^- = 0.0125 mol/L. Before the test the steel wire has undergone a passivation treatment for seven days. The pre-tensile forces used in the tests vary from F = 7.31 N to F = 42.3 N, which are equivalent to the prestress level from $0.162\sigma_{ult}$ to $0.938\sigma_{ult}$.

Figures 8 and 9 show the results of the corrosion starting time (t_s) visibly observed during the tests and the final break time (t_r) of the steel wire, respectively, obtained from 18 immersion tests with different pre-tensile forces applied in the tested steel wire. It can be seen from Figure 8 that, except for one test with F = 26.95 N, all other tests have very similar corrosion starting times with a mean value of 9.24 min and a standard deviation of 1.03 min. This indicates that the prestress of the steel wire has almost no influence on the chemical degradation process of the passivation film. This seems understandable. The time required for corrosion imitation is a chemical process during which the passivation surrounding the steel is gradually destroyed. Unlike the corrosion starting time, however, the break time of the steel wire is found to be closely related to the applied pre-tensile force (see Figure 9). The larger the applied pre-tensile force, the shorter the break time of the steel wire.

This seems to be expected. As the applied pre-tensile force increases, the stress will be close to the ultimate strength of the steel, thus the allowable reduction of the cross-section area becomes small. As demonstrated by Equation (3), when $F \rightarrow F_{max}$ the break time of the steel wire will be very close to the corrosion starting time $t_r \rightarrow t_s$; meaning that any tiny corrosion in the steel wire could lead to a break of the wire.

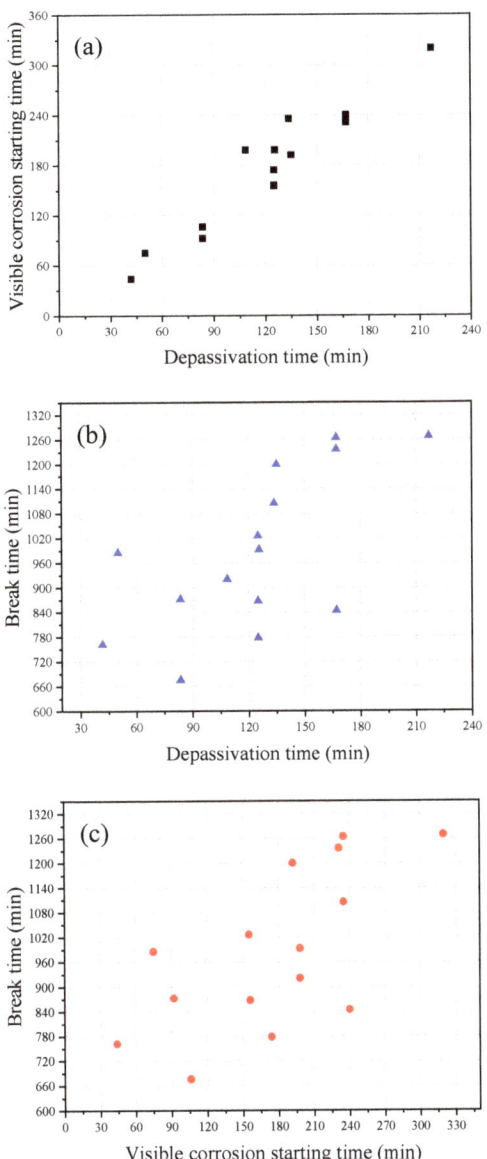

Figure 7. Correlation between (**a**) passivation breakdown time and visible corrosion starting time (correlation coefficient of 0.96), (**b**) between steel wire break time and passivation breakdown time (correlation coefficient of 0.66), (**c**) between steel wire break time and visible corrosion starting time (correlation coefficient of 0.67) obtained from 15 corrosion tests.

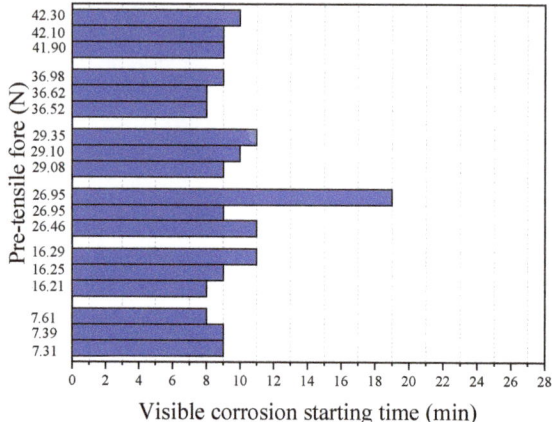

Figure 8. Corrosion starting times at different pre-tensile forces (pH = 7, Cl$^-$ = 0.0125 mol/L).

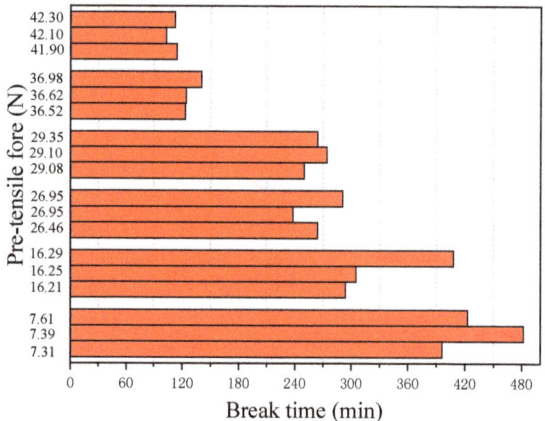

Figure 9. Steel wire break times at different pre-tensile forces (pH = 7, Cl$^-$ = 0.0125 mol/L).

For the convenience of presentation, Equation (3) is rewritten as follows,

$$\Delta F = (\sigma_{ult} v)\Delta t \qquad (5)$$

where $\Delta t = t_r - t_s$ is the time interval between the break time and corrosion starting time of the steel wire and $\Delta F = F_{max} - F$ is the allowable force for the steel wire to break. Figure 10 shows the plot of Δt versus ΔF for the above 18 immersion test data. It is observed from the figure that the relationship between ΔF and Δt is almost linear, indicating that for a given simulated solution the reduction rate of the cross-section area is almost constant. Note that although Δt increases with ΔF, the actual values of Δt for most cases are not very big. For example, for the present 18 immersion tests, the maximum value of Δt is about 420 min, which is only about 7 h. This value, when compared to the design life of a structure that is tens of years, is negligible. This indicates that the break time of the steel wire obtained from the present immersion tests can practically be taken as the corrosion starting time.

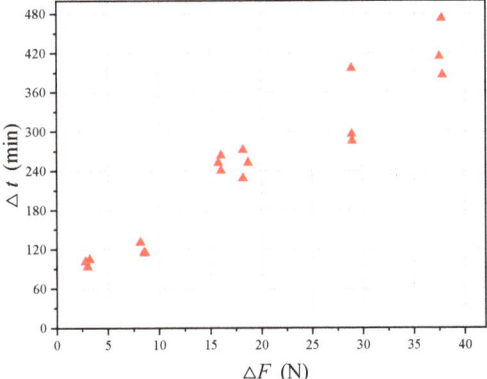

Figure 10. Relationship between Δt and ΔF obtained from 18 corrosion tests.

4.3. Effect of Passivation Film

The effect of the passivation of steel wire on the corrosion-induced break time (t_r) of the steel wire is investigated by using the immersion tests of the bow shaped device in the simulated concrete pore solution of pH = 7.0 and Cl$^-$ = 0.0125 mol/L. A total of 33 (3 × 11) tests are carried out, in which three tests are for the steel wire that has no passivation and the other 3 × 10 tests are for the steel wire that has undergone a passivation treatment in a saturated Ca(OH)$_2$ solution with pH value of 12.5 using different treatment periods. The pre-tensile force used in the tests is F = 30 N. Figure 11 shows the break time of the steel wire obtained from the 33 tests. It can be seen from the figure that the results obtained from any three repeating tests are close, indicating that the test is rather reliable. Interestingly, the longest break time is found in the test where the steel wire has no passivation treatment, followed by the test where the steel wire has only one day passivation treatment. The break times for the steel wire that has passivation treatment between two and 11 days are not very different. While from the tendency of the variation of the curve shown in the figure it seems that the further increase of the passivation treatment period can slightly increase the break time.

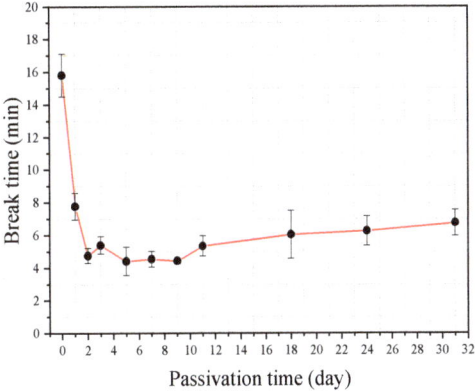

Figure 11. The break time of steel wire with and without passivation treatment (pH = 7, F = 30 N, Cl$^-$ = 0.0125 mol/L).

To further confirm the finding that the passivation of the steel wire does not increase its corrosion break time, the immersion tests of the bow shaped device in the simulated concrete pore solutions of varying Cl$^-$ concentration are also carried out. The steel wire used in the tests includes one without passivation treatment and one with passivation treatment for seven days. The pH value of the simulated solution is kept to be 7.0 for all of the tests. The pre-tensile force used in the tests is F = 30 N. Figure 12 shows the comparison of the break times of the steel wires with and without passivation for eight different Cl$^-$ concentration solutions. It is evident from the figure that the passivation has a negative influence on the break time of steel wire disregarding the Cl$^-$ concentration in the solution. This indicates that, although the passivation film can generally protect the steel from corrosion and thus can postpone the corrosion starting time, it may accelerate the corrosion process after the corrosion starts.

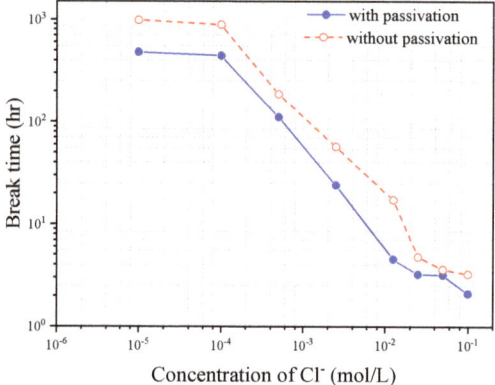

Figure 12. The break time of steel wire with and without passivation treatment under different Cl$^-$ concentration solutions (pH = 7, F = 30 N).

In order to explain why the passivation film has a negative effect on the break time of the steel wire, SEM analysis on the steel wires with and without passivation is carried out. Figure 14 shows the SEM images of the steel wires before they are immersed into the simulated concrete pore solution. It can be seen from the figure that the surface of the steel wire without the passivation is very uniform and smooth; whereas the surface of the steel wire with passivation is not very smooth and it looks like wearing a "knitting cloth". The thickness and mesh density of the "knitting cloth" increase with the passivation time.

Figure 15 shows the SEM images of the corresponding steel wires after three-hour immersion in the simulated concrete pore solution of pH = 7.0 and Cl$^-$ = 0.0125 mol/L. It can be seen from the figure that the corrosion of the steel wire without passivation is almost uniform, whereas the corrosion in other three steel wires that have had passivation treatment exhibits the pitting pattern.

4.4. Effect of Cl$^-$ and OH$^-$ Concentrations

The effect of the simulated concrete pore solution on the corrosion starting time (t_s) and corrosion-induced break time (t_r) of the steel wire is investigated by using the immersion tes t_s of the bow shaped device in the simulated concrete pore solutions of different pH values and different Cl$^-$ concentrations. Before the test the steel wire has undergone a passivation treatment for seven days. The pre-tensile force used in the tes t_s is F = 30 N.

Figure 13 shows the variation of corrosion starting time with Cl$^-$ concentration in three different types of simulated concrete pore solutions of pH = 7, 8.3, and 10.3, respectively. It can be seen from the figure that the corrosion starting time decreases with either the increase of Cl$^-$ concentration or the decrease of pH value in the solution. This kind of results is expected since the higher the Cl$^-$ concentration in the solution or the lower the pH value in the solution, the quicker the depassivation process of the steel will take place.

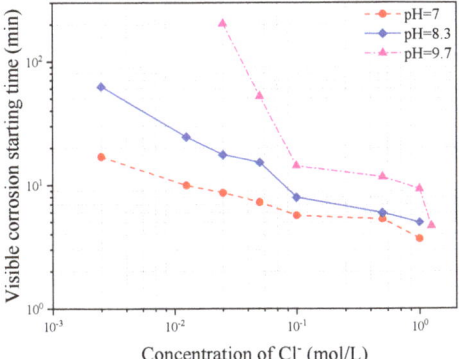

Figure 13. Variation of corrosion starting time with chloride concentration in three different types of simulated concrete pore solutions.

Figure 16 shows the variation of corrosion-induced break time of steel wire with Cl⁻ concentration in the three different types of simulated concrete pore solutions of pH = 7, 8.3 and 10.3, respectively. It can be seen from the figure that the wire break time decreases with the increase of Cl⁻ concentration but turns to be stable or even slightly recovered after the Cl⁻ concentration increases to a certain level. The break time regaining at high Cl⁻ concentrations is probably due to the corrosion product which, at a certain degree, provides a coating function to protect the steel from the chloride attack, and thus reduce the corrosion speed of the steel wire. The decrease of pH value in the solution also leads to a decrease of the wire break time.

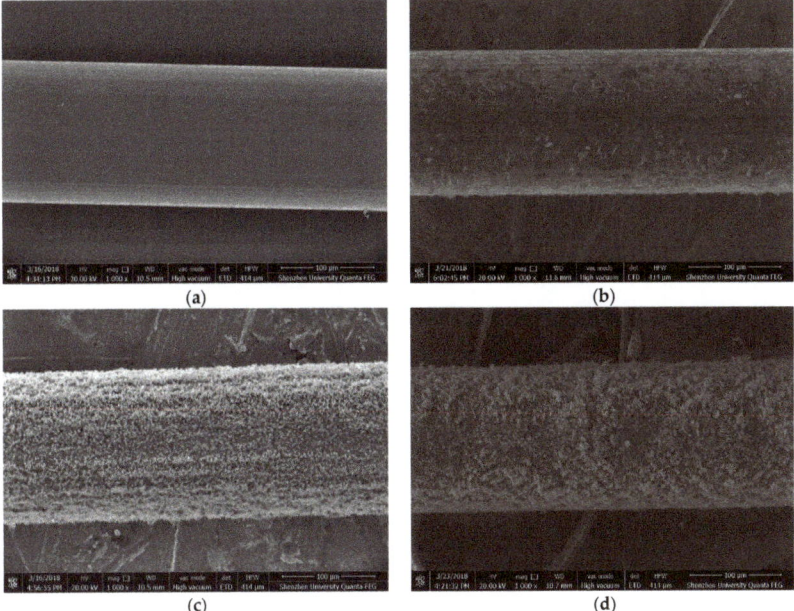

Figure 14. SEM images of steel wires, (**a**) without passivation, (**b**) 1-day passivation, (**c**) 7-day passivation, and (**d**) 11-day passivation.

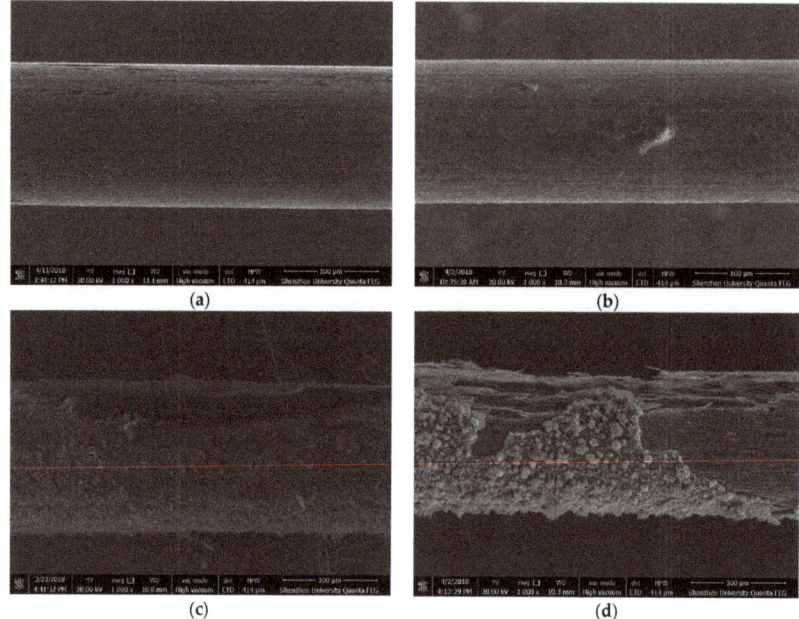

Figure 15. SEM images of corroded steel wires, (**a**) without passivation, (**b**) 1-day passivation, (**c**) 7-day passivation, and (**d**) 11-day passivation.

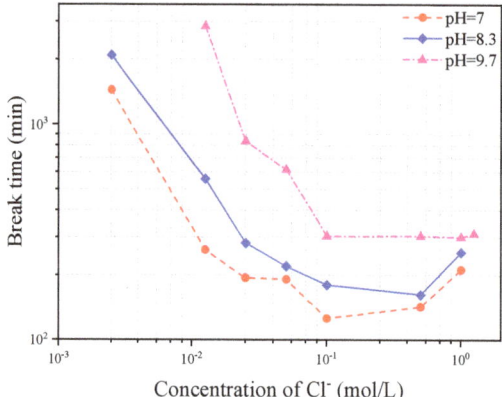

Figure 16. Variation of steel wire break time with chloride concentration in three different types of simulated concrete pore solutions.

5. Conclusions

In this paper a new mechanical-based experimental method has been proposed to investigate the corrosion initiation time and corresponding corrosion behavior of reinforcing steel in simulated concrete pore solutions. The experiment proposed uses a bow shaped device with a prestressed steel wire, which is immersed in a simulated concrete pore solution for corrosion test. The corrosion initiation of the steel wire during the immersion can be observed visibly. From the analysis of the obtained test results, the following conclusions can be drawn:

1. The proposed bow shaped device can be used to determine the corrosion initiation of reinforcing steel under the action of various different simulated concrete pore solutions and the corrosion rate of steel under different corrosion conditions.

2. The corrosion starting time obtained from the present experimental method is comparable with that obtained using the polarization resistance method for depassivation measurement.

3. The passivation film on the steel surface can protect steel from corrosion and thus postpone the corrosion initiation. However, in terms of the corrosion rate, the passivation film has a negative effect. It causes the steel to have pitting corrosion and can accelerate the corrosion process after the corrosion starts.

4. The pre-tensile force in the steel wire can affect the break time of corroded steel but has almost no effect on the corrosion initiation. The closer the pre-tensile force to the ultimate tensile strength of the uncorroded steel wire, the obtained break time of the corroded steel wire will be closer to the corrosion initiation time.

5. Both the chloride concentration and pH value in a simulated solution can affect the process of depassivation of the steel wire. The higher the chloride concentration or the lower the pH value, the quicker the depassivation of the steel wire to take place.

In general, the reduction rate of the cross-section area of the steel wire increases with the chloride concentration or a decrease of pH value in the solution. However, for the case where the chloride concentration is very high and the pH value is small, there is a slight decrease in the corrosion rate due to the coating function of the corrosion products surrounding the wire.

Author Contributions: Conceptualization, D.L. and Y.L. (Yanru Li); validation, J.L., S.X., Z.D., and Y.L. (Yajun Lv); writing—original draft preparation, Y.L. (Yanru Li); writing—review and comment, D.L. and Z.D. All authors have read and agreed to the published version of the manuscript.

Funding: The authors gratefully acknowledge the financial supports provided by the National Natural Science Foundation of China (Grants Nos. 51520105012, 51978406, 51978414) and Shenzhen University.

Conflicts of Interest: The authors declare no conflict of interest.

References

1. Bautista, A.; Paredes, E.; Alvarez, S.; Velasco, F. Welded, sandblasted, stainless steel corrugated bars in non-carbonated and carbonated mortars: A 9-year corrosion study. *Corros. Sci.* **2016**, *102*, 363–372. [CrossRef]
2. Li, D.; Wei, R.; Xing, F.; Sui, L.; Zhou, Y.; Wang, W. Influence of Non-uniform corrosion of steel bars on the seismic behavior of reinforced concrete columns. *Constr. Build. Mater.* **2018**, *167*, 20–32. [CrossRef]
3. Broomfield, J.P. *Corrosion of Steel in Concrete: Understanding, Investigation and Repair*, 2nd ed.; Taylor & Francis: Oxford, UK, 2007.
4. Nawy, E.G. *Concrete Construction Engineering Handbook*; CRC Press: Oxford, UK, 2008.
5. Davis, J.R. *Corrosion: Understanding the Basics*; ASM International: Russell, OH, USA, 2000.
6. Ann, K.Y.; Song, H.-W. Chloride threshold level for corrosion of steel in concrete. *Corros. Sci.* **2007**, *49*, 4113–4133. [CrossRef]
7. Tuutti, K. *Corrosion of Steel in Concrete. CBI Research Report No. 4.82*; Swedish Cement and Concrete Research Institute: Stockholm, Sweden, 1982.
8. Angst, U.; Vennesland, Ø. Critical chloride content in reinforced concrete—State of the art. In *Concrete Repair, Rehabilitation and Retrofitting II, Proceedings of the 2nd International Conference on Concrete Repair, Rehabilitation and Retrofitting; Cape Town, South Africa, 24–26 November 2008*; CRC Press: Oxford, UK, 2009; pp. 311–318. ISBN 978-0-415-46850-3.
9. Angst, U.; Elsener, B.; Larsen, C.K.; Vennesland, Ø. Critical chloride content in reinforced concrete—A review. *Cem. Concr. Res.* **2009**, *39*, 1122–1138. [CrossRef]
10. Green, W.K. Steel reinforcement corrosion in concrete—An overview of some fundamentals. *Corros. Eng. Sci. Technol.* **2020**, *55*, 289–302. [CrossRef]
11. Hobbs, D. Concrete deterioration: Causes, diagnosis, and minimising risk. *Int. Mater. Rev.* **2001**, *46*, 117–144. [CrossRef]

12. Li, L.; Sagüés, A. Chloride Corrosion Threshold of Reinforcing Steel in Alkaline Solutions—Open-Circuit Immersion Tests. *Corrosion* **2001**, *57*, 19–28. [CrossRef]
13. Manera, M.; Vennesland, Ø.; Bertolini, L. Chloride threshold for rebar corrosion in concrete with addition of silica fume. *Corros. Sci.* **2008**, *50*, 554–560. [CrossRef]
14. Figueira, R.; Sadovski, A.; Melo, A.P.; Pereira, E.V. Chloride threshold value to initiate reinforcement corrosion in simulated concrete pore solutions: The influence of surface finishing and pH. *Constr. Build. Mater.* **2017**, *141*, 183–200. [CrossRef]
15. Liu, R.; Jiang, L.; Huang, G.; Zhu, Y.; Liu, X.; Chu, H.; Xiong, C. The effect of carbonate and sulfate ions on chloride threshold level of reinforcement corrosion in mortar with/without fly ash. *Constr. Build. Mater.* **2016**, *113*, 90–95. [CrossRef]
16. Alonso, M.; Castellote, M.; Andrade, C. Chloride threshold dependence of pitting potential of reinforcements. *Electrochim. Acta* **2002**, *47*, 3469–3481. [CrossRef]
17. Mohammed, T.U.; Hamada, H. Corrosion of steel bars in concrete with various steel surface conditions. *ACI Mater. J.* **2006**, *103*, 233–242.
18. Dehghanian, C. Study of surface irregularity on corrosion of steel in alkaline media. *Cem. Concr. Res.* **2003**, *33*, 1963–1966. [CrossRef]
19. Boubitsas, D.; Tang, L. The influence of reinforcement steel surface condition on initiation of chloride induced corrosion. *Mater. Struct.* **2015**, *48*, 2641–2658. [CrossRef]
20. Mammoliti, L.; Brown, L.; Hansson, C.; Hope, B. The influence of surface finish of reinforcing steel and ph of the test solution on the chloride threshold concentration for corrosion initiation in synthetic pore solutions. *Cem. Concr. Res.* **1996**, *26*, 545–550. [CrossRef]
21. Montemor, M.; Simões, A.; Ferreira, M. Chloride-induced corrosion on reinforcing steel: From the fundamentals to the monitoring techniques. *Cem. Concr. Compos.* **2003**, *25*, 491–502. [CrossRef]
22. Andrade, C.; Alonso, C. Test methods for on-site corrosion rate measurement of steel reinforcement in concrete by means of the polarization resistance method. *Mater. Struct.* **2004**, *37*, 623–643. [CrossRef]
23. Song, H.; Saraswathy, V. Corrosion monitoring of reinforced concrete structures—A review. *Int. J. Electrochem. Sci.* **2007**, *2*, 1–28.
24. Elsener, B. Macrocell corrosion of steel in concrete—Implications for corrosion monitoring. *Cem. Concr. Compos.* **2002**, *24*, 65–72. [CrossRef]
25. Leelalerkiet, V.; Kyung, J.-W.; Ohtsu, M.; Yokota, M. Analysis of half-cell potential measurement for corrosion of reinforced concrete. *Constr. Build. Mater.* **2004**, *18*, 155–162. [CrossRef]
26. Law, D.; Cairns, J.; Millard, S.; Bungey, J. Measurement of loss of steel from reinforcing bars in concrete using linear polarisation resistance measurements. *NDT E Int.* **2004**, *37*, 381–388. [CrossRef]
27. Hachani, L.; Fiaud, C.; Triki, E.; Raharinaivo, A. Characterisation of steel/concrete interface by electrochemical impedance spectroscopy. *Br. Corros. J.* **1994**, *29*, 122–127. [CrossRef]
28. Meira, G.; Andrade, C.; Vilar, E.; Nery, K. Analysis of chloride threshold from laboratory and field experiments in marine atmosphere zone. *Constr. Build. Mater.* **2014**, *55*, 289–298. [CrossRef]
29. Zhang, F.; Pan, J.; Lin, C. Localized corrosion behaviour of reinforcement steel in simulated concrete pore solution. *Corros. Sci.* **2009**, *51*, 2130–2138. [CrossRef]
30. Clough, R.W.; Penzien, J.; Griffin, D.S. *Dynamics of Structures*; McGraw-Hill: London, UK, 1993.
31. Huet, B.; L'Hostis, V.; Miserque, F.; Idrissi, H. Electrochemical behavior of mild steel in concrete: Influence of pH and carbonate content of concrete pore solution. *Electrochim. Acta* **2005**, *51*, 172–180. [CrossRef]
32. Chang, C.-F.; Chen, J.-W. The experimental investigation of concrete carbonation depth. *Cem. Concr. Res.* **2006**, *36*, 1760–1767. [CrossRef]

Publisher's Note: MDPI stays neutral with regard to jurisdictional claims in published maps and institutional affiliations.

© 2020 by the authors. Licensee MDPI, Basel, Switzerland. This article is an open access article distributed under the terms and conditions of the Creative Commons Attribution (CC BY) license (http://creativecommons.org/licenses/by/4.0/).

Article

Experimental Study on Corrosion Performance of Oil Tubing Steel in HPHT Flowing Media Containing O_2 and CO_2

Yihua Dou, Zhen Li, Jiarui Cheng * and Yafei Zhang

School of Mechanical Engineering, Xi'an Shiyou University, Xi'an 710065, China; xsyoucjr@163.com (Y.D.); lizhenxsyu@sina.com (Z.L.); effyzhang@126.com (Y.Z.)
* Correspondence: cjr88112@163.com; Tel.: +86-180-9249-2490

Received: 26 October 2020; Accepted: 17 November 2020; Published: 18 November 2020

Abstract: The high pressure and high temperature (HPHT) flow solution containing various gases and Cl^- ions is one of the corrosive environments in the use of oilfield tubing and casing. The changing external environment and complex reaction processes are the main factors restricting research into this type of corrosion. To study the corrosion mechanism in the coexistence of O_2 and CO_2 in a flowing medium, a HPHT flow experiment was used to simulate the corrosion process of N80 steel in a complex downhole environment. After the test, the material corrosion rate, surface morphology, micromorphology, and corrosion product composition were tested. Results showed that corrosion of tubing material in a coexisting environment was significantly affected by temperature and gas concentration. The addition of O_2 changes the structure of the original CO_2 corrosion product and the corrosion process, thereby affecting the corrosion law, especially at high temperatures. Meanwhile, the flowing boundary layer and temperature changed the gas concentration near the wall, which changed the corrosion priority and intermediate products on the metal surface. These high temperature corrosion conclusions can provide references for the anticorrosion construction work of downhole pipe strings.

Keywords: high pressure and high temperature; O_2–CO_2 coexistence environment; flow-induced corrosion

1. Introduction

The thermal recovery technology of heavy oil, which is applied to the production of tight and shale oil in the oil field, has the characteristics of high construction temperature, complex flowing medium, and strong corrosiveness. Generally, the production process contains a variety of anions and cations and multicomponent gases, with temperature higher than 150 °C. As the only channel for oil or gas exploitation, the downhole tubing string is inevitably corroded by the high pressure, high temperature, and complex flowing solution, which cause deformation and fracture damage of tubing string, thereby affecting well site safety [1].

N80 tubing steel, which is commonly used in thermal recovery wells, is damaged by the corrosion of O_2, CO_2, and H_2S in the liquid containing HCO_3^-, Cl^-, SO_4^{2-}, Ca^{2+}, Mg^{2+}, and Na^+ [2]. The corrosion in the high temperature flowing liquid environment where O_2, CO_2, and Cl^- coexist is the focus of intense research for N80 tubing steel. The essence of the CO_2 corrosion for tubing steel is the electrochemical corrosion of metals in aqueous carbonate solution [3]. Thus, many studies have investigated CO_2 corrosion based on these factors. Bai et al. [4] investigated the influence of CO_2 partial pressure on the properties of J55 carbon steel in 30% crude oil/brine at 65 °C. The corrosion rate significantly increases as the CO_2 partial pressure increases from 0 to 1.5 MPa and decreases from 1.5 to 5 MPa, and the surface of J55 carbon steel is covered by $FeCO_3$ and $CaCO_3$. The CO_2 partial pressure changes the system pH and CO_2 solubility in crude oil, which further affects the formation and protection performance of the corrosive film. Zhang et al. [5] studied the corrosion behavior of N80 carbon steel under dynamic supercritical CO_2 (8 MPa)–water environment. The results demonstrated no essential difference in the electrochemical corrosion mechanism between supercritical CO_2 and nonsupercritical CO_2 (5 MPa) environments. The corrosion rate under dynamic condition is higher than that under static condition in the initial time and flowing fluid hinders the formation of corrosive film and then increases the corrosion rate.

De Waard et al. [6] studied the reaction process of the anode and cathode through the point-position mechanical polarization curve, and the anode reaction formula (Fe $\rightarrow$ Fe^{2+} + $2e^-$) containing OH^- was obtained. Nesic et al. [7] divided the pH value of the solution into PH < 4, 4 < PH < 6, and PH > 6; for the CO_2 anode electrochemical corrosion reaction process, an intermediate product $Fe(CO_3)OH$ might be generated. Linter et al. [3] reported that $Fe(OH)_2$ occurs first in the process of CO_2 corrosion, and then $FeCO_3$ would be further generated. The CO_2 corrosion rate is generally controlled by the reaction of the cathode. Scholars locally and internationally have conducted many studies on cathode corrosion. The results showed that the corrosion rate of steel in CO_2 solution is controlled by H evolution kinetics [8].

O_2 is a catalyst for CO_2 corrosion and is also a cathodic depolarizer in the corrosion process of steel. When the protective film is not formed on the carbon steel surface, the corrosion rate will increase with the O_2 content, while the CO_2 content plays a decisive role in the corrosion of carbon steel after the O_2 content reaches saturation in solution. Once the protective film is formed on the surface, the corrosion of carbon steel is not affected by the O_2 content [9]. In the presence of O_2, a double-layer film structure is present in the CO_2 corrosive system, the inner of which is mainly Fe_2O_3, and the outer is the loose and porous corrosion product $FeCO_3$ [10]. The increase in O_2 content will result in a dense outer layer of corrosion product film to protect the steel surface. Another study [11] has found that the corrosion products on the 3Cr steel surface are formed by the accumulation of granular products under high temperature conditions where O_2 and CO_2 coexist, and the outer corrosion scales are mainly $FeCO_3$ and Fe oxides.

The influence of Cl^- is reflected in two aspects, as follows [12]: (1) the possibility of passivation film formation on the sample surface is reduced, or the damage of the passivation film is accelerated; therefore, the local corrosion reaction is promoted. (2) The solubility of CO_2 in the aqueous solution is reduced, which could alleviate the corrosion of carbon steel. The corrosion product film on the surface of N80 steel is denser, and the adhesion is high at a low Cl^- content (5000 mg/L), which indicates that corrosion resistance is improved [13]. When the Cl^- content increases twofold, the protective effect of the corrosive film decreases due to the weak compactness of the corrosive film, thereby resulting in an accelerated corrosion rate of steel. Some studies have compared the effects of Cl^- concentration and CO_2 concentration on the corrosion of carbon steel. The test results of Wang et al. [14] showed that the corrosion rate of N80 steel in 1–3 wt% NaCl solution is the highest. Meanwhile, there is a critical partial pressure of CO_2 in the chloride-containing solution that affects the corrosion rate. Zhang et al. [15] tested the corrosion law of N80 steel under different CO_2 partial pressures in detail. The results show that the increase in CO_2 partial pressure accelerates the corrosion reaction but, at the same, time forms a dense protective layer. However, in a chloride-containing solution, Cl^- is sufficient to penetrate the protective layer and react with the base metal [16].

High-pressure and high-temperature (HPHT) multicomponent media are the most common corrosion environments for N80 steel underground. To study the corrosion law of pipes, we used a high-temperature and high-pressure reactor to create an HPHT flow solution to simulate the corrosion environment. The corrosion rate, surface morphology, surface composition, and corrosion layer profile morphology results were obtained in the test. Meanwhile, the N80 corrosion process in an environment of coexisting CO_2 and O_2 has been discussed in various ways.

2. Experimental

2.1. Experimental System

The HPHT corrosion system was used for the flow-induced corrosion test of the N80 steel, including the HPHT autoclave (Weihai Global Chemical Machinery MFG Co., Ltd., Weihai, China), the lifting platform, the control cabinet, the booster pump, and the air source. As shown in Figure 1, the inside of the autoclave was made of Hastelloy C-276, which can resist corrosion under most conditions. The exterior of the autoclave is made of 304 stainless steel. The samples were installed on the fixture and rotated with the agitator shaft and fixture. Approximately 5 L of mixed fluid (simulated produced fluid in oilfield) was injected into an autoclave and stirred by an agitator. The dissolved oxygen was purged in the solution with injected nitrogen gas for 4 h under a pressure of 0.5 MPa. The autoclave was pressured with pure N_2 to the experimental values (total pressure value and CO_2 partial pressure) with CO_2 gas to the experimental values for 72 h at the flow velocity of 1 m/s. The surface microstructure of the corrosion product scales on the surface of the corroded samples was analyzed by JSM-6390 SEM (JEOL, Tokyo, Japan). The composition of the corroded samples was performed by XRD (LabX XRD-6000, SHIMADZU Co., Ltd., Kyoto, Japan).

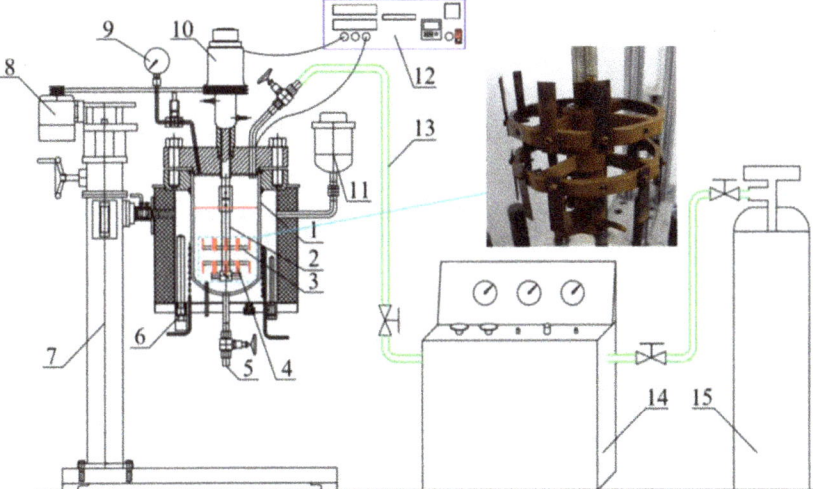

Figure 1. Schematic of experimental setup. 1. Autoclave body; 2. Agitator shaft; 3. Sample fixture; 4. Sample; 5. Outlet; 6. Electric heater; 7. Lifting device; 8. Motor; 9. Pressure gauge; 10. Magnetic stirrer; 11. Expander; 12. Control cabinet; 13. Gas pipeline; 14. Booster pump; 15. CO_2 storage tank.

2.2. Experimental Conditions

The composition of the experimental material, N80 tubing steel, is listed in Table 1. The samples were machined to dimensions of 50 × 13 × 1.5 mm. Before the experiment, the samples were placed in acetone to remove the surface oil and sequentially polished with 500#, 800#, 1200#, and 2000# sandpaper. The samples were scrubbed two times with distilled cotton in distilled water and cleaned for 15 s with distilled water. The samples were dried in cold air and placed in the dryer for 24 h. Finally, the samples were weighed using a digital balance with an accuracy of 0.1 mg.

After the experiment, the corroded samples were photographed to record the surface variations and then immersed in an acid solution (500 mL of HCl and 3.5 g hexamethylenamine diluted with water to 1000 mL) for 10 min to remove product films [4]. The samples were placed on absolute ethanol for washing two times until the acid cleaning solution on the surface was completely removed.

The samples were dried and weighed before the test. When testing XRD, the product film of the sample was peeled off, and the crystal grain was ground to 0.1–10 µm. These grains were filled into the material tray and pressed to make the surface flat.

Table 1. Chemical composition of N80 steel (wt%).

Materials	C	Si	Mn	P	S	Cr	Mo	Ni	Cu	Fe
N80	0.22	0.21	1.77	0.01	0.003	0.036	0.021	0.028	0.019	97.683

The volume of the produced liquid used in each set of experiments was 5 L, and the liquid ingredients are listed in Table 2. The total pressure of HPHT corrosion was 5 MPa. The experimental temperature varied from 50 to 200 °C, and the solution flow velocity was fixed at 1 m/s. The CO_2 partial pressures were 0.25, 0.50, and 0.75 MPa, and the O_2 partial pressures were adjusted to 0.05, 0.10, and 0.15 MPa, respectively.

Table 2. Composition of test solution used in corrosion experiments.

HCO_3^- (mg/L)	Cl^- (mg/L)	Ca^{2+} (mg/L)	Na^+ (mg/L)	SO_4^{2-} (mg/L)	Salinity (mg/L)	pH
800	2000	10	1785	400	4995	7.8

3. Results

3.1. Effects of Solution Temperature and CO_2 Partial Pressure on Corrosion Rate

The corrosion rate of N80 steel measured by changing the CO_2 concentration and temperature in the CO_2–O_2 coexisting environment (0.25 MPa CO_2 and 0.05 MPa O_2) is shown in Figure 2 and Figure 3. The results show that the corrosion rating of N80 steel in this condition is severe or extremely severe according to the NACE SP 0775-2013 standard. When the partial pressure of CO_2 is 0.25 MPa, the corrosion rate increases with increasing temperature. When the partial pressure of CO_2 is greater than 0.5 MPa, the corrosion rate will have a maximum value at 100 °C with the increase in temperature. A comparison of the corrosion rate at different CO_2 partial pressures is shown in Figure 3. As the temperature increases, corrosion rate varies directly with CO_2 partial pressure at a temperature less than 100 °C and inversely at a temperature higher than 100 °C. This result indicates that the increase in CO_2 concentration at low temperature promotes the corrosion reaction, whereas the change in CO_2 concentration at high temperature weakens the effect of corrosion.

3.2. Macroscopic Morphology and Product Components

Figure 4 shows the surface morphology of N80 steel after corrosion at different temperatures in an O_2 and CO_2-containing environment. The results show that thick corrosion products are attached to the surface of the corroded material, which causes the metal matrix to be completely covered by precipitated scales. The product films on the surface are loose and thick, and a portion of them falls off at 50 and 100 °C. As the temperature increases from 100 to 200 °C, the product films become thin and smooth, and the color changes from deep red to black. The morphology of N80 steel after corrosion in a high-temperature environment containing 0.05 MPa O_2 by changing the partial pressure of CO_2 is shown in Figure 5. When the partial pressure of CO_2 is 0.25 MPa, the dark-brown corrosion surface is relatively flat, and a loose corrosion product film is attached. No hard scale is present on the surface of the corroded sample because the solubility of CO_2 will decrease at 200 °C. When the partial pressure of CO_2 reaches 0.75 MPa, the surface oxide of the material increases, and the product film thickens.

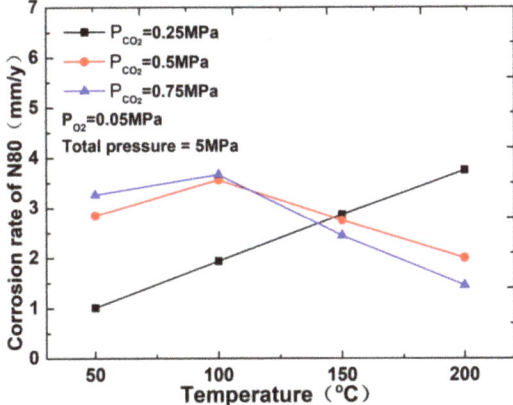

Figure 2. Corrosion rate of N80 steel at different temperatures.

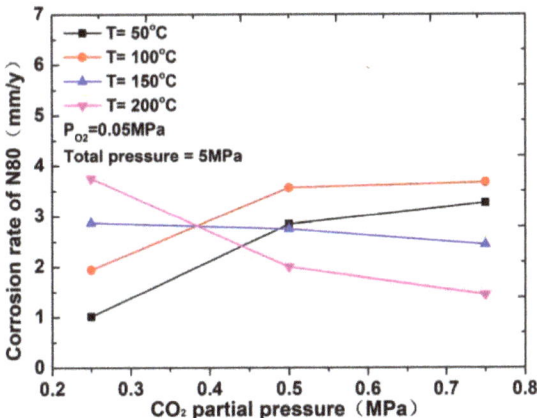

Figure 3. Corrosion rate of N80 steel at different CO_2 partial pressures.

(**a**) 50 °C (**b**) 100 °C (**c**) 150 °C (**d**) 200 °C
0.5 MPa CO_2 + 0.05 MPa O_2

Figure 4. Corrosion macroscopic morphology of N80 steel at different temperatures in O_2 and CO_2 environment.

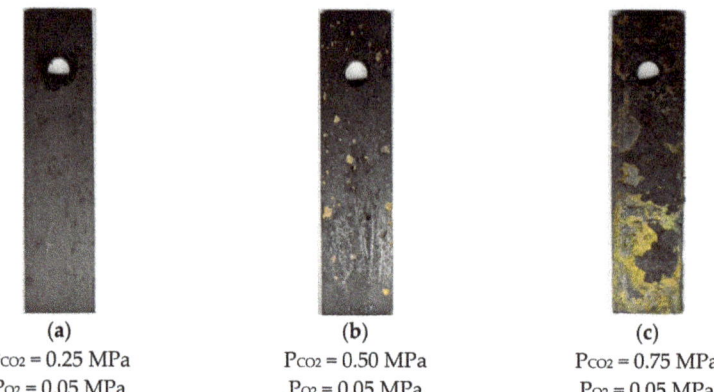

(a) P$_{CO2}$ = 0.25 MPa
P$_{O2}$ = 0.05 MPa

(b) P$_{CO2}$ = 0.50 MPa
P$_{O2}$ = 0.05 MPa

(c) P$_{CO2}$ = 0.75 MPa
P$_{O2}$ = 0.05 MPa

Figure 5. Corrosion macroscopic morphology of N80 steel at different CO_2 partial pressures (200 °C).

The SEM morphology and XRD test results of corrosion products of N80 steel in an environment of coexisting CO_2–O_2 at different temperatures are shown in Figure 6. The results show that the corrosion product film thickness is mainly composed of $FeCO_3$ and Fe_2O_3 at 50 °C. The granular products on the surface are wrapped in the corrosion layer at 100 °C, which causes the surface of the sample to be uneven. The corrosion products are mainly composed of $FeCO_3$, Fe_2O_3, and Fe_3O_4. Fe_2O_3 is oxidized again to Fe_3O_4 at 150 °C, which increases the particle attachment. When the temperature continues to increase to 200 °C, the product structure on the surface of the material is stable and dense.

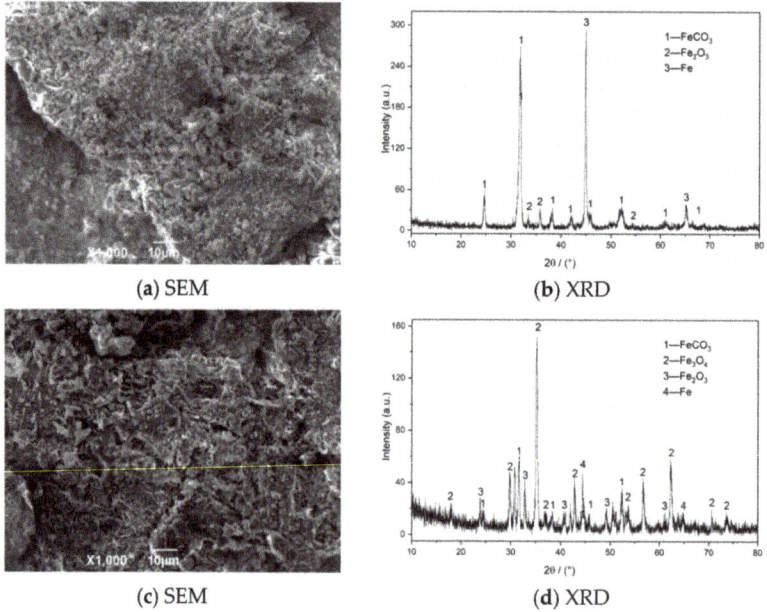

(a) SEM (b) XRD

(c) SEM (d) XRD

Figure 6. Cont.

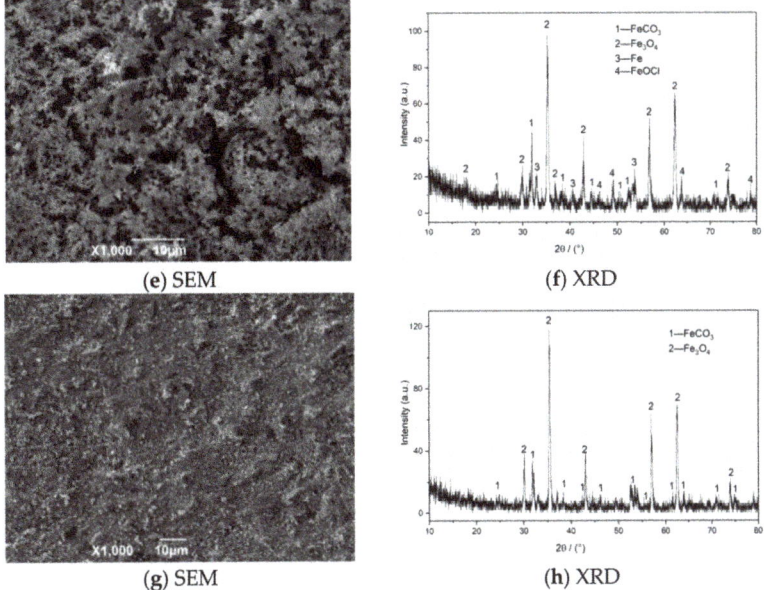

Figure 6. SEM morphology and XRD of N80 steel under different temperatures in environment of coexisting CO_2–O_2, (**a,b**) T = 50 °C, (**c,d**) T = 100 °C, (**e,f**) T = 150 °C, (**g,h**) T = 200 °C. P_{CO2} = 0.5 MPa, P_{O2} = 0.05 MPa.

The SEM morphology and XRD test results of corrosion products of N80 steel at different CO_2 partial pressures are shown in Figure 7. The SEM result with a CO_2 partial pressure of 0.25 MPa shows that the corrosion products on the surface layer are relatively thin and have not been connected to form an overall structure. When the partial pressure of CO_2 increases to 0.5 MPa, the product film falls off significantly, and large corrosion cracks appear on the surface. The iron oxides completely cover the original corrosion products to form a double-layer film structure on the surface in a high-concentration CO_2 environment (P_{CO2} = 0.75 MPa). This kind of membrane structure is not dense or unstable and is easily washed off by the solution, which causes the internal matrix to continue to corrode. The XRD measurement results of the corrosion product of the material show that the surface of the corroded material is covered with $FeCO_3$ grains and iron oxides because when the corrosive environment contains CO_2 and O_2, $FeCO_3$ and $Fe(OH)_2$ corrosion product films will appear on the initial surface, and then O_2 in the environment will oxidize with $Fe(OH)_2$ and $FeCO_3$ to form iron oxides [3,7]. The corrosion products on the final surface are mainly iron oxides, such as $FeCO_3$, Fe_2O_3, and Fe_3O_4. The outer layer of the formed double layer corrosion product film is mainly composed of FeCO3 and Fe oxides, while the inner layer is mainly composed of regular grained $FeCO_3$. Compared with a single CO_2 corrosion product, the appearance of oxides reduces the integrity of the corrosion product film and increases the corrosion rate of the material.

Figure 8 shows the surface structure after removing the outer layer of the attached product film. The corrosion surface is composed of small corrosion pits that are connected to uneven honeycomb areas at 50 °C. At 100 and 150 °C, the corroded surface is spliced by a large area of massive corrosion pits. The original corrosion products are composed of flaky structures due to the gradual expansion of the local fine pits to form a certain area, and the corrosion products gradually accumulate to form a layer. Under the action of liquid shearing force, the flaky corrosion products on the surface of the material fall off to form a large uneven area. Deep corrosion pits appear on the film surface at 150 °C, and a small amount of corrosion product is mixed in the corrosion pits. This result indicates that the flowing liquid penetrated into the material matrix through the cracks, which causes serious pitting corrosion in the internal surface. The surface of the material is relatively regular and smooth without evident pitting after corrosion at 200 °C, and the corrosion products are mainly flakes.

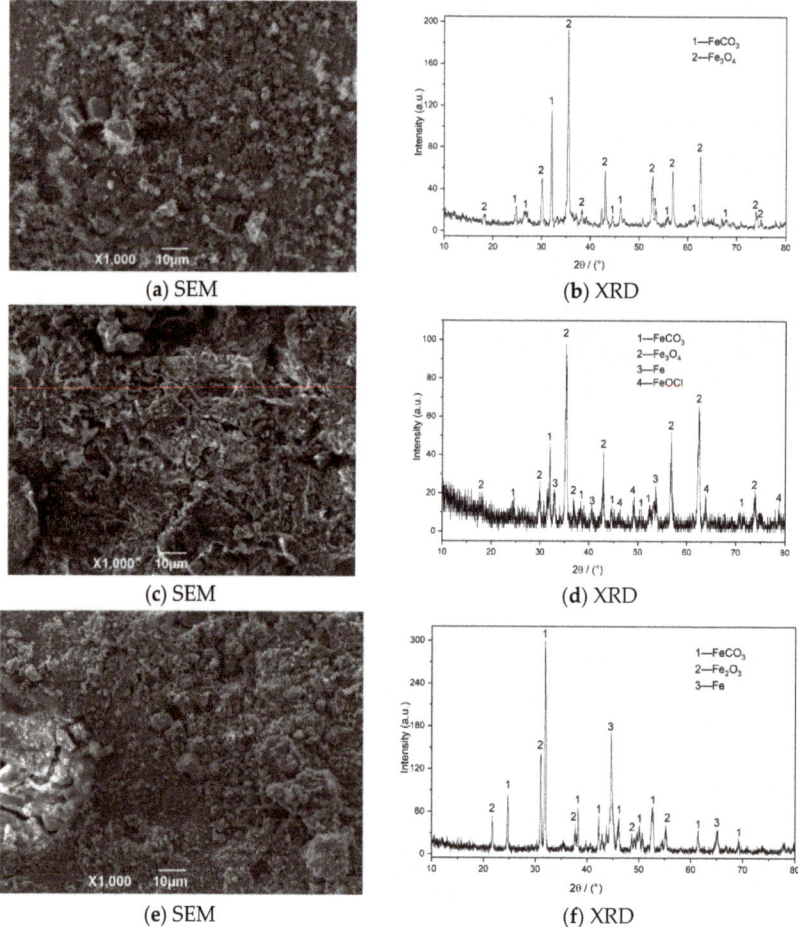

Figure 7. SEM morphology and XRD of N80 steel under different CO_2 partial pressures in environment of coexisting CO_2-O_2; (**a,b**) P_{CO2} = 0.25 MPa, (**c,d**) P_{CO2} = 0.50 MPa, (**e,f**) P_{CO2} = 0.75 MPa. P_{O2} = 0.05 MPa, T = 100 °C.

Figure 9 shows the cross-sectional morphology of the corrosion sample. According to different positions, the sample surface can be divided into an adhesion layer, corrosion layer, and matrix. Among these layers, the adhesion layer is the loosest and consists of crystalline particles or oxide. The corrosion layer is mainly formed by the accumulation of reaction intermediate products and dense corrosion product. The corrosion layer is not enough to prevent macromolecules and ions from contacting the metal surface, especially in flowing liquids, because it contains large pores. For the same reason, the corrosion layer is a porous medium, which cannot prevent Cl ions from contacting the substrate. Hence, when the temperature is in the range of 100 to 150 °C, serious continuous corrosion pits appear in the corrosion layer and extend to the depths of the matrix.

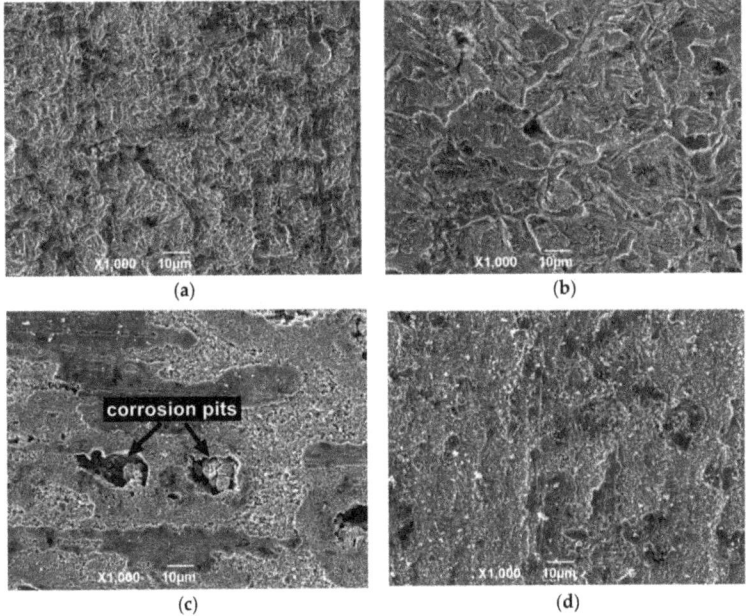

Figure 8. SEM morphology of N80 steel after removing corrosion product film; (**a**) T = 50 °C, (**b**) T = 100 °C, (**c**) T = 150 °C, (**d**) T = 200 °C, P_{CO2} = 0.50 MPa, and P_{O2} = 0.05 MPa.

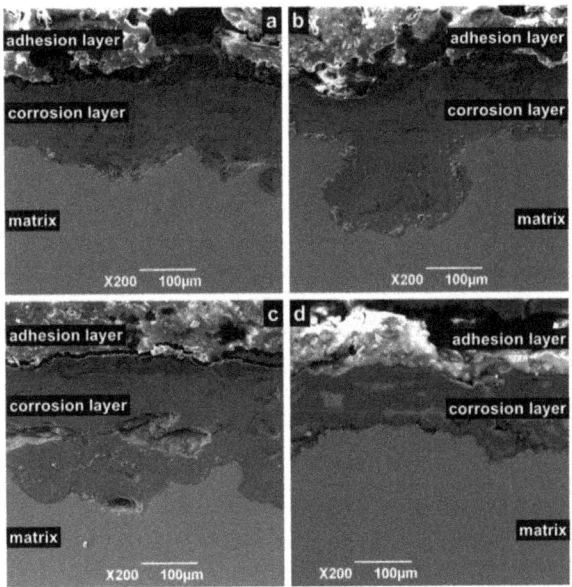

Figure 9. Cross-sectional SEM morphology of N80 steel at different temperatures in environment of coexisting CO_2–O_2; (**a**) T = 50 °C, (**b**) T = 100 °C, (**c**) T = 150 °C, (**d**) T = 200 °C, P_{CO2} = 0.50 MPa, and P_{O2} = 0.05 MPa.

4. Discussion

Given that the total pressure in this experiment is 5 MPa, according to the law of O_2 and CO_2 gas saturation pressure changing with temperature, the fluid in the kettle is in the form of gas–liquid coexistence at this experimental temperature. During the rotation of the sample, the surface of the material is subjected to the fluid shear force, which accelerates the rupture of the corrosion product film. Meanwhile, the rapid fluid flow accelerates the convective mass transfer of the gas in the boundary layer of the sample surface, thereby changing the corrosion process. According to the characteristics shown by the experimental results, the main factor affecting the corrosion of N80 in the high-temperature and high-pressure flowing media is the change in the reaction of O_2 or CO_2 under different flow velocities and temperatures.

4.1. Formation of Corrosive Environment Containing CO_2

The essence of the corrosion of CO_2 and steel materials is the electrochemical reaction of metals in carbonic acid solution. Although the carbonic acid produced by dissolving CO_2 in water is a dibasic acid, the acidity of carbonic acid is higher than that of hydrochloric acid under the same pH value, which will cause serious corrosion to steel materials. Nesic [7] divided the pH value of the solution as follows: pH < 4, 4 < pH < 6, and pH > 6. Nesic believes that $Fe(CO_3)OH$ intermediate products can be generated during the electrochemical reaction of the anode. Thus, the anode reaction is as follows:

$$\text{pH} < 4$$
$$H_3O^+ + e^- \rightarrow H + H_2O, \tag{1}$$

$$4 < \text{pH} < 6$$
$$H_2CO_3 + e^- \rightarrow H + HCO_3^-, \tag{2}$$

$$\text{pH} > 6$$
$$2HCO_3^- + 2e^- \rightarrow 2H_2 + 2CO_3^{2-}, \tag{3}$$

Moreover, the overall reaction equation is as follows:

$$Fe + CO_2 + H_2O \rightarrow FeCO_3 + H_2 \tag{4}$$

In this experiment, the pH of several groups of media is between 4 and 6. As shown in Figure 10, the CO_2 corrosion rate decreases with increasing temperature. Therefore, when the temperature is less than 100 °C, a H depolarization process exists in the electrochemical reaction. The process is completed by HCO_3^- and H^+, which are decomposed by H_2CO_3 in the solution. The higher the partial pressure of CO_2 is, the higher the solubility of H_2CO_3 is, and the higher the H^+ concentration produced by the hydrolysis of H_2CO_3 in the solution will be, thereby resulting in fast corrosion of the material [17,18]. However, in a high-temperature environment (T > 150 °C), the decrease in the solubility of CO_2 and increase in pH of the solution reduce the anode reaction rate. Therefore, the corrosion rate increases with the increase in CO_2 concentration at low temperature but decreases at high temperature.

4.2. Diffusion and Reaction of O_2 in CO_2 Solution

The results of comparing the corrosion rate of N80 steel with and without O_2 are shown in Figure 11. In an environment containing only CO_2 gas, the surface corrosion rate of N80 shows a law that first increases and then decreases. When the temperature reaches 200 °C, the corrosion rate decreases to <0.5 mm/a. In an environment containing only O_2 gas, the corrosion rate increases with the increase in temperature, and the maximum corrosion rate occurs at 200 °C. When O_2 is added to the CO_2 corrosive medium, the corrosion rate is higher than that of a single gas environment. The corrosion ratio obtained by comparison is shown in Figure 12. When the CO_2 partial pressure is less than 0.5 MPa, the corrosion rate change law is similar to that of a single CO_2 environment.

However, when the partial pressure of CO_2 is 0.75 MPa, the corrosion rate continues to increase with the increase in temperature, which indicates that the addition of O_2 significantly changes the CO_2 corrosion process in a high-concentration and high-temperature environment. Given that the flow velocity does not change, the forced convection flow in the boundary layer of the same phase medium is close. The results show that the corrosion contrast at different temperatures is significantly different, which indicates that the O_2 mass transfer law in the corrosion product film affects the electrochemical reaction of the substrate.

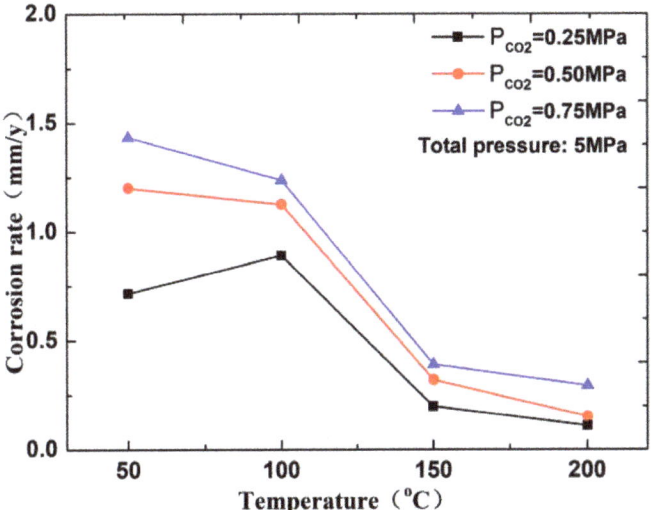

Figure 10. Corrosion rate of N80 steel at different temperatures in CO_2 environment.

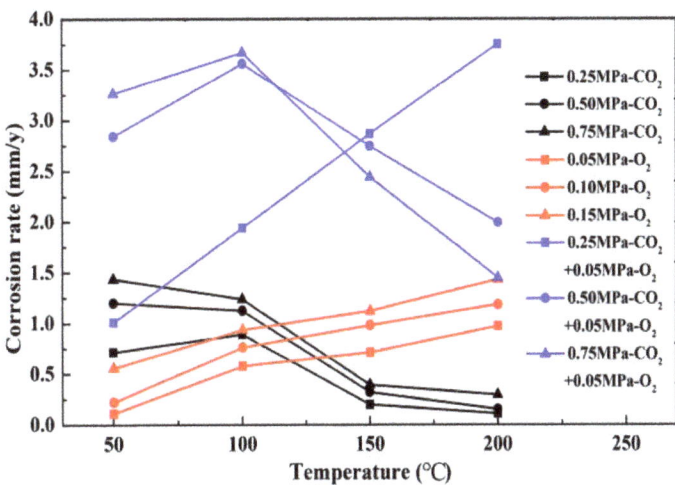

Figure 11. Comparison of corrosion rates of N80 steel in different media.

The diffusion and mass transfer of O_2 in the film restricts the rate of the depolarization reaction because the N80 surface forms a $FeCO_3$ film on the metal surface in the H_2CO_3 solution. In a steady-state mass transfer unit, $dC/dt = 0$, the total molar flux is defined by the Nernst–Plank equation, as follows [18]:

$$J = -D\frac{dC}{dx} - \frac{nFDC}{RT}\frac{df}{dx} + Cv = J_d + J_m + J_c. \tag{5}$$

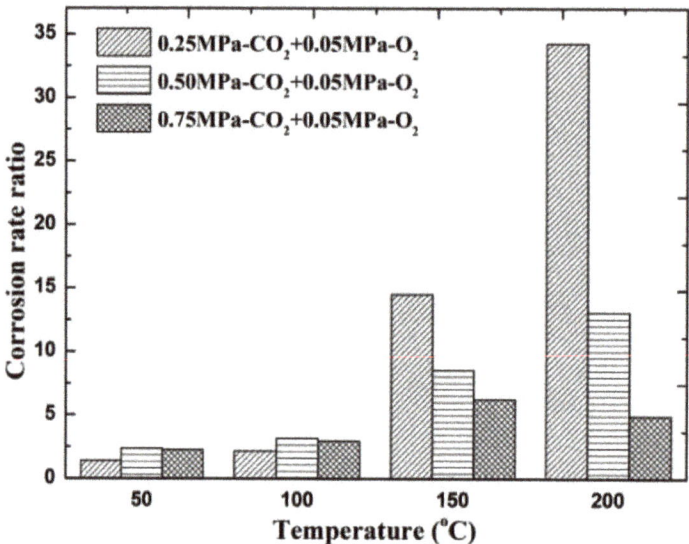

Figure 12. Comparison of corrosion rates of O_2 and O_2–CO_2 coexistence conditions.

Take the corrosion of N80 steel in water at 150 °C and 5 MPa pressure as an example. According to the first boundary condition of Fick's second law [19], the expression of reactant concentration at any position in the boundary layer is obtained as follows:

$$\frac{C^y - C^a}{C^w - C^a} = 1 - \mathrm{erf}\left(\frac{y}{\sqrt{4Dt}}\right), \tag{6}$$

$$C^y = C^w - (C^w - C^a)\,\mathrm{erf}\left(\frac{y}{\sqrt{4Dt}}\right), \tag{7}$$

where the erf(y) is the error function, and the O_2 concentration in the water is approximately equal to 0.044 mol/L [20]. According to the diffusion layer thickness formula $\delta = (\pi Dt)^{0.5}$, the calculated boundary layer thickness when O diffuses for 1 s is 0.08 mm. Therefore, three positions near the wall—that is, y = 0.05 mm, diffusion layer interface y = 0.1 mm, and outside the diffusion layer y = 0.15 mm—are taken to calculate the change in O concentration C^y. As shown in Figure 13, when the wall surface O concentration was 1/10 of the diffusion outer layer concentration, the O molecules participating in the electrochemical cathode reaction accounted for 90% of the total molecules. The O concentration in the diffusion layer (x = 0.05 mm) significantly decreases with time—that is, approximately 38% in 1 s. Meanwhile, the diffusion layer interface (x = 0.1 mm) only decreases by 10%. This result shows that in the film, within a short time of the formation of the electrochemical system, the concentration of the reactants forms a larger concentration gradient in the diffusion layer because the oxygen content that is involved in the reaction accounts for 90%, the electrochemical reaction is rapid, and the system is controlled by the reactant concentration.

According to the solubility of O_2 and CO_2 in water at different temperatures tested by Duan and Broden [20,21], as shown in Table 3, the solubility of CO_2 in water is 10- to 20-fold higher than that of O_2 at temperatures ranging from 50 to 200 °C. This result shows that CO_2 derived more reactants than O_2 in a high-temperature environment. However, under the coexistence of O_2 and CO_2, O_2 first combines with Fe^{2+} or Fe^{3+} to form $Fe(OH)_3$ (Equations (8) and (9)) because O_2 oxidizes more than H_2CO_3 and H^+. A part of Fe^{2+} can combine with HCO^{3-} to form $FeCO_3$ crystals (Equation (10)). In a high-temperature environment, $Fe(OH)_3$ and $FeCO_3$ are decomposed into Fe_2O_3, which has poor adhesion and poor ability to form a film, as shown in Equations (11)–(13).

$$\text{First reaction stage} \\ 4Fe^{2+} + 4H^+ + O_2 \rightarrow 4Fe^{3+} + 2H_2O, \tag{8}$$

$$\text{First reaction stage} \\ Fe^{3+} + 3H_2O \rightarrow Fe(OH)_3 + 3H^+, \tag{9}$$

$$\text{First reaction stage} \\ Fe^{2+} + HCO_3^- \rightarrow FeCO_3 + H^+ + 2e, \tag{10}$$

$$\text{Second reaction stage} \\ Fe(OH)_3 \rightarrow FeO(OH) + H_2O, \tag{11}$$

$$\text{Second reaction stage} \\ 2FeO(OH) \rightarrow Fe_2O_3 + H_2O, \tag{12}$$

$$\text{Second reaction stage} \\ 4FeCO_3 + O_2 \rightarrow 2Fe_2O_3 + 4CO_2. \tag{13}$$

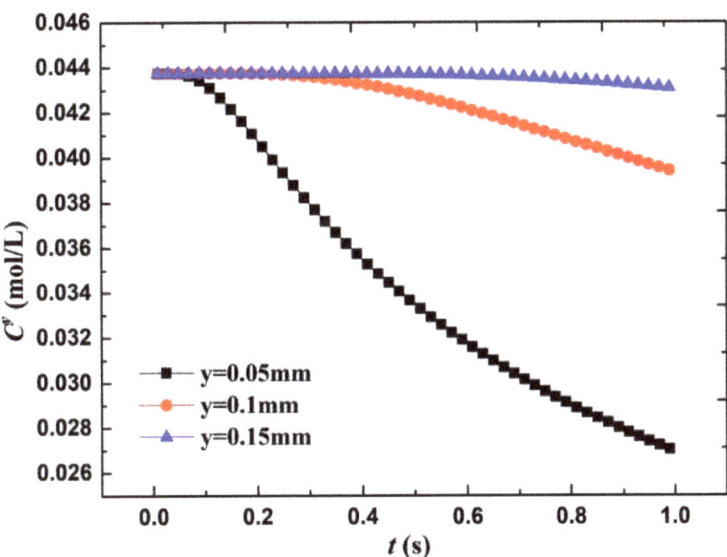

Figure 13. Change in O_2 concentration on material surface with reaction time.

Table 3. Solubility of O_2 and CO_2 in water under 5 MPa pressure (mol/L) [20,21].

Gas	T (°C) 50	60	90	100	120	130	150	210
O_2	0.044	—	—	0.036	—	0.038	0.044	—
CO_2	—	0.669	0.495	—	0.415	—	0.377	0.299

In addition to the effect on the priority of the reaction on the metal surface, the penetration of O_2 into the product film will also affect the surface structure. As shown in Figure 14, local reactions occurred inside the product film because the $FeCO_3$ film was not enough to isolate O_2 molecules, and the liquid flow made the product film thin. When O_2 molecules enter the film, they will react with $FeCO_3$ to form Fe_2O_3 and CO_2, and a small amount of H_2 will be generated by the partial H evolution reaction. When gas accumulates inside, it will swell inside the product film and eventually break the film. The flow of external liquid will also accelerate the rupture of the corrosion film.

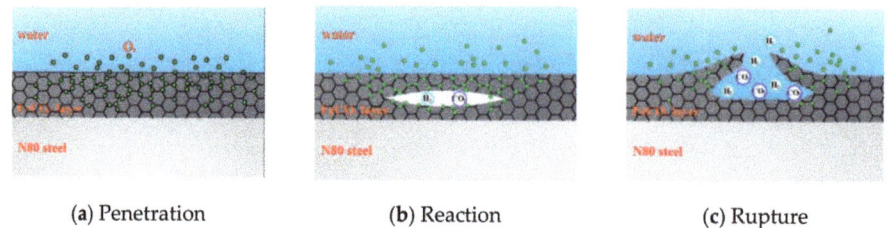

(a) Penetration (b) Reaction (c) Rupture

Figure 14. Schematic of corrosion in high-temperature environment of coexisting O_2 and CO_2.

4.3. Influence of Temperature

In addition to the influence of mass transfer, reaction priority, and reaction position on the corrosion of N80 in the coexisting environment, temperature also changes the progress and rate of several reactions. First, temperature will affect the corrosion process dominated by chemical reactions and the dissolution and diffusion of CO_2/O_2 in the solution. Second, temperature will affect the oxidation of Fe^{2+}, the deposition of corrosion product $FeCO_3$, and the formation of corrosion intermediate products. These factors increase the complexity of the corrosion process and surface characteristics of the material.

Temperature can affect the gas dissolution and diffusion rate and further affect the composition, compactness, and corrosion rate of the corrosion product film of N80 steel in the CO_2–O_2 environment. As shown in Figure 2, when the partial pressures of CO_2 are 0.50 and 0.75 MPa, and the temperature is less than 50 °C, the slow diffusion rate of O_2 causes the surface corrosion of the material to be dominated by CO_2 corrosion. When the temperature continues to increase and is less than 150 °C, part of the $FeCO_3$ film will be transformed into Fe_2O_3. At this time, the corrosion rate increases because Fe_2O_3 is loose and easily falls off. Meanwhile, due to changes in the structure of the corrosion product film, a part of the corrosion product film is easily penetrated by O_2 and Cl^-. Fe ions can also diffuse into the solution through the microporous structure on the corrosion product film, which accelerates the dissolution of iron. When the temperature is higher than 150 °C, several iron oxides become dense and protective, and the corrosion rate will decrease.

5. Conclusions

To study the corrosion law of tubing materials in HPHT flowing media, corrosion experiments were carried out in a test liquid containing O_2 and CO_2. According to the corrosion rate, corrosion morphology, and product composition obtained by the experiments, the following conclusions can be drawn.

(1) When the CO_2 concentration was low in the coexistence environment, the corrosion rate of N80 was affected by O_2 and continued to increase with increasing temperature. When the CO_2 concentration increased (when the volume fraction was higher than 10%), the corrosion rate had a maximum value near 100 °C.

(2) O_2 and temperature affected the changes in intermediate products and film structure of the CO_2 corrosion. The oxide formed by the reaction of O_2 and Fe will cover the surface of regular $FeCO_3$

grains, which may destroy the original single structure of the product film and cause the surface film to rupture or form cracks. The temperature increase will also accelerate the rupture process.

Author Contributions: Funding acquisition, Y.D.; Investigation, Y.D. and J.C.; Writing—original draft preparation, Z.L. and J.C.; Experiment performing, J.C. and Y.Z.; Writing—review and editing, J.C. All authors have read and agreed to the published version of the manuscript.

Funding: This research was funded by National Natural Science Foundation of China (grant no. 51674199 and 51974246). The APC was funded by and supported by the Basic Research Program of Natural Science of Shaanxi Province Effect (grant no. 2019JQ-809).

Acknowledgments: This work was supported by the Institute of Safety Evaluation and Control of Completion Test System.

Conflicts of Interest: The authors declare no conflict of interest.

References

1. Singh, A.; Lin, Y.H.; Ebenso, E.E.; Liu, W.Y.; Pan, J.; Huang, B. Gingko biloba fruit extract as an eco-friendly corrosion inhibitor for J55 steel in CO_2 saturated 3.5% NaCl solution. *J. Ind. Eng. Chem.* **2015**, *24*, 219–228. [CrossRef]
2. Fu, A.Q.; Geng, L.Y.; Li, G.; Cai, R.; Li, G.S.; Li, G.F.; Bai, Z.Q. Corrosion failure analysis of an oil tube used in a western oilfield. *Corros. Prot.* **2013**, *34*, 645–648.
3. Linter, B.R.; Burstein, G.T. Reactions of pipeline steels in carbon dioxide solutions. *Corros. Sci.* **1999**, *41*, 117–139. [CrossRef]
4. Bai, H.T.; Wang, Y.Q.; Ma, Y.; Zhang, Q.B.; Zhang, N.S. Effect of CO_2 partial pressure on the corrosion behavior of J55 carbon steel in 30% crude oil/brine mixture. *Materials* **2018**, *11*, 1765–1773. [CrossRef] [PubMed]
5. Zhang, G.A.; Liu, D.; Li, Y.Z.; Guo, X.P. Corrosion behavior of N80 carbon steel in formation water under dynamic supercritical CO_2 condition. *Corros. Sci.* **2017**, *120*, 107–120. [CrossRef]
6. De Waard, C.; Milliams, D.E. Carbonic acid corrosion of steel. *Corrosion* **1975**, *31*, 177–181. [CrossRef]
7. Nesic, S.; Postlethwaite, J.; Olsen, S. An electrochemical model for prediction of corrosion of mild steel in aqueous carbon dioxide solutions. *Corrosion* **1996**, *52*, 280–294. [CrossRef]
8. Ha, H.M.; Gadala, I.M.; Alfantazi, A. Hydrogen evolution and absorption in an API-X100 line pipe steel exposed to near-neutral pH solutions. *Electrochim. Acta* **2016**, *204*, 18–30. [CrossRef]
9. Choi, Y.S.; Farelas, F.; Nesic, S.; Magalhaes, C.D.; Azevedo, A. Corrosion behavior of deep water oil production tubing material under supercritical CO_2 environment: Part 1-effect of pressure and temperature. *Corrosion* **2013**, *70*, 38–47. [CrossRef]
10. Patrick, J.M.; Paul, M.N. *Corrosion and Corrosion Control*; Wiley Press: Hoboken, NJ, USA, 1985; pp. 105–116.
11. Huang, X.B.; Yin, Z.F.; Li, H.L.; Bai, Z.Q.; Zhao, W.Z. Corrosion of N80 tubing steel in brine at 1.2MPa CO_2 containing trace amounts of H_2S. *Corros. Eng. Sci. Technol.* **2013**, *47*, 78–83. [CrossRef]
12. Nesic, S.; Postlethwaite, J.; Vrhovac, M. CO_2 corrosion of carbon steel—From mechanistic to empirical modelling. *Corros. Rev.* **1997**, *15*, 1–2. [CrossRef]
13. Bai, Z.Q.; Li, H.L.; Liu, D.X.; Wang, X.F. Corrosion factors of N80 steel in simulated H_2S/CO_2 environment. *Mater. Prot.* **2003**, *36*, 32–34.
14. Wang, G.J.; Liu, J.S.; Zhang, J. Corrosion behaviors of three metal materials in autoclave containing saline medium. *Corros. Prot.* **2010**, *31*, 167–169.
15. Zhang, Q.; Li, A.Q.; Wen, J.B.; Bai, Z.Q. Effect of CO_2 Partial Pressure on CO_2/H_2S Corrosion of Oil Tube Steel. *J. Iron Steel Res.* **2004**, *16*, 72–74.
16. He, W.; Knudsen, O.O.; Diplas, S. Corrosion of stainless steel 316L in simulated formation water environment with CO_2-H_2S-Cl. *Corros. Sci.* **2009**, *51*, 2811–2819. [CrossRef]
17. Jiang, X.; Zheng, Y.G.; Ke, W. Effect of flow velocity and entrained sand on inhibition performances of two inhibitors for CO_2 corrosion of N80 steel in 3% NaCl solution. *Corros. Sci.* **2005**, *47*, 2636–2658. [CrossRef]
18. Hamann, C.H.; Hamnett, A.; Vielstich, W. *Electrochemistry*; Wiley-VCH: Weinheim, German, 2007.
19. Kays, W.M.; Crawford, M.E.; Weigand, B. *Convective Heat and Mass Transfer*; McGraw-Hill: New York, NY, USA, 2007.

20. Duan, Z.; Sun, R. An improved model calculating CO_2 solubility in pure water and aqueous NaCl solutions from 273 to 533 K and from 0 to 2000 bar. *Chem. Geol.* **2003**, *193*, 257–271. [CrossRef]
21. Kolev, N. Solubility of O_2, N_2, H_2 and CO_2 in water. In *Multiphase Flow Dynamics 3*; Springer: Berlin/Heidelberg, Germany, 2011.

Publisher's Note: MDPI stays neutral with regard to jurisdictional claims in published maps and institutional affiliations.

 © 2020 by the authors. Licensee MDPI, Basel, Switzerland. This article is an open access article distributed under the terms and conditions of the Creative Commons Attribution (CC BY) license (http://creativecommons.org/licenses/by/4.0/).

Article

The Corrosion Features of Q235B Steel under Immersion Test and Electrochemical Measurements in Desulfurization Solution

Peng Gong, Guangxu Zhang and Jian Chen *

School of Chemistry, Chemical Engineering and Life Science, Wuhan University of Technology, Wuhan 430070, China; gpeng2018@163.com (P.G.); Zhanggx2002@163.com (G.Z.)
* Correspondence: chenjian0501@whut.edu.cn

Received: 9 July 2020; Accepted: 20 August 2020; Published: 27 August 2020

Abstract: With the continuous tightening marine diesel engines emission standards, removing sulfur oxides (SO_X) by sodium hydroxide solution absorption is a highly efficiency and economic method, which has been a hot area of research. The ensuing desulfurization solution is a new corrosive system, the aim of this paper is to ascertain the corrosion feature of Q235B steel in desulfurization solution, which lays a theoretical foundation for industrialization. For this purpose, mass loss, electrochemical techniques and surface analyses were applied. The results of mass loss highlight a reduction in the corrosion rate with 35 days of immersion. Higher exposure time increased the compactness of the corrosion product layer and changed phase composition. These conclusions are supported by surface analyses, such as X-ray diffraction and scanning electron microscope. However, electrochemical results showed that the polarization resistance R_p was fluctuant. Both of R_p and charge transfer resistance R_t reach a maximum after immersing 21 days. In addition, although the sediments attached to the steel surface could inhibit corrosion, pitting corrosion aggravated by hydrolyzation of $FeSO_4$ should be given more attention.

Keywords: Q235B steel; desulfurization solution; corrosion mechanism; pitting corrosion

1. Introduction

As human beings taking the environment into consideration severely, the effective disposal of sulfur oxides (SO_X) in off-gas from marine diesel engine has been a worldwide puzzle and hot spots of research. Moreover, international Maritime Organization (IMO) had announced a stricter emission standard for SO_X in the regulation 14 [1]. In order to minimize the costs of tail gas up-to-standard discharge, it is an alternative method of reducing SO_X emissions by sodium hydroxide solution absorption [2]. Further, the technique has many advantages, such as high desulfurization efficiency (≥98%), no secondary pollution, sodium sulfate by-product as an industrial chemical, etc. At present, the chemical absorption mechanism, transfer mechanism and theoretical calculation approaches have been explored [3–5]. The desulfurization solution is a new corrosion system, therefore, the erosion problems of steel in desulfurization solution need to be investigated systematically for industrialization application early.

Ordinary carbon steels typical such as Q235B steels are being used as one of the main materials in China due to the shortage of resources and the consideration of economy. Therefore, it should give preference to Q235B steel in sodium hydroxide desulfurization system. However, it tends to rust when exposed to wet air, saltwater and other corrosive substances [6]. Corrosion would result in uneven steel surface, decreased thickness and deteriorated mechanical properties, further leading to perforations of pipelines and equipment failure. Until now, many studies have issued the corrosion mechanism and impacting factors of Q235B steel in different corrosive mediators. Cheng et al. [7] investigated the corrosion behavior of Q235B carbon steel in sediment from crude oil and found that corrosion pits were initiated under the scale deposits. Yu et al. [8] researched the atmospheric corrosion of Q235 steel in Turpan, indicating that the corrosion rate was 20 g·m^{-2}·a^{-1} and the corrosion products were composed of α-FeOOH, γ-FeOOH, Fe_3O_4, $Fe(OH)_3$. Sulfate ions is the most common ionic forms in desulfurization solution. Liu et al. [9] and Boah et al. [10] obtained the consistent conclusion that sulfate ion was even more corrosive than chloride ion. Xu et al. [11] demonstrated that sodium sulfate was harmful to the stability of the passive film. Interestingly, the presence of a larger amount of sulfate ions even inhibited the nucleation of the pitting of the steel. Whereas tail gas composition is complex, it is not hard to fathom that desulfurization solution would contain large amounts of metals, non-metals ions and organic substance. Great important should therefore be attached to give more insights into the corrosion feature of Q235B steel in desulfurization solution. Considering the requirement of industrialized application, the main objective of this paper was to research the corrosion behavior of Q235B steel in the desulfurization solution by open circuit potential (OCP), electrochemical impedance spectroscopy (EIS), polarization curves, scanning electron microscopy (SEM), X-ray diffraction (XRD), and mass loss. In addition, the corrosion mechanism was also discussed.

2. Experiment

2.1. Desulfurization Solution Analysis

Desulfurization solution was collected from an outdoor tank in Shanghai. pH was measured by a pH meter (PHS-3C, Shanghai Inesa Scientific Instrument Co., Ltd., Shanghai, China). Afterwards, desulfurization solution stored in a polystyrene vessel was sent to the lab for testing and chemical component analyses as soon as possible, the whole process was consistent with GB/T 5750-2006. An inductively coupled plasma-optical emission spectrometer (Prodigy 7) and an ion chromatograph (ICS-6000) were used to analyze its composition, in accordance with GB/T 8538-2016.

2.2. Immersion Test

Fifteen specimens with dimensions of 20 mm × 20 mm × 2 mm, were prepared for the immersion test at an ambient temperature. Prior to test, the specimens were ground smooth with emery papers ranging from grades 600 to 3000 to be up to the mustard of the corrosion tests. The specimens were then degreased with ethanol and dried in cool air. The initial mass and surface area of each specimen were measured by an electronic balance (BS110S) and a slide caliper (TM004), severally. The specimens were immersed in desulfurization solution for 7, 14, 21, 28 and 35 days. The specimens were retrieved at a scheduled time and immersed in a mixed solution (500 mL deionized water +3.5 g hexamethylenetetramine +500 mL 36% hydrochloric acid) vigorously for 30 s, followed by rinsing with water, dried with ethanol, and then weighted. Triplicate samples were taken for the measurements of the final mass after pickling. The corrosion rate of the specimen was calculated as follows:

$$V_{\text{corr}} = \frac{(M_0 - M_1) \times 3650}{\rho s t} \quad (1)$$

where M_0 is the initial mass of the specimen; M_1 is the final mass of the specimen after pickling; ρ is the density of Q235B steel; s is the surface area; and t is the immersion time. The element composition of Q235B steel manufactured by Sougang Mine Co (Qianan, China) is shown in Table 1.

Table 1. Chemical element composition of Q235B steel (wt%).

C	Al	Si	P	S	Mn	Fe
0.15	0.184	0.128	0.017	0.013	0.218	bal

2.3. Electrochemical Measurements

The measurements of the open circuit potential (OCP), electrochemical impedance spectroscopy (EIS), and polarization curves were performed on an electrochemical workstation (CS350H, Wuhan Corrtest Instrument Co., Ltd., Wuhan, China) conducted with a classical three-electrode cell (250 mL) at ambient temperature. The counter electrode was platinum wire electrode, and the reference electrode was Hg/Hg$_2$SO$_4$ electrode (MSE) connected to the cell via a Luggin capillary, which was filled with saturated potassium sulfate solution and 2% pure agar, all potentials were referred to it. The working electrode was Q235B steel, embedded in tetrafluoroethylene with an exposed area of 0.196 cm^2. The working electrode was disposed the same as immersion test samples before the test itself began in order to ensure reliability of figures.

The EIS measurement was carried out with a perturbation signal of 10 mV AC potential versus the OCP in a frequency range from 10^5 Hz to 10^{-2} Hz. The expectant data of EIS were obtained with a stabilized OCP and fitted with a suitable circuit model by a fit software named Zview. The polarization curves measurement was taken by changing the electrode potential automatically at a range from −200 mV to 150 mV vs. OCP with scan rate of 0.167 mV·s^{-1}. In addition, three parallel tests were carried out and the representative value was reported.

2.4. Morphologies and Component Analysis

The surface morphology of the samples was carried out using scanning electron microscopy (SEM, JSM-IT300, Japan Electronics Co. LTD, Tokyo, Japan). The phase composition of corrosion products was analyzed by X-ray diffraction (XRD), XRD was carried out using RU-200B (Rigaku Corporation, Tokyo, Japan), with Cu-target, a tube voltage of 40 kV, a tube current of 30 mA, scanning range from 5° to 70°, and scanning step size of 2°/min.

3. Results and Discussion

3.1. Composition of the Desulfurization Solution

The analysis result of the desulfurization solution was shown in Table 2. The sample was analyzed for component cations and anions, such as SO$_4^{2-}$, SO$_3^{2-}$, Na$^+$, K$^+$, Zn^{2+} and so on. They are clearly believed to influence the corrosion process of metal, usually bringing about serious damage of the pipeline and equipment [12]. SO$_4^{2-}$ ions and Na$^+$ ions were the highest concentrations of anion and cation, respectively, at 15,300.3 mg L^{-1} and 2785.5 mg L^{-1}. The desulfurization solution was a mildly acidic medium with pH 6.95 ± 0.08. Additionally, there were some organics and insoluble solids in desulfurization solution, and the organic phase (PAHs) also has an effect on corrosion but only slightly.

Table 2. The composition and physicochemical properties of the desulfurization solution.

Content	Chemical Formula	Value
	Na$^+$	2785.5
	K$^+$	540.7
	Mg^{2+}	4.7
	Ca^{2+}	26.2
	Zn^{2+}	90.2
Ionic concentration (mg L^{-1})	SO$_4^{2-}$	15,300.3
	SO$_3^{2-}$	9580.5
	NH$_4^+$	224.6
	NO$_3^-$	3.4
	NO$_2^-$	0.3
Organics concentration (mg L^{-1})	PAHs	158.7
Suspended solids concentration (mg L^{-1})		2519.5
pH		6.95 ± 0.08

3.2. Corrosion Morphology

After samples were immersed for 7 and 35 days, its morphologies showed obvious difference. The metallic luster of the samples surface was gradually lost and visible corrosion became more serious. As shown in Figure 1, for 7 days of immersion, general corrosion occurred on the steel surface, flocculent laurel-green precipitates were concatenated and scattered sporadically on the samples surface. After 35 days of immersion, the precipitates cover areas enlarged and thickened. The corresponding SEM photographs showed in Figure 1 after the rust was removed by acid pickling. It is obvious that the pit corrosion was slight in 7 days and subsequently performed increasingly serious with later 28 days of immersion.

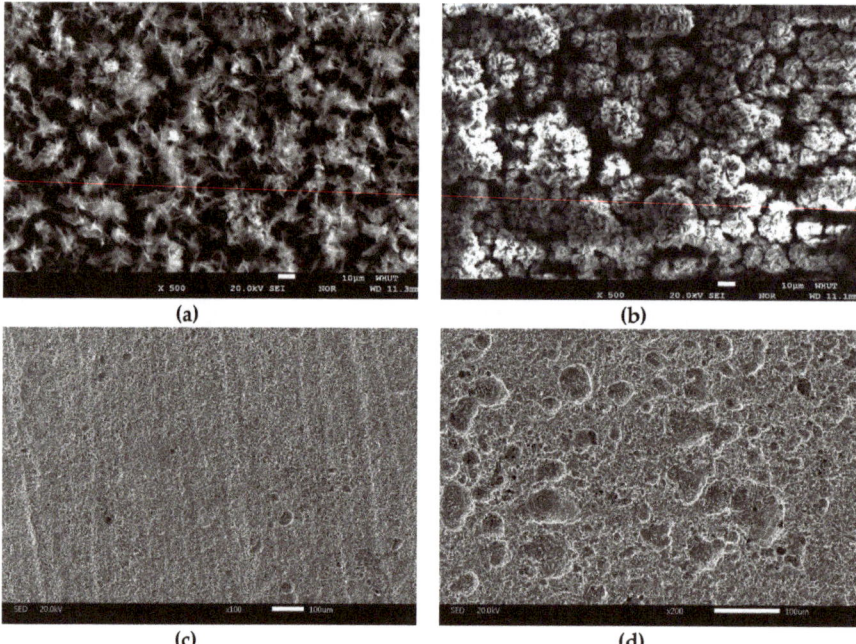

Figure 1. SEM macro-morphologies characteristics of Q235B steel in desulfurization solution with 7 days (**a**) and 35 days (**b**); The corresponding SEM macro-morphologies characteristics after the corrosion products were removed by acid pickling for 7 days (**c**) and 35 days (**d**).

3.3. Corrosion Products Analysis

The XRD spectra of precipitates on the sample surface for 7 days and 35 days of immersion were shown in Figure 2. The results reveal that the primary corrosion products mainly consisted $Fe(OH)_2$ (JCPDS3-903), $Fe(OH)_3$ (JCPDS38-32) and $Fe_2O_3 \cdot H_2O$ (JCPDS13-92). After immersing for 35 days, corrosion film became thicker and more compact (Figure 1a,b), the end corrosion products were mainly composed of $Fe_2O_3 \cdot H_2O$ (JCPDS13-92), $FeSO_4 \cdot 4H_2O$ (JCPDS81-19) and Fe_3O_4 (JCPDS3-862). The phases changed enormously and are different from that of the steel in pure Na_2SO_4 solution [13].

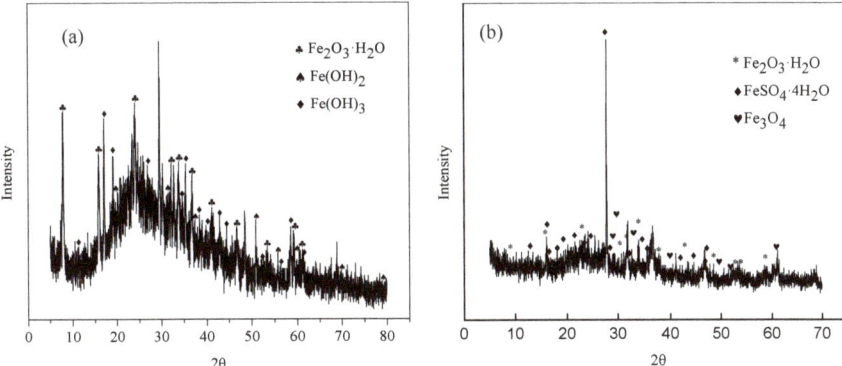

Figure 2. X-ray diffraction (XRD) analysis of corrosion products of Q235B steel in desulfurization solution with different immersion time (**a**) 7 days and (**b**) 35 days.

3.4. Mass Loss and Corrosion Rate

The relationship between mass loss and immersion time was presented in Figure 3. The mass loss of the samples increased continuously with prolonging the immersion time. It indicated that the samples were incessantly subjected to corrosion. Table 3 listed the time dependence of the corrosion rate (ΔV_{corr}) reckoned from the mass loss for 35 days. The (ΔV_{corr}) reduced sharply in the first 21 days of immersion. It had been reported that compact corrosion products film such as Fe_3O_4 on the matrix surface could form an effective anticorrosive film [14]. Therefore, the corrosive ion in desulfurization solution would not contact accessibly with the matrix and the (ΔV_{corr}) decreased. However, corrosion pits were found on the matrix surface after the corrosion products were removed (Figure 1c,d). There were a few pitting holes dispersing on the surface after 7 days of immersion, while the local corrosion became severe with time. The variation of (ΔV_{corr}) back up the conclusion. In the meantime, pH of the corrosion electrolyte decreased with time. It was positively associated with extent local corrosion and negatively correlated with mass loss. It could be inferred that local corrosion was dominant in the later stage of immersion, and there were corrosion reactions forming hydrogen ions.

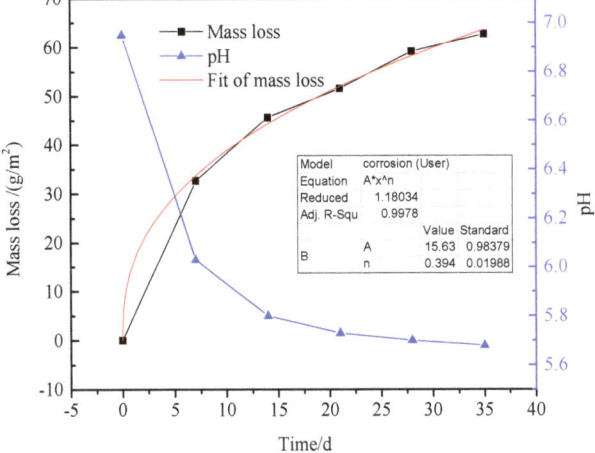

Figure 3. Curve of mass loss and pH with time of Q235B steel in desulfurization solution.

Table 3. Corrosion rate of Q235B steel in desulfurization solution.

Time/d	7	14	21	28	35
V_{corr}/(mm/a)	0.22	0.15	0.11	0.10	0.08
ΔV_{corr}/(mm/a)	—	0.07	0.04	0.01	0.02

3.5. Open Circuit Potential Measurements

The relationships between OCP and immersion time of Q235B steel in desulfurization solution were shown in Figure 4. The OCP decreased sharply from −0.018 V to −1.111 V with time at the first day. Such a reduction was caused by an accelerated anodic reaction rate, according to mixed potential theory [15]. Three days later, the OCP augment a little and then renewed to reduce. The augment of the OCP attributed to the suppression of anodic reaction which might be induced by the accumulation of precipitates (Figure 1a,b) [16]. Whereas corrosion products began to form on the steel surface as the OCP continued to decline monotonically. It indicated that the gradually decreased of the OCP was caused by the adsorption for anion and an accelerated anodic reaction, due to the local corrosion (Figure 1c,d) [17]. Additionally, on subsequent days, the OCP reach a comparatively steady-state value (−1.127 V). This phenomenon manifested that cathodic reaction and anodic reaction achieved a balance [18].

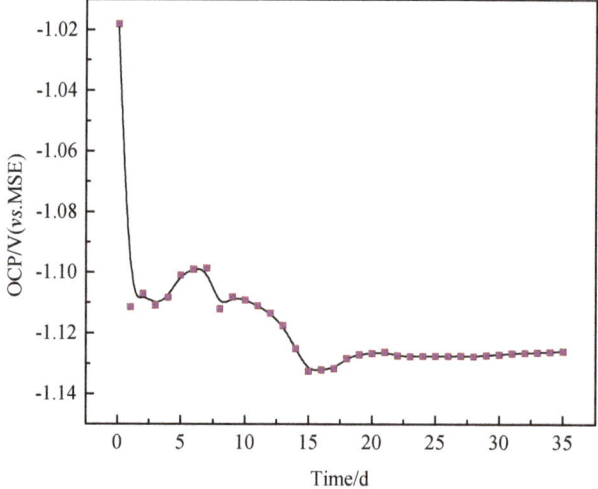

Figure 4. The trend of open circuit potential (OCP) for Q235B steel in desulfurization solution.

3.6. Electrochemical Impedance Spectroscopy

The EIS was utilized to research interface reaction ability and the electrons migration in the corrosion products film [19]. The results of EIS measurements of Q235B steel at OCP in desulfurization solution with different immersion time were presented in Figure 5, containing Nyquist plots and Bode plots. The Nyquist spectrum (Figure 5a) indicated a single narrow capacitance loop for all specimens, which manifested that the electrode was not a pure capacitor [20,21]. It was likely related to the compactness and distribution of precipitates. Additionally, the diameter of the capacitive semicircle increased with time, which indicated the improvement of corrosion resistance. In Bode plots (Figure 5c), the maximum phase angle values approached about 60°, which demonstrated that the corrosion products film was porous [22]. Figure 5b revealed that there was only one time constant under the range of the frequency measured. Consequently, the one-time constant equivalent electrical circuits $R_s(R_tQ_t)$, as shown in Figure 6, is suitable to fit the experimental data. In the electrical analog circuits, R_s represents the resistance of solution, and R_t corresponds to the charge transfer resistance, linked to the resistivity properties of the passive film. Moreover, Q_t is a constant phase element (CPE) related to the dispersion of a double layer capacitance of the corrosion product layer [23], which is used to compensate for non-homogeneity in the electrochemical system.

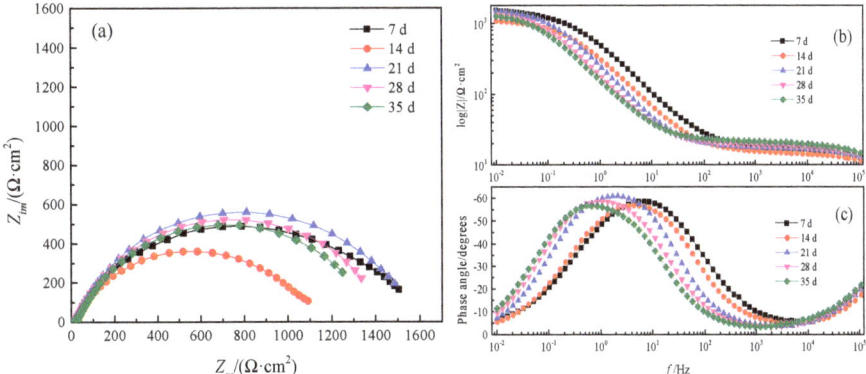

Figure 5. Electrochemical impedance spectra (EIS) of Q235B steel in desulfurization solution with 7 days, 14 days, 21 days, 28 days and 35 days, respectively; (**a**): Nyquist plot, (**b**): Bode magnitude, (**c**): Phase angle plot.

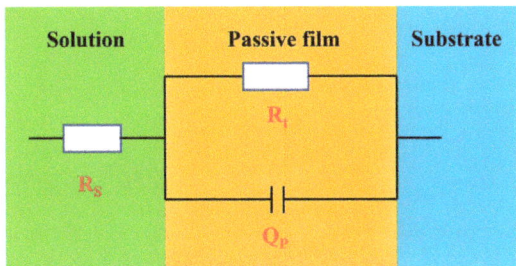

Figure 6. Equivalent circuits $R_s(R_tQ_t)$ used in the fitting procedure of the EIS experimental data.

The corresponding fitting results are listed in Table 4. R_s continuously increased with immersion time, it indicated that there was slight change for solution property in different immersion stage [24]. A decrease in pH value represents an increase of ion concentration (Figure 3), which leads to increased conductivity of solution. R_t value increased first and then decreased, reaching the maximum value on the 21st day. Theoretically, R_t was only determined by the charge-transfer-controlled corrosion and inversely proportional to V_{corr} [25]. Thus, the change of R_t demonstrated that V_{corr} decreased with time for 21 days and then increased after 21 days. It is probably because corrosion products with higher electrochemical activity might participate in cathodic reaction [26], weakening the protection of the corrosion products film, or corrosion products might fall off from the matrix surface.

Table 4. Parameters of equivalent circuits obtained by fitting the experimental results of EIS.

Time/d	$R_s/(\Omega\cdot cm^2)$	$Q_{dl}/(F\cdot cm^{-2})$	n	$R_t/(\Omega\cdot cm^2)$
7	14.03 ± 0.15	(4.48 ± 0.03) × 10^{-4}	0.7474 ± 0.008	1092 ± 2.6
14	15.68 ± 0.13	(7.82 ± 0.05) × 10^{-4}	0.7588 ± 0.005	1224 ± 3.1
21	16.63 ± 0.11	(9.01 ± 0.07) × 10^{-4}	0.7988 ± 0.003	1545 ± 1.8
28	17.77 ± 0.13	(1.29 ± 0.08) × 10^{-4}	0.7784 ± 0.005	1469 ± 2.5
35	20.01 ± 0.09	(1.61 ± 0.05) × 10^{-4}	0.7619 ± 0.006	1430 ± 2.7

3.7. Polarization Curves

Figure 7 showed the polarization curves obtained from Q235B steel in desulfurization solution with different immersion time. As shown in Figure 7, the shape of the anodic and cathodic was similar over the potential domain tested for both samples. In general, a continuous augment in the current density of the cathodic branches was noted as the potential decreased. Because the cathodic branches represented the hydrogen evolution, such as the augment of hydroxide ion [27]. Anodic branches shifted to a great current region, appearing the similar tendency as that observed in cathodic branches. This was due to the cause that the declined pH of the examined solution (Figure 2) caused more severe acidic environments and, with increasing the measurement time, so either transformation of the corrosion products composition (Figure 2) or physical structure of the passive film changed (Figure 1) [28].

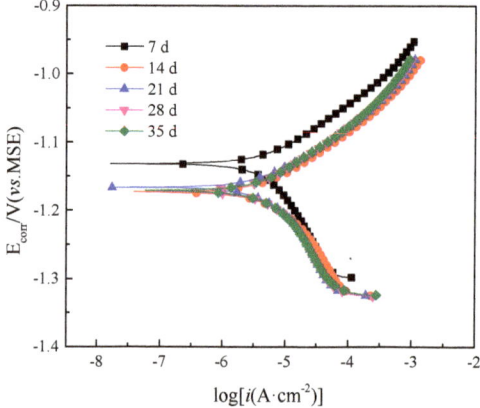

Figure 7. Polarization curves of Q235B steel in desulfurization solution with different immersion time.

The corresponding electrochemical parameters are presented in Table 5, where E_{corr} is the corrosion potential, i_{corr} (reckoned from intercept of anodic and cathodic Tafel curves) is the corrosion current density, R_p is the polarization resistance. From this table, the E_{corr} decreased rapidly with time in the first 14 days of immersion and then slightly increased from 14 days but subsequently decreased from 21 days to 35 days. The i_{corr} was approximately correlated with the V_{corr} and had inverse change with the E_{corr}. The fact illustrated that corrosion products film had no protective effect in the first 14 days. However, it decreased the corrosion rate from 14 days to 21 days. The increase of i_{corr} also indicated that the steel matrix was suffered from severe corrosion at time from 21 days to 35 days.

Table 5. Parameters of polarization curves with different immersion time.

	7 days	14 days	21 days	28 days	35 days
E_{corr} V	−1.133	−1.174	−1.167	−1.172	−1.171
i_{corr} μA/cm²	7.231	8.074	6.690	7.532	7.459
R_p Ω·cm²	3607	3231	3899	3497	3463

On comparative evaluation, the change law of the corrosion rate obtaining from polarization curves was quite different from that of mass loss. As was shown in Table 3, the V_{corr} reduced gradually with 35 days of immersion. This was due to the different principle of measurement. For immersion test, the calculated V_{corr} was in fact average value, which should be regarded as uniform corrosion rate. On the basis of NACE Standard RP0775-2005, the V_{corr} of immersing 35 days was 0.08 mm·a^{-1}, which could be accepted. Nevertheless, the V_{corr} presented in electrochemical measurement was a state variable, which was close to pitting corrosion rate. It could cause more serious damage and thus should be given more attention.

3.8. The Corrosion Mechanism of Q235B Steel

In oxygenated electrolyte solution, anodes and cathodes would separate [29]. The initial anodic and cathodic reactions of Q235B steel in desulfurization solution are primarily presented as follows:

$$Fe \rightarrow Fe^{2+} + 2e^- \tag{R1}$$

$$O_2 + 4e^- + 2H_2O \rightarrow 4OH^- \tag{R2}$$

Reactions (R1) and (R2) refer to the anodic and cathodic reaction, respectively. $Fe(OH)_2$ was formed in the process of ionic migration according to chemical reaction (R3). Further, $Fe(OH)_2$ could be produced from the hydrolysis of Fe^{2+} ions in the anodic region, reaction (R4). Thermodynamic and $\varphi_{(MSE)}$-pH diagrams predicted that $Fe(OH)_2$ was not stable under such conditions. It is easy to be oxidized to $Fe(OH)_3$ according to chemical reaction (R5).

$$Fe^{2+} + 2OH^- \rightarrow Fe(OH)_2 \downarrow \tag{R3}$$

$$Fe^{2+} + 2H_2O \rightarrow Fe(OH)_2 + 2H^+ \tag{R4}$$

$$4Fe(OH)_2 + O_2 + 2H_2O \rightarrow 4Fe(OH)_3 \tag{R5}$$

Based on metal electrochemical corrosion of thermodynamic principles, the initial corrosion products (Figure 2, 7 days) could spontaneously transform into stabilized phase (Figure 2, 35 days) as shown in reactions (R6) and (R7). In addition, SO_3^{2-} ions suffered from oxidation into SO_4^{2-} ions by reaction (R8) increasing the concentration of SO_4^{2-} ions.

$$2Fe(OH)_3 \rightarrow Fe_2O_3 \cdot H_2O + 2H_2O \tag{R6}$$

$$2Fe(OH)_3 + Fe^{2+} \rightarrow Fe_3O_4 + 2H_2O + 2H^+ \tag{R7}$$

$$2SO_3^{2-} + O_2 \rightarrow 2SO_4^{2-} \tag{R8}$$

The concentration of H^+ ions slowly rise as the reactions (R4) and (R7) progress, reducing the pH down from 6.95 to 5.68 (Figure 3). In general, steel would corrode severely in solutions comprising SO_4^{2-} ions. The V_{corr} and i_{corr} slowed down generally in the presence of sediments because $Fe_2O_3 \cdot H_2O$ and Fe_3O_4 improved the compactness of corrosion product films, which compose a barrier from the corrosive ions toward the matrix surface. Moreover, O_2 was the main cathode depolarizer in this corrosion environment, therefore limit diffusion of O_2 naturally turn into rate-control step of cathode reaction [30,31]. It is possible that the decrease of oxygen content in the solution would cause the same effect on V_{corr}. When oxygen content in the solution declined to a certain extent, it is easy to bring out local O_2 concentration nonuniform underneath the corrosion products film, causing more localized corrosion (Figure 1c,d). Additionally, iron oxides could be analogous to a kind of cathode depolarizer, which resulted in scattered pit cavity [32]. The growth of corrosion pits can produce Fe^{2+} ions, and the insoluble corrosion products might attract anions [33]. The hydrolysis of Fe^{2+} ions could acidize the local region of the pitting hole and induce the migration of anions, leading to the acceleration of pitting corrosion, reaction (R9) [34,35]. If the ratio of Fe^{2+} ions and SO_4^{2-} ions exceeded the threshold in the local area, to form insoluble corrosion products $FeSO_4 \cdot 4H_2O$, reaction (R10) [36]. The corrosion products films might fall off from the steel surface with the increase of sediments, which explained that the decrease in quantity of R_p and R_t with time from 28 days of immersion (Tables 4 and 5).

$$4Fe^{2+} + 4SO_4^{2-} + 6H_2O + O_2 \rightarrow 2\,Fe_2O_3 \cdot H_2O + 4H_2SO_4 \tag{R9}$$

$$\text{Fe}^{2+} + \text{SO}_4^{2-} + 4\text{H}_2\text{O} \rightarrow \text{FeSO}_4 \cdot 4\text{H}_2\text{O} \tag{R10}$$

4. Conclusions

(1) There was scale sediments attached to the surface of the steel. For 7 days of immersion, the sediments patches were connected and distributed on the matrix surface. The compactness of the sediments increased with time. The initial corrosion products was composed of Fe(OH)_2, Fe(OH)_3, and $\text{Fe}_2\text{O}_3 \cdot \text{H}_2\text{O}$, the end corrosion products consisted of $\text{Fe}_2\text{O}_3 \cdot \text{H}_2\text{O}$, $\text{FeSO}_4 \cdot 4\text{H}_2\text{O}$, and Fe_3O_4.

(2) For immersion test, the V_{corr} reduced gradually with 35 days of immersion. However, the results of electrochemical measurement showed that the V_{corr} was fluctuant in reality.

(3) In the initial immersion stage (7 days), the primary corrosion type was general corrosion, pitting corrosion was slight and dispersed under the sediments. In the later stage of corrosion (35 days), the cyclic regeneration mechanism of acid, induced by oxidation hydrolysis of FeSO_4, aggravated the pitting corrosion.

(4) Though the sediments attached to the steel surface could inhibit corrosion, pitting corrosion under the sediments would bring about more serious damage (leak of pipeline and increase of equipment fault rate) thus should be given more attention.

Author Contributions: Funding acquisition, G.Z.; Investigation, P.G. and G.Z.; Writing—original draft preparation, G.Z. and J.C.; Experiment designing, P.G.; Experiment performing, P.G.; Data analysis, P.G. and G.Z.; Writing—review and editing, P.G. All authors have read and agreed to the published version of the manuscript.

Funding: This research was funded by the National Science and Technology Program of China grant number CDGC01-KT16.

Acknowledgments: The authors wish to acknowledge the direction of Yang Miao (School of Chemistry, Chemical Engineering and Life Science, Wuhan University of Technology, Wuhan) and Hao Song (Sinopec Research Institute of Petroleum Processing, Beijing) in revision.

Conflicts of Interest: The authors declare no conflict of interest.

References

1. Martínez, A.H. Study of exhaust gascleaning systems for vessels to fulfill IMO III in 2016. *Univ. Politcnica Catalunya* **2011**, *5*, 37.
2. Anttila, M.; Hämäläinen, R.; Tuominiemi, S. Method and an Equipment for Reducing the Sulphur Dioxide Emissions of a Marine Engine. U.S. 20070798720, 16 May 2007.
3. Zidar, M. Gas-liquid equilibrium-operational diagram: Graphical presentation of absorption of SO_2 in the $NaOH-SO_2-H_2O$ system taking place within a laboratory absorber. *Ind. Eng. Chem. Res.* **2000**, *39*, 3042. [CrossRef]
4. Bandyopadhyay, A.; Biswas, M.N. Modeling of SO_2 scrubbing in spray towers. *Sci. Total Environ.* **2007**, *383*, 25. [CrossRef] [PubMed]
5. Liu, C.F.; Shih, S.M. Effects of flue gas components on the reaction of $Ca(OH)_2$ with SO_2. *Ind. Eng. Chem. Res.* **2006**, *45*, 8765. [CrossRef]
6. Machmudah, S.; Zulhijah, R.; Setyawan, H.; Kanda, H.; Goto, M. Magnetite thin film on mild steel formed by hydrothermal electrolysis for corrosion prevention. *Chem. Eng. J.* **2015**, *268*, 76–85. [CrossRef]
7. Cheng, Q.; Tao, B.; Liu, S.; Zhang, W.; Liu, X.; Li, W.; Liu, Q. Corrosion behavior of Q235B steel in sediments water from crude oil. *Corros. Sci.* **2016**, *111*, 61–71. [CrossRef]
8. Yu, Q.; Dong, C.; Fang, Y.; Fang, Y.; Xiao, K.; Guo, C.; He, G.; Li, X. Atmospheric corrosion of Q235 carbon steel and Q450 weathering steel in Turpan, China. *J. Iron Steel Res. Int.* **2016**, *23*, 1061–1070. [CrossRef]
9. Liu, G.; Zhang, Y.; Ni, Z.; Huang, R. Corrosion behavior of steel submitted to chloride and sulphate ions in simulated concrete pore solution. *Constr. Build. Mater.* **2016**, *115*, 1–5. [CrossRef]
10. Boah, J.K.; Somuah, S.K.; LeBlanc, P. Electrochemical behavior of steel in saturated calcium hydroxide solution containing Cl-, SO42-, and CO32- Ions. *Corrosion* **1990**, *46*, 153–158. [CrossRef]

11. Xu, P.; Jiang, L.; Guo, M.; Zha, J.; Chen, L.; Chen, C.; Xu, N. Influence of sulfate salt type on passive film of steel in simulated concrete pore solution. *Constr. Build. Mater.* **2019**, *223*, 352–359. [CrossRef]
12. Huang, B.; Yang, G.H. Research progress of ship tail gas gas cleaning desulfurization denitration and PM removal equipment. *Chem. Ind. Eng. Prog.* **2013**, *32*, 2826.
13. Persaud, S.Y.; Carcea, A.G.; Newman, R.C. An electrochemical study assisting the interpretation of acid sulfate stress corrosion cracking of NiCrFe alloys. *Corros. Sci.* **2015**, *90*, 383–391. [CrossRef]
14. Revie, R.W.; Uhlig, H.H. Treatment of water and steam systems. In *Corrosion and Corrosion Control: An Introduction to Corrosion Science and Engineering*, 4th ed.; John Wiley & Sons Inc.: New Jersey, NY, USA, 2009; pp. 317–332.
15. Kuang, W.; Mathews, J.A.; Macdonald, D.D. The effect of Anodamine on the corrosion behavior of 1018 mild steel in deionized water: I. Immersion and polarization tests. *Electrochim. Acta* **2014**, *127*, 79–85. [CrossRef]
16. Oliveira, N.; Guastaldi, A. Electrochemical behavior of Ti–Mo alloys applied as biomaterial. *Corros. Sci.* **2008**, *50*, 938–945. [CrossRef]
17. Cai, B.P.; Liu, Y.H.; Tian, X.J.; Wang, F.; Li, H.; Ji, R. An experimental study of crevice corrosion behaviour of 316L stainless steel in artificial seawater. *Corros. Sci.* **2010**, *52*, 3235–3242. [CrossRef]
18. Zheng, L.; Neville, A. Corrosion Behavior of Type 316L Stainless Steel in Hydraulic Fluid and Hydraulic Fluid/Seawater for Subsea Applications. *Corrosion* **2009**, *65*, 145–153. [CrossRef]
19. Cheng, Q.; Song, S.; Song, L.; Hou, B. Effect of Relative Humidity on the Initial Atmospheric Corrosion Behavior of Zinc during Drying. *J. Electrochem. Soc.* **2013**, *160*, C380–C389. [CrossRef]
20. Cheng, Q.; Chen, Z. The cause analysis of the incomplete semi-circle observed in high frequency region of EIS obtained from TEL-covered pure copper. *Int. J. Electrochem.* **2013**, *8*, 8282–8290.
21. Wu, Y.H.; Liu, T.M.; Sun, C.; Xu, J.; Yu, C.K. Effects of simulated acid rain on corrosion behaviour of Q235 steel in acidic soil. *Corros. Eng. Sci. Technol.* **2010**, *45*, 136–141. [CrossRef]
22. Gonzalez, J.E.G.; Mirza-Rosca, J.C. Study of the corrosion behavior of titanium and some of its alloys for biomedical and dental implant applications. *J. Electroanal. Chem.* **1999**, *471*, 109–115. [CrossRef]
23. Sheng, X.; Ting, Y.P.; Pehkonen, S.O. The influence of sulphate-reducing bacteria biofilm on the corrosion of stainless steel AISI 316. *Corros. Sci.* **2007**, *49*, 2159–2176. [CrossRef]
24. Mohammadloo, H.E.; Sarabi, A.A.; Sabbagh, A.A.; Salimi, R.; Sameie, H. The effect of solution temperature and pH on corrosion performance and morphology of nanoceramic-based conversion thin film. *Mater. Corros.* **2014**, *64*, 535–543. [CrossRef]
25. Medhashree, H.; Shetty, A.N. Electrochemical investigation on the effects of sulfate ion concentration, temperature and medium pH on the corrosion behavior of Mg–Al–Zn–Mn alloy in aqueous ethylene glycol. *J. Magnes. Alloys* **2017**, *5*, 64–73. [CrossRef]
26. Zou, Y.; Wang, J.; Zheng, Y. Electrochemical techniques for determining corrosion rate of rusted steel in seawater. *Corros. Sci.* **2011**, *53*, 208–216. [CrossRef]
27. Sánchez-Tovar, R.; Montañés, M.T.; García-Antón, J. The effect of temperature on the galvanic corrosion of the copper/AISI 304 pair in LiBr solutions under hydrodynamic conditions. *Corros. Sci.* **2010**, *52*, 722–733. [CrossRef]
28. Li, D.G.; Wang, J.D.; Chen, D.R. Influence of pH value on the structure and electronic property of the passive film on 316L SS in the simulated cathodic environment of proton exchange membrane fuel cell (PEMFC). *Int. J. Hydrog. Energy* **2014**, *39*, 20105–20115. [CrossRef]
29. Dong, Z.H.; Shi, W.; Guo, X.P. Initiation and repassivation of pitting corrosion of steel in carbonated concrete pore solution. *Corros. Sci.* **2011**, *53*, 1322–1330. [CrossRef]
30. Hubbard, A. *Corrosion and Corrosion Control: An Introduction to Corrosion Science and Engineering*, 4th ed.; Revie, R.W., Uhlig, H.H., Eds.; John Wiley & Sons: Hoboken, NJ, USA, 2008; 490p, Journal of Colloid & Interface Science, **2008**, *328*, 463–463.
31. Stratmann, M.; Streckel, H. On the atmospheric corrosion of metals which are covered with thin electrolyte layers—I. Verification of the experimental technique. *Corros. Sci.* **1990**, *30*, 715–734. [CrossRef]
32. Yan, M.C.; Sun, C.; Xu, J.; Dong, J.; Ke, W. Role of Fe oxides in corrosion of pipeline steel in a red clay soil. *Corros. Sci.* **2014**, *80*, 309–317. [CrossRef]
33. Reffass, M.; Sabot, R.; Jeannin, M.; Berziou, C.; Refait, P. Effects of NO2− ions on localised corrosion of steel in NaHCO3 + NaCl electrolytes. *Electrochim. Acta* **2007**, *52*, 7599–7606. [CrossRef]

34. Hao, L.; Zhang, S.; Dong, J.; Ke, W. Evolution of corrosion of MnCuP weathering steel submitted to wet/dry cyclic tests in a simulated coastal atmosphere. *Corros. Sci.* **2012**, *58*, 175–180. [CrossRef]
35. Wang, J.H.; Wei, F.I.; Chang, Y.S.; Shih, H. The corrosion mechanisms of steel and weathering steel in SO_2 polluted atmospheres. *Mater. Chem. Phys.* **1997**, *47*, 1–8. [CrossRef]
36. González, J.; Miranda, J.; Otero, E.; Feliu, S. Effect of electrochemically reactive rust layers on the corrosion of steel in a $Ca(OH)_2$ solution. *Corros. Sci.* **2007**, *49*, 436–448. [CrossRef]

© 2020 by the authors. Licensee MDPI, Basel, Switzerland. This article is an open access article distributed under the terms and conditions of the Creative Commons Attribution (CC BY) license (http://creativecommons.org/licenses/by/4.0/).

MDPI
St. Alban-Anlage 66
4052 Basel
Switzerland
Tel. +41 61 683 77 34
Fax +41 61 302 89 18
www.mdpi.com

Materials Editorial Office
E-mail: materials@mdpi.com
www.mdpi.com/journal/materials

www.ingramcontent.com/pod-product-compliance
Lightning Source LLC
LaVergne TN
LVHW070359100526
838202LV00014B/1349